Climate-Smart Technologies for Sustainable Agricultural Production

Dillip Kumar Swain · Geetha Mohan ·
Poonam Biswal · Ahmad Faisal · Naoya Wada
Editors

Climate-Smart Technologies for Sustainable Agricultural Production

Innovations and Practices from India and Japan

Springer

Editors
Dillip Kumar Swain
Agricultural and Food Engineering Department
Indian Institute of Technology Kharagpur
Kharagpur, West Bengal, India

Geetha Mohan
Global Research Centre for Advanced Sustainability Science
University of Toyama
Toyama, Japan

Poonam Biswal
Agricultural and Food Engineering Department
Indian Institute of Technology Kharagpur
Kharagpur, West Bengal, India

Ahmad Faisal
Agricultural and Food Engineering Department
Indian Institute of Technology Kharagpur
Kharagpur, West Bengal, India

Naoya Wada
Global Research Centre for Advanced Sustainability Science
University of Toyama
Toyama, Japan

ISBN 978-3-032-25665-2 ISBN 978-3-032-25666-9 (eBook)
https://doi.org/10.1007/978-3-032-25666-9

This Springer imprint is published by the registered company Springer Nature Switzerland AG
The registered company address is: Gewerbestrasse 11, 6330 Cham, Switzerland

Preface

Climate change is fundamentally transforming agricultural systems globally, thereby increasing risks to food security, rural livelihoods, and natural resources. The increasing variability of climate patterns, extreme weather events, land degradation, and resource constraints necessitate prompt and systemic interventions that go beyond incremental adaptation. In this context, climate-smart agriculture (CSA) has been introduced as a comprehensive framework that integrates efforts to enhance productivity, strengthen resilience, and ensure environmental sustainability across agri-food systems.

This edited volume, *Climate-Smart Technologies for Sustainable Agricultural Production: Innovations and Practices from India and Japan*, presents the outcomes of a JSPS–ICSSR bilateral research initiative and its associated international workshop on *Climate-Smart Farming Resilient Technologies and Practices in Japan and Opportunities for Indian Farmers and Agri-tech Start-up Entrepreneurs.* The volume reflects a shared commitment to advancing science-based, policy-relevant, and practice-oriented solutions through India–Japan collaboration, bringing together two countries that face distinct yet converging challenges in climate adaptation, rural transformation, and sustainable agricultural development.

The bilateral project was jointly led by Dillip Kumar Swain, Professor, Agricultural and Food Engineering Department, Indian Institute of Technology Kharagpur, India, and Geetha Mohan, Professor and Vice-Dean, Graduate School of Sustainability Studies, and Professor at the Global Research Centre for Advanced Sustainability Science (GRASS), University of Toyama, Japan. Their leadership enabled a transdisciplinary platform that brought together researchers, policymakers, industry practitioners, extension systems, and agri-tech entrepreneurs through a brainstorming session. The output presented in this volume, which reflects an integrated vision that combines technological innovation, institutional analysis, and real-world implementation experiences.

The book is structured into three thematic sections. The first addresses climate-smart farming systems and enabling technologies, including AI-based soil carbon monitoring, agrometeorological advisory services, region-specific climate-smart practices, agricultural policy reform, and ICT-enabled resource circulation models.

These chapters demonstrate how data, digital infrastructure, and policy alignment can reduce climate risk while improving farm-level decision-making. The second section focuses on conservation agriculture and organic farming, highlighting resource recycling, carbon and energy footprints, watershed-based planning, and organic farming as a climate-resilient strategy. Together, these chapters provide empirical evidence for nature-based and low-input solutions that deliver environmental and economic co-benefits. The third section explores innovative adaptation and transformation pathways, including farmer-centric digital platforms, experimental and simulation-based assessments of climate-adaptive technologies, rural revitalization strategies, and circular bioeconomy models such as agricultural waste-to-energy systems.

A central contribution of this volume lies in its alignment with the United Nations Sustainable Development Goals (SDGs). Collectively, the chapters advance SDG 2 (Zero Hunger) through resilient and productive agricultural systems; SDG 12 (Responsible Consumption and Production) via resource efficiency and circular economy approaches; SDG 13 (Climate Action) by strengthening adaptation and mitigation in agriculture; and SDG 15 (Life on Land) through sustainable land management, soil health, and biodiversity conservation. By grounding these global goals in applied case studies from India and Japan, the book translates international sustainability commitments into actionable strategies.

This volume is intended for policymakers, researchers, development practitioners, and agri-tech innovators seeking evidence-based pathways for climate-smart agricultural transitions. By bridging policy, technology, and practice, it aims to inform decision-making, stimulate innovation, and support inclusive and sustainable agri-food systems. The editors gratefully acknowledge the support of JSPS and ICSSR, the contributions of all authors, and Springer Nature for providing a global platform to disseminate this collaborative work.

Kharagpur, India — Dillip Kumar Swain
Toyama, Japan — Geetha Mohan
Kharagpur, India — Poonam Biswal
Kharagpur, India — Ahmad Faisal
Toyama, Japan — Naoya Wada

Contents

Chapter 1
Introduction: Climate-Smart Agricultural Transformation Through Science, Policy, and Practice in India and Japan

Mohan Geetha, Dillip Kumar Swain, Poonam Biswal, Ahmad Faisal, and Naoya Wada

Abstract Agriculture stands at a critical crossroads in the Anthropocene, confronting climate variability, ecological degradation, and demographic change. This introductory chapter conceptualizes climate-smart agriculture (CSA) as a systems-based transformation that integrates ecological restoration, digital innovation, institutional reform, and participatory governance. Drawing comparative insights from India and Japan, the chapter examines how smallholder vulnerability and farmland abandonment represent distinct yet converging challenges for sustainable agricultural futures. By synthesizing themes of soil organic carbon management, conservation and organic practices, climate advisory systems, digital agriculture platforms, and policy alignment, the chapter positions CSA within broader global sustainability discourse. It argues that resilient agri-food systems require coordinated transformation across biophysical, informational, and institutional domains.

Keywords Climate-smart agriculture (CSA) · Soil organic carbon · Digital agriculture · Institutional reform · Agricultural resilience

---

M. Geetha · N. Wada
Graduate School of Sustainability Studies, University of Toyama, Toyama, Japan

M. Geetha (✉) · D. K. Swain · N. Wada
Global Research Centre for Advanced Sustainability Science (GRASS), University of Toyama, Toyama, Japan
e-mail: geetha@eco.u-toyama.ac.jp

D. K. Swain · P. Biswal · A. Faisal
Agricultural and Food Engineering Department, Indian Institute of Technology Kharagpur, Kharagpur, India

D. Kumar Swain et al. (eds.), *Climate-Smart Technologies for Sustainable Agricultural Production*, https://doi.org/10.1007/978-3-032-25666-9_1

1.1 Agriculture at the Crossroads of Climate, Ecology, and Societal Transition

Agriculture currently stands at a significant inflection point in history. Across continents, food production systems face increasing challenges due to climatic variability, land degradation, biodiversity loss, and widening socioeconomic disparities. Rising temperatures, unpredictable precipitation patterns, prolonged droughts, extreme rainfall events, and shifts in seasonal cycles are progressively compromising the stability of agricultural yields and rural livelihoods (FAO 2012; IPCC 2014;). Concurrently, agriculture continues to both contribute to and mitigate climate change through land-use planning, soil carbon management, livestock systems, and energy consumption practices. This dual role of agriculture, as both susceptible to and influential within the climate system, underscores its central importance in the discourse on global sustainability. The urgency of transforming agricultural systems is particularly apparent in countries such as India and Japan, where climate-related stresses intersect with demographic shifts, institutional complexities, and evolving market structures. India's predominantly smallholder farming system remains highly vulnerable to monsoon variability and heat stress, and yields of major crops are projected to decline under future climate scenarios unless appropriate adaptation strategies are implemented (Naresh Kumar et al. 2013, 2014, 2015). Conversely, Japan faces a distinct yet equally pressing structural transformation: the abandonment of agricultural land, driven by rapid aging, depopulation, and declining farm profitability, particularly in rural prefectures such as Toyama (Su et al. 2018; Shoyama et al. 2021). These contrasting national contexts share a common challenge: maintaining agricultural productivity while enhancing resilience, mitigating environmental degradation, and revitalizing rural communities.

This volume arises from the recognition that climate-smart agriculture (CSA) should be understood not merely as a single technological intervention but as a multidimensional transformation encompassing ecological restoration, digital innovation, institutional reform, and community engagement. Throughout its chapters, the publication consolidates empirical case studies, methodological innovations, and policy analyses from India and Japan to exemplify how climate-smart technologies and practices are being implemented across diverse agro-ecological and socio-economic contexts. Instead of portraying climate-smart agriculture as a universal blueprint, the volume illustrates that resilience-building must be locally rooted, scientifically substantiated, and institutionally endorsed. The conceptual foundation of Climate-Smart Agriculture (CSA) is anchored in three interconnected pillars: enhancing productivity, strengthening resilience, and diminishing greenhouse gas emissions (FAO 2012; Campbell et al., 2014). However, the practical implementation of these pillars necessitates more than mere agronomic modifications. It requires the development of advanced sensing technologies for evaluating soil carbon, adaptive advisory systems, conservation-oriented cultivation practices, institutional reforms concerning price support and market accessibility, and innovative land-use strategies for aging rural landscapes. By integrating these dimensions, the publication positions climate-smart agriculture within a broader transformation of agri-food systems.

1.2 Soil, Carbon, and the Biophysical Foundations of Climate Resilience

Central to sustainable agricultural transformation is soil. Soil organic carbon (SOC) is increasingly acknowledged as a crucial indicator of soil health, crop productivity, and climate mitigation potential. Traditional laboratory-based methods for measuring SOC, while scientifically reliable, are often time-consuming and limited in spatial scope. Recent advancements in digital soil mapping (DSM), hyperspectral remote sensing, and machine learning enable more scalable and dynamic monitoring of SOC, as discussed in this chapter. The integration of satellite imagery, proximal sensors, and generative artificial intelligence models improves predictive accuracy and reduces monitoring costs. These innovations represent a shift toward data-driven soil stewardship, in which the potential for carbon sequestration can be quantified and linked to climate mitigation strategies. The significance of soil extends beyond merely accounting for carbon. Conservation agriculture practices, such as minimal tillage, residue retention, and crop diversification, improve soil structure, reduce erosion, and stabilize yields under climatic stress, as elaborated in Chap. 7. Empirical assessments of carbon and energy footprints indicate that, although short-term transitions may entail trade-offs, long-term conservation systems ultimately enhance soil carbon stocks and reduce net emissions. These findings concur with broader global evidence suggesting that soil-centered interventions are among the most effective strategies for climate mitigation in the agricultural sector (IPCC 2014).

Organic farming systems, as discussed in Chap. 8, further exemplify the potential of ecological nutrient cycling, biological nitrogen fixation, and compost-based soil management to decrease dependence on synthetic inputs. By eliminating synthetic nitrogen fertilizers and advocating diversified crop rotations, organic systems substantially reduce greenhouse gas emissions and enhance soil resilience. Significantly, organic farming is not characterized as a nostalgic return to pre-industrial agricultural practices but is instead recognized as a rigorously scientific, climate-resilient production strategy that incorporates soil biology, biodiversity, and farmer expertise. In Japan, regenerative models such as synecoculture and resource-recycling organic systems (Chap. 6) exemplify how biodiversity-rich cropping systems can stabilize soil moisture, increase microarthropod populations, and mitigate climatic extremes. These systems highlight a vital insight: climate-smart agriculture should focus on restoring ecological processes rather than simply compensating for their degradation through external inputs.

1.3 Climate Risk, Advisory Systems, and Adaptive Decision-Making

Climate resilience is not solely a function of soil management; it equally relies on anticipatory capacity and information systems. The agrometeorological advisory services in India (Chap. 2) exemplify how climate forecasts, downscaled weather models, and decision-support systems can inform farmers' planting schedules, irrigation timing, and pest control strategies. The integration of numerical weather prediction models with local dissemination channels, including mobile applications and community radio, bridges the divide between climate science and actionable steps at the farm level. The Climate-Smart Village (CSV) model implemented by ITC Limited (Chap. 13) extends this logic by integrating climate risk assessment, participatory technology prioritization, and institutional linkages. The employment of NDVI-based yield estimation and rainfall variability indices to categorize villages into resilience–yield quadrants exemplifies an innovative approach to targeted adaptation planning. These methodologies exemplify the synthesis of remote sensing, statistical analysis, and community engagement (BISA-ITC, 2021).

The convergence of climate data and farmer experience underscores a recurring theme throughout the volume: the necessity of integrating technological tools within participatory frameworks to promote adoption and ensure sustainability. Statistical prioritization methods, such as the Climate-Smart Adoption Prioritisation (CSAP) toolkit (Khatri-Chhetri et al. 2019), offer structured decision-making mechanisms; however, their effectiveness depends on stakeholder involvement and institutional support.

1.4 Digital Agriculture and the Reconfiguration of Knowledge Systems

The transformation of agriculture under climate stress increasingly depends on digital infrastructures that improve monitoring accuracy, predictive capabilities, and connectivity among farmers. In both India and Japan, digital technologies are transforming the ways in which agricultural knowledge is generated, disseminated, and implemented. The advent of hyperspectral imaging, machine learning algorithms, Internet of Things (IoT) sensors, and cloud-based analytical platforms signifies not merely technological advancement but a fundamental shift towards data-driven resilience. The digital soil mapping frameworks discussed earlier are complemented by broader digital ecosystems examined in this volume. Farmer-centric digital platforms (Chap. 10) demonstrate how climate information services, AI-driven plant diagnostics, soil health dashboards, and digital market linkages can unify within integrated "phygital" architectures. Such systems connect physical extension services with digital analytics, enabling smallholders to access precision advisories previously available only to high-capital farms. The incorporation of mobile-based climate advisories, soil nutrient recommendations, and pest alerts enhances adaptive capacity by reducing uncertainty and facilitating timely decision-making.

Artificial intelligence assumes a particularly vital role in addressing issues related to data scarcity and heterogeneity. Generative adversarial networks and data augmentation methods improve predictive performance when field-level observations are scarce. These advancements underpin yield forecasting, stress detection, and carbon accounting, crucial elements of climate-smart systems. Significantly, the volume demonstrates that digital innovation is not an end in itself but an instrument for democratizing agricultural knowledge. When integrated with farmer-producer organizations (FPOs), agri-business centers, and institutional credit mechanisms, digital systems promote collective action and enhance market integration.

Japan's experience further underscores the importance of remote sensing and UAV-based monitoring in maintaining multifunctional landscapes. In regenerative organic systems in Toyama (Chap. 6), multispectral NDVI analyses demonstrate seasonal stability patterns associated with soil management practices. These findings highlight that digital monitoring can authenticate ecological assertions and facilitate evidence-based policy development. Across various contexts, digital agriculture is not positioned as a substitute for traditional knowledge but rather as a complementary approach that enhances accuracy, transparency, and scalability. However, digital transformation also presents governance challenges. Issues such as data ownership, privacy, algorithmic transparency, and disparities in technological access necessitate careful institutional planning. Absent inclusive frameworks, digital divides may intensify inequalities between large-scale producers and smallholders. The cases exemplified in this volume consistently highlight the significance of participatory governance and public–private partnerships in promoting equitable technological dissemination.

1.5 Institutional Architecture and Policy Alignment in Climate-Smart Transitions

Technological innovation alone cannot guarantee sustainable agricultural transitions. Institutions, both formal and informal, shape incentives, mechanisms for sharing risks, and patterns of resource allocation. Several chapters in this volume examine the policy framework underpinning agricultural systems, particularly in India. The analysis of agricultural price support reform (Chap. 4) demonstrates how fiscal sustainability, trade obligations under the WTO Agreement on Agriculture, and farmer income stability converge within the discourse on climate adaptation. Minimum Support Price (MSP) regimes, although historically crucial in stabilizing grain production, may also distort cropping patterns and hinder diversification toward climate-resilient crops. Fiscal simulations comparing procurement-based models with price deficiency payments reveal intricate trade-offs between income protection and market efficiency. These policy considerations are directly pertinent to climate-smart agriculture, as price incentives influence farmers' adoption of diversified cropping systems, water-saving technologies, and low-emission practices.

Institutional convergence consistently emerges as a prominent theme. The integration of Climate-Smart Agriculture (CSA) interventions with national initiatives, such as crop insurance programs, irrigation missions, and rural employment guarantees, demonstrates that climate resilience should be embedded in mainstream policies rather than confined to isolated pilot projects. The ITC CSV model exemplifies this integrated approach by aligning climate-smart practices with government schemes, extension services, and financial instruments. Similarly, agrometeorological advisory systems depend on coordinated collaboration among meteorological departments, agricultural universities, and local extension networks.

In Japan, institutional dynamics are exhibited differently. The challenge of farmland abandonment, particularly in aging prefectures such as Toyama (Chap. 12), emphasizes the significance of governance mechanisms for land consolidation, leasing, and cooperative management. When demographic decline diminishes household labor capacity, institutional innovations, such as corporate land leasing arrangements, livestock-sharing systems, and community-based cooperatives, become essential to landscape stewardship. Policy frameworks that facilitate transparent leasing contracts, participatory decision-making, and local employment creation are vital to ensuring that corporate involvement enhances rather than diminishes community resilience. Institutional innovation also relates to climate finance and carbon markets. As soil carbon monitoring improves accuracy, new opportunities emerge to link regenerative practices with carbon credit systems. However, these mechanisms require reliable measurement, reporting, and verification (MRV) protocols, as well as safeguards to protect smallholders' interests. The volume underscores that climate-smart agriculture should be integrated into accountable institutional systems that balance productivity, equity, and environmental sustainability.

1.6 Demographic Change and Landscape Reconfiguration

While India's agricultural challenge primarily focuses on climate variability and the vulnerability of smallholder farmers, Japan's experience highlights demographic transformation and land-use restructuring. Rapid population aging, declining birth rates, and rural depopulation have significantly altered the social fabric of agricultural communities. As documented in Chap. 12, Toyama Prefecture has experienced a consistent population decline and an increase in elderly residents, resulting in widespread abandonment of farmland. GIS-based spatial analyses and vegetation surveys demonstrate clustering of abandoned fields in municipalities characterized by high demographic vulnerability.

Land abandonment entails ecological, economic, and cultural implications. Unmanaged fields experience ecological succession, which modifies biodiversity patterns and disrupts irrigation systems. However, the volume also reinterprets abandonment as a potential space for innovation. Examples include corporate biomass production, specialty crop cultivation, livestock grazing systems modeled after Tajima in Hyogo Prefecture, and vineyard conversion inspired by Tsukuba in Ibaraki Prefecture, which illustrate alternative land-use strategies. These instances demonstrate that climate-smart transformation may encompass not only intensification but also strategic reconfiguration of land utilization. The Tajima livestock model demonstrates how marginal rice fields can be repurposed as pasture, thereby optimizing labor demands and restoring biodiversity (Nakahama et al. 2024). The conversion of vineyards in Tsukuba underscores the significance of market orientation and rural branding in maintaining the viability of specialty crops. These instances emphasize that climate-smart agriculture must adapt to demographic realities and economic considerations. In societies characterized by aging populations, labor-efficient and multifunctional land-use systems emerge as vital elements of resilience.

India's rural transformation follows a distinct trajectory, characterized by the fragmentation of landholdings, migration, and pressures for livelihood diversification. The CSV framework responds to these dynamics by advocating on-farm diversification, agroforestry, livestock integration, and collective marketing mechanisms. In both national contexts, climate-smart strategies have been shown to be inherently linked to the restructuring of rural socio-economic systems.

1.7 Energy Transitions and Multifunctional Agricultural Landscapes

Agriculture's role in climate mitigation extends beyond soil carbon management. Energy footprints, biomass use, renewable energy integration, and agroforestry systems contribute to emission-reduction pathways. Chap. 7's analysis of carbon and energy footprints in conservation agriculture shows that residue retention and reduced tillage lower energy inputs while enhancing carbon sequestration. These findings align with broader global efforts to align agricultural production with net-zero targets.

Bioenergy production on marginal lands, as examined in Toyama's corporate leasing initiatives (Chap. 12), demonstrates the dual potential of land reutilization for both climate mitigation and rural income enhancement. Fast-growing willow systems and biomass feedstocks incorporate agricultural landscapes into regional low-carbon economies. Nevertheless, as highlighted in academic discussions (Mitsui et al. 2010; Timilsina 2012), the expansion of bioenergy must be carefully balanced with food security concerns. The emphasis is placed on multifunctionality: land-use strategies should aim to optimize ecological restoration, food production, energy generation, and community welfare concurrently. Solar sharing systems, in which photovoltaic panels are installed over cultivated land, serve as a paradigmatic example of integrated food–energy systems. By generating revenue from renewable energy while sustaining agricultural production, such systems diversify income sources and bolster rural economic stability. These initiatives indicate a transition from monocultural agriculture toward diversified, climate-resilient landscapes.

1.8 Toward a Systems-Based Climate-Smart Paradigm

The preceding sections collectively demonstrate that climate-smart agriculture cannot be limited to agronomic modifications or isolated technological interventions. Instead, it signifies a comprehensive reorientation of agricultural production systems towards ecological integrity, adaptive capacity, and socio-economic resilience. The chapters compiled in this volume present a consistent and persuasive insight: sustainable agricultural transformation arises from the integration of biophysical science, digital innovation, institutional reform, and community agency. At the biophysical level, the importance of soil health, carbon sequestration, biodiversity, and water management is unequivocal. Soil organic carbon mapping (Chap. 1), conservation agriculture (Chap. 7), organic nutrient cycling (Chap. 8), and regenerative systems (Chap. 6) collectively emphasize that soil functions both as a driver of productivity and a regulator of climate. Absent the restoration of soil biological processes and carbon reserves, efforts toward intensification remain ecologically precarious. This aligns with broader ecological research indicating that semi-managed agricultural systems provide essential ecosystem services, such as hydrological regulation and biodiversity support (Uchida and Ushimaru 2014; Takeuchi et al. 2016).

At the informational level, climate risk forecasting and digital advisory systems (Chaps. 2 and 10) facilitate anticipatory adaptation strategies. The integration of climate projections, NDVI-based yield estimation, and participatory prioritization tools within the Climate Smart Village framework (Chap. 13) exemplifies how predictive analytics can be translated into context-specific actions. Importantly, the evidence indicates that resilience enhancement occurs not solely through technological adoption but also via institutional embedding, such as linkages to insurance schemes, collective machinery pools, and farmer-producer organizations. Institutional architecture, therefore, emerges as a vital determinant of sustainability outcomes. The fiscal and policy analysis of agricultural price support systems (Chap. 4) highlights how macroeconomic frameworks influence cropping patterns, resource utilization, and environmental externalities. When policy incentives are aligned with climate-smart objectives, such as promoting diversification, efficient irrigation, and emission-reducing practices, structural transformation becomes achievable. Conversely, misaligned incentives pose a risk of entrenching agricultural systems in unsustainable trajectories.

Demographic and spatial restructuring further complicates the nexus between climate and agriculture. Japan's experience with farmland abandonment (Chap. 12) exemplifies that agricultural sustainability is intrinsically linked to the vitality of rural demographics. Land-use transformation, including livestock conversion, biomass production, vineyard specialization, and solar sharing, reflects adaptive responses to aging populations and labor shortages. These models demonstrate that climate-smart agriculture must be attentive to demographic realities and that multifunctional landscapes can develop from structural decline when supported by institutional innovation.

The comparison between India and Japan presented in this volume enhances the conceptual understanding of climate-smart transformation. India faces challenges related to climatic volatility and vulnerability of smallholders within densely populated agricultural regions. Conversely, Japan faces depopulation and underutilization of land in its technologically advanced yet aging rural systems. Despite these structural contrasts, climatic vulnerability in densely populated smallholder systems and demographic contraction in technologically advanced rural landscapes, the foundational principles of resilience, diversification, ecological restoration, and participatory governance converge. The cross-national dialogue underscores that climate-smart agriculture is not a rigid template but a flexible framework adaptable to diverse socio-ecological conditions.

1.9 Agriculture in the Anthropocene: Beyond Incremental Adaptation

Agriculture in the Anthropocene operates within planetary boundaries, constrained by climate change, biodiversity loss, and resource limitations. The necessary transformation is not merely incremental but fundamentally structural. This volume advances that transformation by illustrating how innovation, encompassing technological, institutional, and social dimensions, can be effectively mobilized across various scales.

Several overarching lessons emerge. First, resilience must be multidimensional. Yield stability, income security, carbon mitigation, and ecological restoration must be pursued simultaneously rather than sequentially. The CSV model's yield–resilience quadrants (Chap. 13) provide a compelling visualization of this multidimensional objective. Secondly, it is imperative that technology be democratized. AI-driven diagnostics, remote sensing platforms, and digital MRV systems present significant potential; however, this potential can only be realized when access is equitable, capacity building is prioritized, and institutional trust is established. Farmer field schools, participatory workshops, and cooperative governance structures serve as essential links facilitating the transition from innovation to adoption. Third, land should be regarded as multifunctional. The reconfiguration of abandoned land in Toyama (Chap. 12), the integration of livestock and crop systems (Chap. 6), and the blending of food and energy production exemplify how sustainable landscapes can optimize multiple ecosystem services concurrently. This multifunctionality enhances resilience by diversifying income sources and ecological functions. Fourth, policy coherence is essential. Harmonizing agricultural, climate, water, and rural development policies improves efficiency and mitigates conflicting incentives. The integration of Climate-Smart Agriculture (CSA) initiatives with national schemes in India exemplifies this coherence, while Japan's land-leasing reforms illustrate governance adaptation in aging contexts.

Ultimately, climate-smart agriculture should be regarded as a constantly evolving paradigm. As projections of climate change become more severe, adaptation strategies must be flexible and responsive. Ongoing monitoring, iterative learning processes, and collaboration across various sectors are imperative. The interdisciplinary nature of this volume, encompassing soil science, hydrology, economics, digital technology, rural sociology, and policy analysis, mirrors the multifaceted complexity of the challenge.

1.10 Positioning this Volume within the Global Sustainability Discourse

The analyses presented in this volume go beyond geographically confined case studies to engage more deeply into international debates on sustainable intensification, agroecology, carbon management, digital agriculture, and rural revitalization. Based on empirical evidence from India and Japan, the contributions illustrate that climate-smart agriculture is not merely a standardized technological package, but rather a transformation sensitive to contextual factors shaped by ecological conditions, institutional frameworks, and socio-economic realities. By integrating experiences from both the Global South and the Global North, this volume demonstrates that although pathways may vary, the fundamental drivers of resilience, such as soil regeneration, water stewardship, diversified livelihoods, and institutional coherence, remain universally pertinent. The evidence closely aligns with broader commitments to sustainability concerning food security, water governance, responsible production, and climate mitigation. Enhancements in soil organic carbon stocks, reductions in greenhouse gas emissions, strengthened adaptive capacity, and improved resource-use efficiency are not isolated achievements; rather, they represent interconnected outcomes of systemic transformation. Simultaneously, the case studies emphasize that climate-smart agriculture cannot be solely attributed to technological optimism. Sustainable progress relies on participatory governance, long-term institutional support, transparent market linkages, and the fostering of trust among rural communities and external stakeholders.

Collectively, the contributions of this volume affirm that climate-smart technologies and practices form essential foundations for resilient agricultural futures. When ecological restoration, digital innovation, institutional collaboration, and policy alignment function harmoniously, agriculture transitions from climate vulnerability toward adaptive stability. The subsequent analyses examine these interconnections in empirical detail, fostering a comprehensive understanding of how sustainable agricultural production can be re-envisioned amid rapid environmental and socio-economic change.

References

BISA–ITC (2021) Adaptation strategies for managing climatic risks in Madhya Pradesh agriculture. Borlaug Institute for South Asia and ITC Limited

Campbell BM, Thornton PK, Zougmoré RB, Van Asten P, Lipper L (2014) Sustainable intensification: what is its role in climate-smart agriculture? Curr Opin Environ Sustain 8:39–43. https://doi.org/10.1016/j.cosust.2014.07.002

FAO (2012) The state of food and agriculture. Food and Agriculture Organization of the United Nations

IPCC (2014) Climate change 2014: impacts, adaptation, and vulnerability. Cambridge University Press

Khatri-Chhetri A, Pant A, Aggarwal PK, Vasireddy VV, Yadav A (2019) Stakeholders' prioritization of climate-smart agriculture interventions: evaluation of a framework. Agric Syst 174:23–31. https://doi.org/10.1016/j.agsy.2019.03.002

Mitsui Y, Seto S, Nishio M, Minato K, Ishizawa K, Satoh S (2010) Willow clones with high biomass yield in short rotation coppice in the southern region of Tohoku district (Japan). Biomass Bioenergy 34(4):467–473

Nakahama N, Hamano T, Fujimoto M, Eto A (2024) Utilization of abandoned land as cattle grazing restores butterfly and flowering plant diversities in Japan. Ecol Res 39(5):730–743

Naresh Kumar S, Aggarwal PK, Saxena R, Rani S, Jain S, Chauhan N (2013) An assessment of regional vulnerability of rice to climate change in India. Clim Change 118(3–4):683–699

Naresh Kumar S, Aggarwal PK, Kumar U, Jain S, Chauhan N (2014) Vulnerability of wheat production to climate change in India. Clim Res 59:173–187

Naresh Kumar S, Govindakrishnan PM, Swarooparani DN, Nitin C, Jain S, Aggarwal PK (2015) Assessment of impact of climate change on potato and potential adaptation gains in the indo-Gangetic Plains of India. Int J Plant Prod 9(1):151–170

Shoyama K, Nishi M, Hashimoto S, Saito O (2021) Outcome-based assessment of the payment for mountain agriculture: a community-based approach to countering land abandonment in Japan. Environ Manag 68(3):353–365

Su G, Okahashi H, Chen L (2018) Spatial pattern of farmland abandonment in Japan: identification and determinants. Sustainability 10(10):3676

Takeuchi K, Ichikawa K, Elmqvist T (2016) Satoyama landscape as social–ecological system: historical changes and future perspective. Curr Opin Environ Sustain 19:30–39

Timilsina GR (2012) Biofuels: the food versus fuel debate. CABI Rev 1–8

Uchida K, Ushimaru A (2014) Biodiversity declines due to abandonment and intensification of agricultural lands: patterns and mechanisms. Ecol Monogr 84:637–658

Chapter 2
Advanced Sensing and AI for Soil Organic Carbon Monitoring and Mapping

Anshuman Nayak, Rachna Singh, Ushasi Dam, Mouli Sarkar, Dillip Kumar Swain, and Somsubhra Chakraborty

Abstract Soil organic carbon (SOC) is essential for understanding the global carbon cycle and promoting sustainable land management. However, traditional methods for SOC mapping are often costly, time-consuming, and limited to small-scale applications. Digital soil mapping (DSM) now offers a transformative approach, integrating remote sensing techniques and machine learning algorithms to enable efficient SOC monitoring. This study highlights hyperspectral remote sensing, combining data from spaceborne, airborne, and UAV platforms with field and laboratory data to produce SOC maps of higher spatial, spectral, and temporal resolutions. Advanced analytical methods such as multiple linear regression (MLR), partial least squares regression (PLSR), regression-kriging, and machine learning models, including random forest (RF), support vector machines (SVM), and artificial neural networks (ANN), enhance SOC prediction accuracy and reduce reliance on conventional methods. Additionally, using proximal sensors and spectroradiometers provides flexibility, allowing applications across diverse agroecological settings. This multimodal approach not only refines SOC mapping but also supports sustainable carbon sequestration and adaptive land management practices. As a result, the approach strengthens SOC monitoring and supports broader environmental planning. Future trends in this field emphasise integrating sensor fusion and mobile applications, making SOC monitoring more accessible for real-time decision-making and adequate policy support. Ultimately, the study demonstrates that DSM-based SOC mapping offers a scalable, adaptable solution for environmental sustainability in an increasingly climate-sensitive world.

Keywords Soil organic carbon · Digital soil mapping · Hyperspectral remote sensing · Machine learning · Carbon sequestration

A. Nayak · R. Singh · U. Dam · M. Sarkar · D. K. Swain · S. Chakraborty (✉)
Agricultural and Food Engineering Department, Indian Institute of Technology Kharagpur, Kharagpur, India
e-mail: somsubhra@agfe.iitkgp.ac.in

D. Kumar Swain et al. (eds.), *Climate-Smart Technologies for Sustainable Agricultural Production*, https://doi.org/10.1007/978-3-032-25666-9_2

2.1 Introduction

Soil Organic Carbon (SOC) is vital in maintaining ecological balance, supporting agricultural productivity, and mitigating climate change. It is a significant component of Soil Organic Matter (SOM), which consists of materials at various stages of decomposition, including plant and animal residues, fine organic particles, and a wide range of soil organisms. Since SOM contains roughly 55–60% carbon by mass, SOC represents a significant fraction of it and is often used as a key indicator of soil health. Beyond its role in soil fertility and structure, SOC is central to the global carbon cycle, the continuous exchange of carbon among the soil, vegetation, oceans, and atmosphere. The SOC pool stores an estimated 1500 PgC in the top 1 m of soil, which is more than is held in the atmosphere (about 800 PgC) and terrestrial vegetation (500 PgC) combined (FAO and ITPS 2015). This vast reservoir makes SOC a critical link between soil health and broader ecological and climatic processes. By regulating carbon fluxes, SOC helps offset greenhouse gas emissions and mitigates climate change. At the same time, adequate SOC levels improve soil structure, increase water-holding capacity, and enhance nutrient availability. Hence, SOC is a critical parameter that needs to be monitored regularly to assess the health of the agricultural ecosystem (Weindorf and Acosta Martinez 2025). This chapter will discuss the procedures, algorithms, and technologies used for estimating soil organic carbon. While some widely used gold-standard laboratory techniques are discussed, the major focus will be on remote sensing technologies for estimating SOC.

2.1.1 Laboratory Methodologies

Laboratory methods, such as dry combustion using elemental analysers and wet oxidation techniques, are widely used to quantify soil organic carbon. Among these, the Walkley-Black method, developed by A. Walkley and I. A. Black, is one of the most widely used chemical approaches, particularly in agricultural studies (Walkley and Black 1934). This method involves the wet oxidation of organic carbon with potassium dichromate ($K_2Cr_2O_7$) in the presence of concentrated sulphuric acid (H_2SO_4). The underlying chemical principle involves the oxidation of organic carbon by dichromate ions, as shown in Eq. (2.1).

$$2Cr_2O_7^{(2-)} + 3C + 16H^+ \rightarrow 4Cr^{(3+)} + 3CO_2 + 8H_2O \tag{2.1}$$

Followed by a back-titration of the excess dichromate with a ferrous ammonium sulphate solution Eq. (2.2)

$$Cr_2O_7^{(2-)} + 6Fe^{(2+)} + 14H^+ \rightarrow 2Cr^{(3+)} + 6Fe^{(3+)} + 7H_2O \tag{2.2}$$

The Walkley-Black method is favoured for its straightforwardness, cost-effectiveness, and practicality, especially in regions where advanced analytical instruments are not readily available. Despite its simplicity, one major issue is its incomplete oxidation of all organic carbon; a correction factor of 1.3 is often applied to account for this, assuming an average carbon recovery of 77%, which can introduce variability. Further drawbacks include interference from other reducing agents in the soil, such as chlorides and ferrous iron, which can lead to overestimating carbon content. Furthermore, the method involves hazardous chemicals, such as concentrated acids and carcinogenic hexavalent chromium, and requires strict safety protocols and proper waste disposal, making it impractical for real-time estimation. A detailed step-by-step guide to this procedure can be found in the analytical flowchart referenced in Fig. 2.1. Hence, due to these limitations, there is a pressing demand for rapid, non-invasive, and more cost-effective methods to measure SOC.

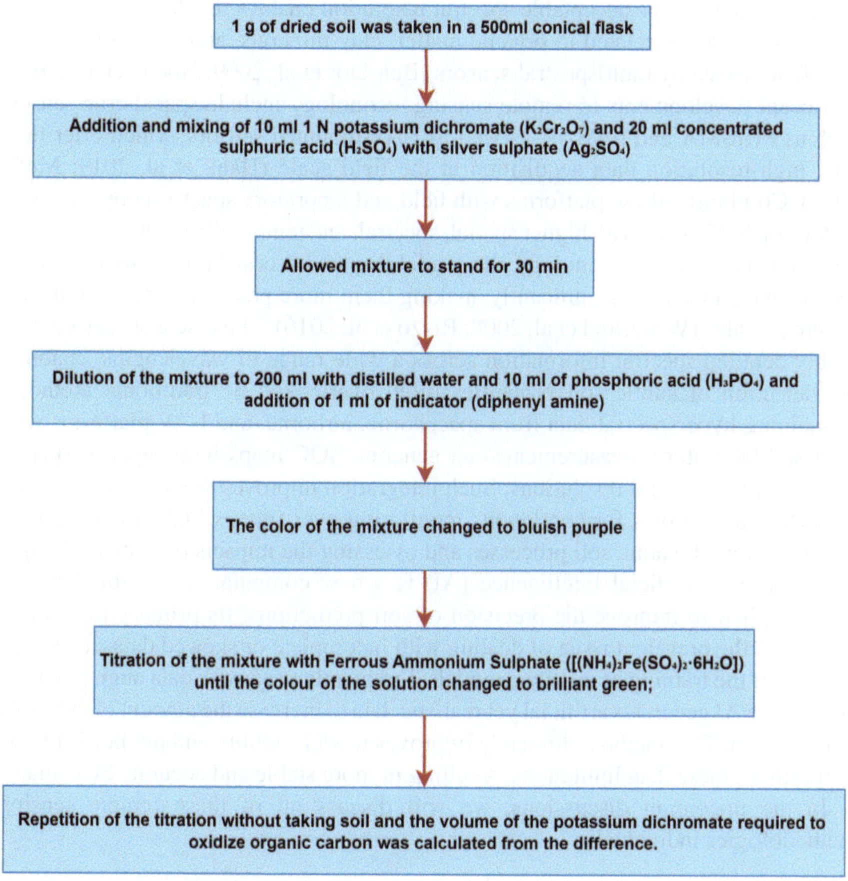

Fig. 2.1 A step-by-step depiction of the Walkley-Black method of soil organic carbon estimation

2.1.2 Remote Sensing Technologies

Modern techniques such as visible and near-infrared (vis–NIR) spectroscopy and machine learning provide a fast, non-destructive, and eco-friendly way to estimate SOC. Spectroscopy investigates how light interacts with matter, while spectrometry measures the light absorbed or emitted at different wavelengths. By analysing these patterns, we can determine the types and amounts of substances in soil, providing valuable information about its composition and properties. SOC prediction often involves complex, nonlinear relationships between soil reflectance and carbon content. For example, spectral absorptions at ~1730 nm and ~ 2100–2300 nm are linked to organic carbon functional groups, while indirect effects such as soil texture and moisture also influence reflectance (Baumgardner et al. 1986; Lobel and Ashner 2002; Curcio et al. 2013). Hyperspectral remote sensing, on the other hand, provides a unique opportunity to disentangle these interactions and quantify SOC more accurately. This acceptable spectral resolution enables the detection of subtle absorption features related to organic matter, clay minerals, and soil moisture that are often missed by multispectral sensors (Ben-Dor et al. 2009; Nocita et al. 2015).

Recent developments in remote sensing technology include spaceborne sensors such as PRISMA and EnMAP, as well as UAV-mounted sensors, which offer flexible, high-resolution data acquisition at the field scale (Hank et al. 2019; Mulla 2021). Combining these platforms with field and laboratory spectroscopy allows us to develop SOC maps with higher spatial, spectral, and temporal resolutions. With the rapid advancement of technology, the availability and accessibility of hyperspectral sensors have increased significantly, making them more practical for soil studies at different scales (Wetterlind et al. 2008; Rizzo et al. 2016). These sensors can capture highly detailed spectral information across a wide range of wavelengths, enabling the detection of subtle soil properties often overlooked by traditional methods. Combining hyperspectral data from spaceborne, airborne, and UAV platforms with field and laboratory measurements can generate SOC maps with superior spatial, spectral, and temporal resolutions. Such integration improves the precision of SOC prediction and allows for continuous monitoring over time, which is essential for understanding dynamic soil processes and assessing the impacts of climate change.

Generative Artificial Intelligence (AI) is a new computational method that is being applied to improve the precision of soil predictions. Its primary function is to address the prevalent issue of dealing with incomplete or skewed datasets, which can hinder the training of effective models. Frequently used with data augmentation, Generative AI generates artificial yet realistic data to increase the amount of available training data. This method ultimately improves model stability and predictability by overcoming these data limitations, resulting in more stable and accurate SOC maps.

In the upcoming discussions, we will discuss all of these remote sensing methodologies individually.

Fig. 2.2 Nix colour sensor

2.2 Remote Sensing Technologies

2.2.1 NixPro Colour Sensors

Colour sensors are advanced optical devices that detect and quantify colours by measuring the intensity of reflected or transmitted light across different wavelengths, often converting this information into standardised colour spaces such as RGB, LAB, HSV, or CMYK. They are widely used across industries, from printing, textiles, and paints to food quality assessment and environmental monitoring, where objective, reproducible colour evaluation is essential. The Nix Pro Colour Sensor is a compact, handheld tool designed for precise and consistent colour measurement (Fig. 2.2). Unlike human vision, which can be influenced by ambient light and perception biases, the Nix Pro uses an internal calibration system. It blocks external light to capture surfaces' true colours with high accuracy. It connects via Bluetooth to a smartphone or tablet and provides instant colour values in multiple colour models through the Nix app. This makes it particularly valuable for tasks such as paint matching, design, material analysis, and scientific applications like soil colour measurement, where accurate colour quantification can be linked to soil organic matter, texture, and mineral composition (Stiglitz et al. 2016; Mukhopadhyay et al. 2020; Swetha and Chakraborty 2021; Jha et al. 2021; Swetha et al. 2022; Carvalho et al. 2024; Cean et al. 2024; Alam et al. 2025). Its portability, user-friendliness, and reliability make it an indispensable colour-sensing tool for professionals and researchers.

2.2.2 Mobile Camera-Based RGB Sensors

A mobile camera-based RGB sensor uses a smartphone's built-in camera to capture and analyse colour information from objects or surfaces in terms of the three primary colour components—red, green, and blue (RGB). Modern smartphone cameras are equipped with highly sensitive image sensors and advanced image-processing algorithms, allowing them to record subtle variations in colour under different lighting conditions. Using specialised mobile applications, the raw pixel data from the camera can be processed to extract precise RGB values, which can then be converted into other colour spaces such as LAB, HSV, or HEX for more detailed analysis. This approach is cost-effective, portable, and widely accessible, making it popular in applications like digital design, agriculture, soil analysis, food quality assessment, and remote diagnostics (Han et al. 2016; Swetha et al. 2020; Gorthi et al. 2021; Lavanya et al. 2023; Nayak et al. 2023; Gain et al. 2024; Lavanya et al. 2024; Pate et al. 2025;). However, the accuracy of mobile camera-based RGB sensing can be influenced by factors such as illumination, camera quality, and white balance settings. Despite these limitations, with proper calibration and standardised lighting conditions, mobile cameras provide a practical alternative to dedicated colourimeters or spectrometers, enabling quick, on-the-go colour measurements directly from a device that most people already carry.

2.2.3 Hyperspectral Sensors

A hyperspectral sensor is a specialised optical device designed to capture detailed information from across the electromagnetic spectrum by dividing light into many narrow, continuous wavelength bands. Unlike standard RGB sensors, which measure only three broad bands (red, green, and blue), hyperspectral sensors can record hundreds of bands, often covering regions from the visible (400–700 nm) to near-infrared (700–1000 nm), shortwave infrared (1000–2500 nm), and sometimes even thermal regions. Each pixel captured by a hyperspectral sensor contains a complete reflectance spectrum, known as a "spectral signature," which provides unique information about the chemical composition, physical structure, and condition of materials.

2.2.3.1 Spectroradiometers

Spectroradiometers are high-precision optical devices used to measure the intensity of electromagnetic radiation over a continuous spectrum as a function of wavelength. It works by gathering light reflected or emitted from a surface, passing it through a monochromator that separates the incoming radiation into its constituent wavelengths, and then using extremely sensitive detectors to measure the intensity. A detailed spectral fingerprint of the target material is provided by spectroradiometers, which, in contrast to broadband sensors, record reflectance data in high-resolution continuous spectral bands (Fig. 2.3) (typically 1–10 nm wide).

When applied to soil samples, spectroradiometers are typically used in the visible (Vis, 350–700 nm), near-infrared (NIR, 700–1000 nm), and shortwave infrared

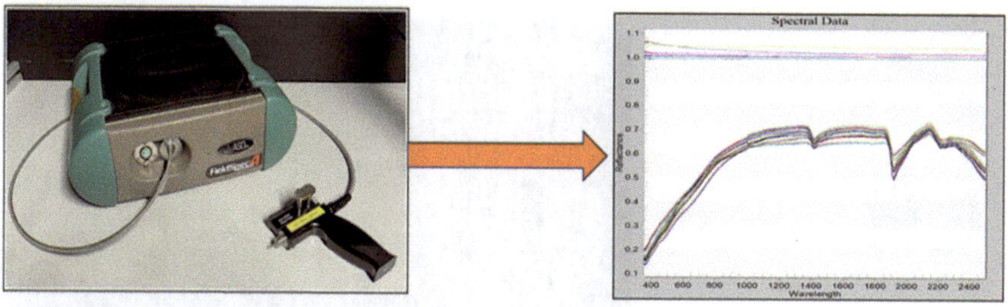

Fig. 2.3 Spectroradiometer

(SWIR, 1000–2500 nm) regions. These regions contain diagnostic absorption features related to key soil attributes:

- Organic matter (broad absorptions around 1900–2100 nm),
- Moisture (clear absorption between 1400 and 1900 nm),
- Clay minerals (Al–OH, Mg–OH, and Si–OH vibrations at 1400–2200 nm),
- Iron oxides (strong absorptions around 400–1000 nm),

Spectroradiometers allow the identification of chemical bonds, mineralogical composition, and indicators of soil properties. DSM, precision agriculture, soil science, and environmental monitoring all make extensive use of measurements due to their speed and non-destructive nature.

Essentially, spectroradiometers are hyperspectral sensors that are based on the ground. They bridge the gap between laboratory spectroscopy and field/airborne imaging, making them an effective tool for precise and economical soil characterisation and fertility evaluation.

2.2.3.2 Hyperspectral Cameras

Hyperspectral imaging (HSI) refers to the process of collecting and analysing this spectral information to identify and characterise materials based on their unique spectral signatures. It combines the advantages of imaging (spatial information) with spectroscopy (spectral information), enabling both the visualisation of objects and the identification of their composition. In soil science, hyperspectral imaging can help to estimate soil organic carbon, moisture, and mineral content by analysing subtle variations in reflectance across different wavelengths (Tahmasbian et al. 2018; Malmir et al. 2019; Xu et al. 2021). The strength of HSI lies in its ability to detect what is not visible to the human eye, making it a powerful non-destructive tool for research, monitoring, and decision-making across scientific and industrial fields.

It can be performed using different types of sensors depending on the scale and purpose of the study, ranging from ground-based systems to airborne and satellite platforms (Fig. 2.4).

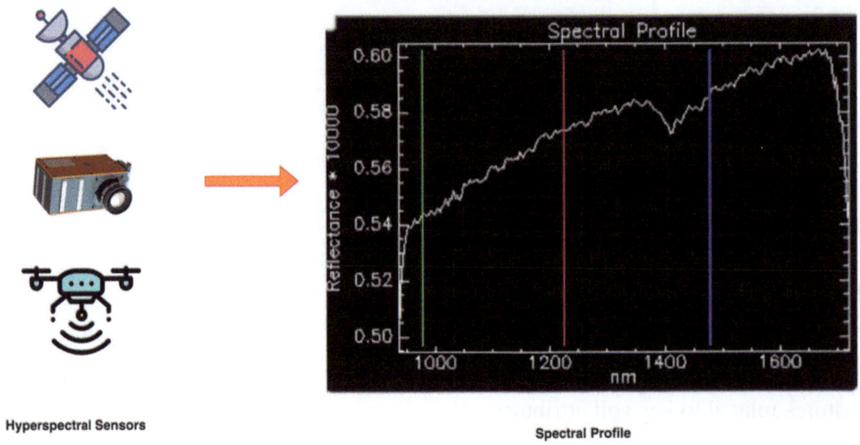

Fig. 2.4 Hyperspectral imaging camera

Ground-Based Hyperspectral Imaging

Ground-based hyperspectral imaging is typically conducted in laboratories or in the field with handheld or mounted hyperspectral cameras and spectroradiometers. These setups provide high spatial and spectral resolution under controlled conditions, making them ideal for soil analysis, crop monitoring, and material studies. Ground-based systems minimise atmospheric effects and allow precise calibration, which is why they are widely used in soil science to measure soil organic carbon, moisture, or nutrient status (Tahmasbian et al. 2018; Malmir et al. 2019).

Airborne Hyperspectral Imaging

Airborne hyperspectral imaging uses sensors mounted on drones or aircraft. These systems cover larger areas than ground-based sensors while maintaining fine resolution. They are widely applied in precision agriculture, forestry, mineral exploration, and environmental monitoring (Hively et al. 2011; Hbirkou et al. 2012; Tan et al. 2020). UAV-mounted hyperspectral cameras, in particular, are flexible and cost-effective for localised studies.

Satellite-Based Hyperspectral Imaging

Satellite-based hyperspectral imaging operates at regional to global scales. Equipped with advanced spaceborne hyperspectral sensors, satellites provide broad coverage and enable long-term monitoring of Earth's surface. Although their spatial resolution is often lower than ground or airborne systems, they are invaluable for large-scale applications such as climate studies, land-use change detection, water quality monitoring, and global soil and vegetation mapping.

2.2.4 Fourier Transform Infrared (FTIR) Spectroscopy

A fast, analytical technique used to identify and quantify chemical substances by measuring how infrared light is absorbed by a sample. FTIR spectroscopy utilises mid-infrared radiation (4000–400 cm^{-1}) to interrogate soil samples (Fig. 2.5), revealing specific absorption bands associated with organic matter, clay minerals, and functional groups (Margenot et al. 2023).

It relies on the fact that molecular bonds absorb infrared radiation at specific vibrational frequencies. When infrared light is scanned over a sample, specific wavelengths are absorbed, and molecular bonds are stretched, bent, or twisted. Instead of measuring each wavelength individually, FTIR uses an interferometer, a standard Michelson interferometer, to modulate a broad range of IR wavelengths simultaneously, producing an interferogram. This interferogram is a sum signal containing information on all the frequencies of IR light absorbed by the sample. The Fourier Transform is subsequently applied to the interferogram to convert it into a standard IR spectrum, which presents absorbance (or transmittance) as a function of wavenumber (cm^{-1}). The Fourier Transform used in FTIR spectroscopy to convert the time-domain interferogram into a frequency-domain spectrum is given by Eq. (2.3):

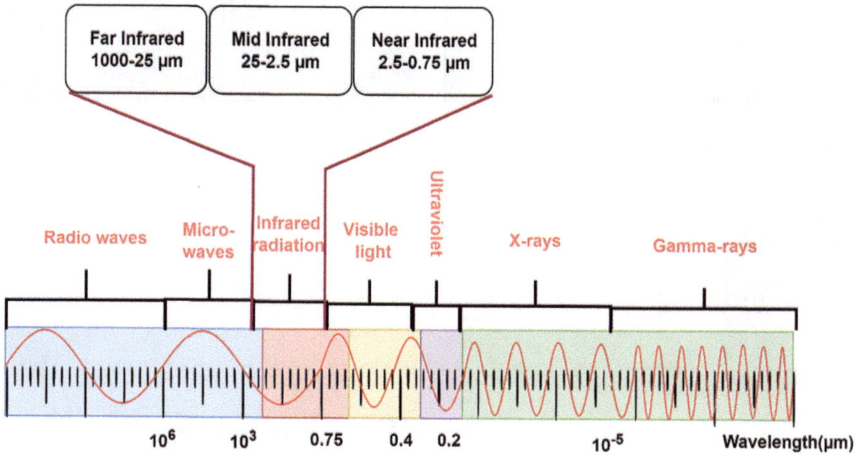

Fig. 2.5 Electromagnetic spectrum highlighting the infrared region

$$F(v) = \int_{-\infty}^{\infty} f(t).e^{-2\pi ivt}.dt \tag{2.3}$$

Where,

$F(v)$	Frequency-domain spectrum (wavenumber domain)
$f(t)$	Time-domain signal (interferogram)
v	Frequency or wavenumber
t	Time or optical path difference
$e^{-2\pi ivt}$	Exponential kernel function for thc transformation

A FTIR spectrometer equipped with a transmittance accessory was used to record spectral data (Fig. 2.6) over the mid-IR range (4000–400 cm^{-1}). Each spectrum was measured by averaging 24 scans at 4 cm^{-1} resolution in order to reduce random noise. Soil samples were mixed with potassium bromide (KBr) at a 1:9 ratio and pressed into pellets at high pressure using a hydraulic press prior to analysis. The resulting spectral signatures were acquired in the software, and baseline correction was applied to improve spectral quality (Fig. 2.7). Steps for collecting soil sample data using FTIR Spectroscopy are outlined in Fig. 2.8.

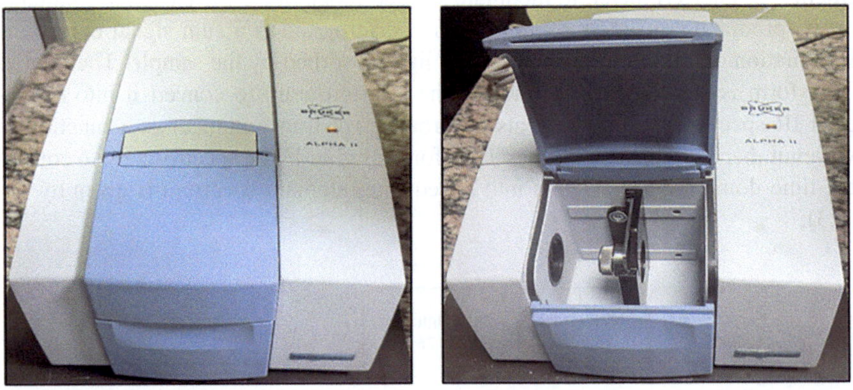

Fig. 2.6 FTIR spectroscopy

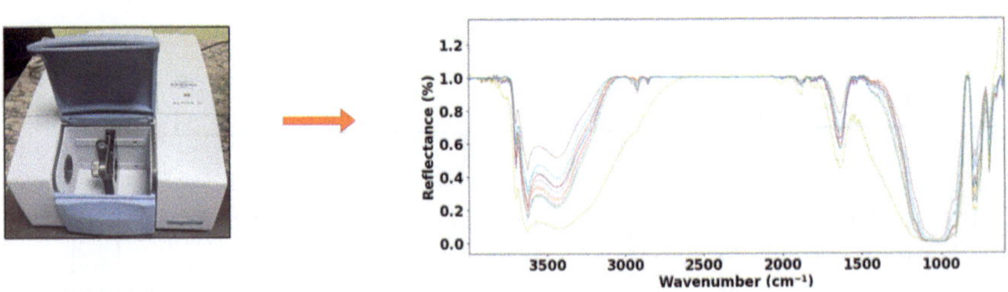

Fig. 2.7 Illustration of FTIR spectra for 10 different soil samples

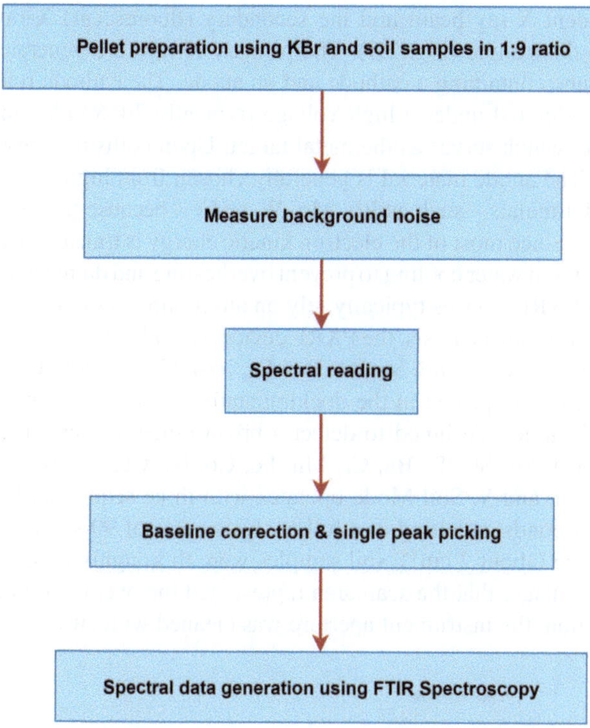

Fig. 2.8 Flowchart explaining the sample analysis procedure with FTIR

2.2.5 Portable X-Ray Fluorescence (PXRF) Spectrometry

X-ray fluorescence (XRF) spectrometry has been applied in soil studies since the 1930s (Jones 1982). With technological improvements, portable X-ray fluorescence (PXRF) devices have emerged, offering key advantages such as minimal sample preparation, rapid and non-destructive analysis, high throughput, accurate elemental detection, low operational costs, and in-situ applicability (Zhu et al. 2011; Ulmanu et al. 2011; Weindorf et al. 2014; Cean et al. 2024; Da Costa et al. 2024; Dasgupta et al. 2024; Alam et al. 2025). PXRF has been extensively employed to evaluate heavy metal contamination (Weindorf et al. 2012a; Weindorf et al. 2013b; Parsons et al. 2013; Hu et al. 2014; Paulette et al. 2015). Moreover, PXRF-derived elemental data have served as proxies for various soil properties, including pH (Sharma et al. 2014a), cation exchange capacity (Sharma et al. 2014b), calcium and gypsum contents (Zhu and Weindorf 2009; Weindorf et al. 2013a), soil texture (Zhu et al. 2011; Mancini et al. 2024), salinity (Swanhart et al. 2014), and soil horizonation (Weindorf et al. 2012b; Weindorf et al. 2012c).

PXRF spectrometry enables the detection and quantification of numerous elements, ranging from magnesium (Mg) to uranium (U), based on the interaction between an incident X-ray beam and the secondary (fluorescent) X-rays emitted by atoms within the sample (Fig. 2.9). The primary X-rays are generated inside a vacuum-sealed tube containing a cathode and an anode. The cathode releases electrons that are accelerated under a high voltage (typically 20–60 kV) and directed toward the anode, which serves as the metal target. Upon collision, these electrons produce X-rays. The anode material is generally chosen from high atomic number, high melting point metals—such as Rh, Mo, W, or Cr—because they yield intense primary radiation. Since most of the electron kinetic energy is transformed into heat, older systems required water cooling to prevent overheating and damage to the anode, whereas modern PXRF devices typically rely on air-cooling systems.

Before conducting soil analysis, the PXRF device was calibrated using a 316 alloy clip containing 16.13% Cr, 1.78% Mn, 68.76% Fe, 10.42% Ni, 0.20% Cu, and 2.10% Mo, which was securely placed in the docking station (Fig. 2.10). The instrument was run in "Soil Mode," designed to detect a broad range of elements, including Sr, Zr, Mo, Ag, Cd, Sn, Sb, Ti, Ba, Cr, Mn, Fe, Co, Ni, Cu, Zn, Hg, As, Se, Pb, Rb, P, S, Cl, K, Ca, and V. Soil Mode operates with three sequential beams, each scanning for 30 seconds, resulting in a full analysis cycle of 90 s. Given the small PXRF window size (about 2 cm^2), soil samples were thoroughly mixed to achieve homogeneity and ensure that the scan area represented the overall composition. To avoid contamination, the instrument aperture was cleaned with air after every scan.

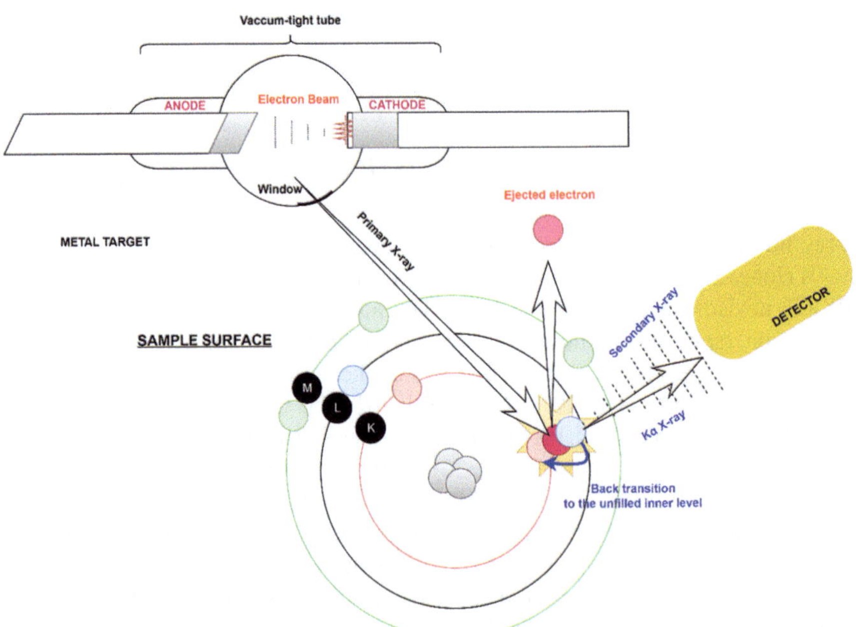

Fig. 2.9 Fundamentals of X-ray fluorescence: interaction of primary X-rays with sample surface leading to electron excitation and emission of secondary X-rays (fluorescence)

Fig. 2.10 Sample scanning using PXRF in the field

The recorded elemental data were stored in the onboard system and subsequently exported to MS Excel for further processing.

2.2.6 Sensor Fusion

The integration of multiple data sources, commonly referred to as sensor fusion, has emerged as a promising frontier in SOC mapping. By combining hyperspectral imagery, proximal sensing, terrain attributes, and climatic variables, sensor fusion enables the capture of both direct and indirect indicators of SOC (Mulder et al. 2011; Mahmood et al. 2012; Veum et al. 2017). For example, UAV-based hyperspectral data provide detailed field-level insights, while spaceborne observations extend spatial coverage; when fused together, these multi-scale datasets produce more accurate and generalizable SOC predictions. The effectiveness of sensor fusion is further enhanced through advanced machine learning algorithms capable of processing heterogeneous datasets. Machine learning and Neural Network algorithms can integrate spectral, environmental, and spatial covariates simultaneously, refining SOC predictions beyond what single-source datasets can achieve (Kalopesa et al. 2025; Mundada and Jain 2025). Ultimately, this multimodal approach not only improves prediction accuracy but also supports sustainable carbon sequestration strategies, adaptive land management practices, and broader environmental planning initiatives.

2.3 Analytical Techniques for Remote Sensing

The raw spectral and elemental data collected from remote and proximal sensors do not directly correspond to SOC values. To turn all of this data into accurate and useful SOC estimates, you need strong analytical methods. These methods are meant to model the complicated and nonlinear connections that often exist between the sensor readings and the real soil properties. Choosing the right analytical model is very important for making accurate SOC maps that can be used for environmental monitoring and land management, which is good for the environment.

2.3.1 Algorithms

The accurate translation of spectral data into SOC estimates requires robust analytical models. Traditionally, MLR has been applied, but its assumptions of linearity limit its performance with complex spectral data. A more widely adopted method is PLSR, which effectively handles collinear spectral bands and has been extensively used for SOC estimation. To further enhance mapping, regression-kriging integrates regression models with geostatistical interpolation, capturing both deterministic and spatially autocorrelated variability.

However, because SOC-reflectance relationships are rarely linear, advanced machine learning models have shown superior performance. Models such as RF, SVM, and ANN excel at handling high-dimensional, nonlinear spectral data. RF provides robust predictions while reducing overfitting, SVM is effective in high-dimensional feature spaces, and ANNs can capture complex hierarchical interactions. Numerous studies confirm that these machine learning models consistently outperform conventional regression approaches in SOC prediction accuracy.

2.3.2 Global to Local Models

A key strategy in modern digital soil mapping involves leveraging both global and local models to balance broad applicability with high local accuracy. This dual approach allows for a scalable yet precise method for creating detailed soil maps, particularly for crucial properties like SOC.

Global models are developed using large, harmonised datasets, such as extensive soil spectral libraries, to generate SOC predictions that are valid across diverse geographical regions. These models are foundational for large-scale environmental monitoring and climate change studies, providing a broad understanding of soil characteristics over vast areas.

However, a significant drawback of global models is that they often yield biased and inaccurate predictions at a local or field-specific scale. This occurs because the immense diversity in soil mineralogy, land management practices, and environmental conditions within a global library can introduce significant variability that doesn't accurately reflect the more uniform conditions of a specific site. As a result, their direct application for precision tasks is often limited.

Local models, in contrast, are tailored to specific regions or even individual fields. By training or fine-tuning a model with site-specific data—often gathered from sources like UAV-mounted sensors or ground-based spectroscopy—much higher prediction accuracy can be achieved. This approach is crucial for applications like precision agriculture, where detailed, field-level information is essential for optimising resource management, improving crop yields, and making informed decisions. The primary limitation is the cost and time required to collect enough local data to build a robust model from scratch.

To bridge the gap between the broad scope of global models and the precision of local ones, innovative methods like the GLOBAL-LOCAL approach have been developed. This technique leverages the extensive information within large soil spectral libraries to create spectroscopic models that can be accurately applied at a local level. The process, often called "spiking," enhances a global library with a small amount of local data to create a more accurate, localised model.

The process typically involves these steps:

1. **Representative Local Data**: A small, representative set of soil samples (often called a "Lab" set) is collected from the target local site and analysed using conventional methods.
2. **Identifying Spectral Neighbours**: Using statistical techniques like Principal Component Analysis (PCA) and k-nearest neighbours, the most spectrally similar samples ("neighbours") to the local "Lab" set are identified from the large global library. This step effectively mines the global database for soils that are most like the target site.
3. **Model Calibration**: A new, hybrid calibration model is developed using these selected neighbours from the global library. The model's accuracy is initially tested by predicting the SOC of the "Lab" set.
4. **Final Model Creation**: The local "Lab" set is then added to (or "spiked" into) the set of selected neighbours to create the final, refined GLOBAL-LOCAL model. This model can then be used to make highly accurate predictions for the remaining unknown samples at the local site.

This hybrid approach has been shown to be significantly more accurate than using a global model alone and can perform as well as, or even better than, models built exclusively with a large number of site-specific samples. A key advantage of the GLOBAL-LOCAL method is that it drastically reduces the need for extensive and costly local soil analysis, as only a small number of representative samples are required. By effectively mining rich information from large spectral libraries, this method provides a cost-effective, powerful tool for improving the accuracy of local soil predictions (Fig. 2.11).

2.3.3 Generative AI

Generative Artificial Intelligence (AI) represents an emergent computational method for improving the precision of soil predictions. It addresses a perennial challenge in research: incomplete, limited, or skewed datasets, which can significantly hinder model training and performance. To mitigate this issue, Generative AI uses a process called data augmentation. This technique involves training an algorithm to learn the

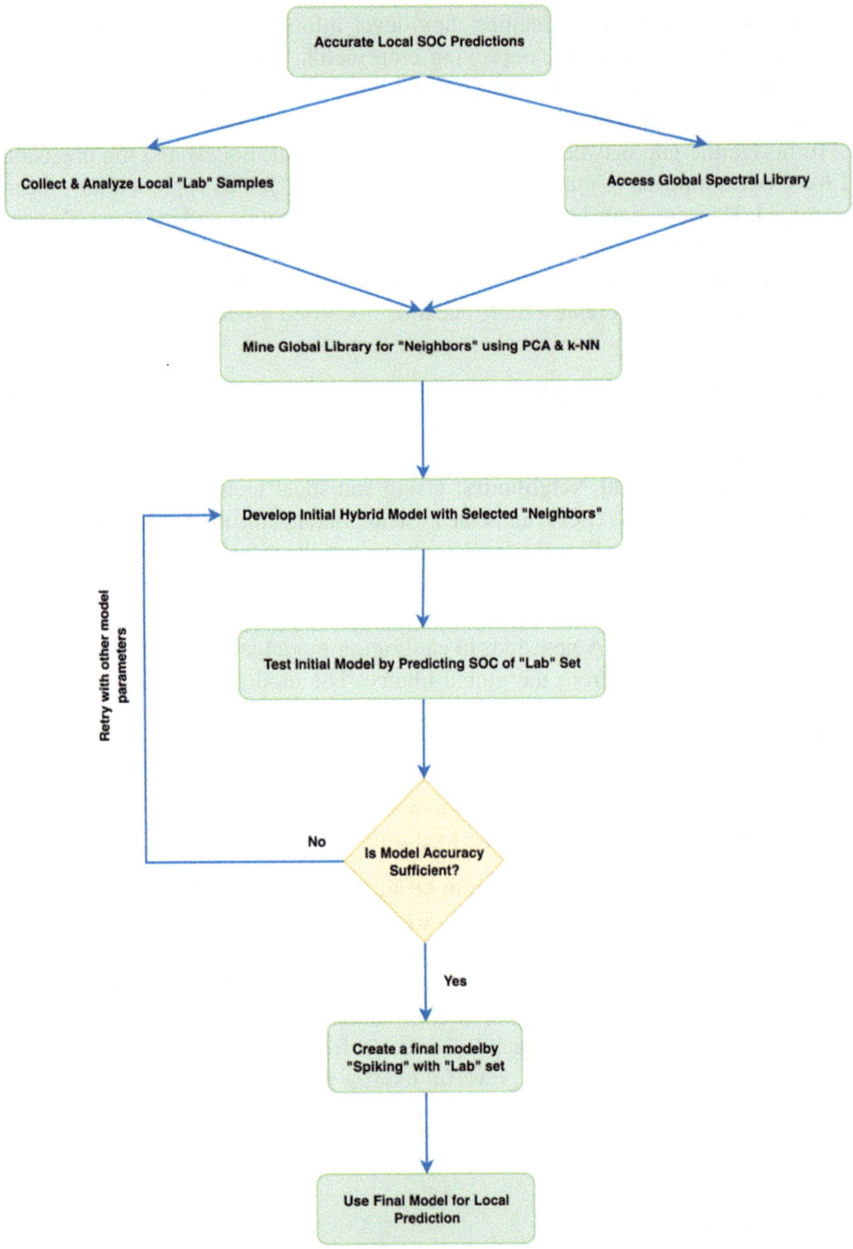

Fig. 2.11 Flowchart illustrating the detailed steps of the global-to-local approach

underlying distribution of an existing dataset and then generating new, synthetic data points that are statistically consistent with the original samples.

A prominent example of this is the Triplet Variational Autoencoder (TVAE), an algorithm specifically designed to model the complex joint distribution of multivariate tabular data, such as soil spectral measurements. The synthetic data generated by TVAE maintains the statistical characteristics of the original dataset. By supplementing the training set with synthetic data, the models can be trained on a more comprehensive and balanced dataset, an effective strategy for addressing data scarcity and ultimately improving model robustness and generalizability.

Beyond VAEs, Generative Adversarial Networks (GANs) offer another powerful approach. The Conditional Tabular GAN (CTGAN), for example, is tailored to handle the complexities inherent in soil datasets, including mixed data types and non-Gaussian distributions. CTGAN utilises an adversarial training process with two neural networks: a generator that produces synthetic data and a discriminator that learns to distinguish between real and synthetic samples. This iterative competition compels the generator to produce high-fidelity synthetic data that preserves complex correlations between variables. While VAEs typically offer more stable training, GANs may generate more realistic samples, and the selection between them is contingent upon the specific characteristics of the dataset. The application of these generative models to augment datasets thus signifies a substantial advancement in predictive soil science.

The empirical evidence underscores the efficacy of this approach. Tables 2.1 and 2.2 present a comparative analysis of model performance before and after data augmentation.

Table 2.1 Model performance before data augmentation

Model	R^2	MSE (%)	RMSE (%)
Random forest	0.529	0.078	0.280
SVM	0.560	0.073	0.271
XGBoost	0.502	0.083	0.288
Gradient boosting	0.545	0.076	0.275
Neural network	0.550	0.075	0.273
Stacking	0.570	0.071	0.267

Table 2.2 Model performance after augmentation (20% synthetic data)

Model	R^2	MSE (%)	RMSE (%)
Random forest	0.730	0.043	0.206
SVM	0.650	0.054	0.233
XGBoost	0.690	0.048	0.220
Gradient boosting	0.654	0.054	0.232
Neural network	0.712	0.045	0.211
Stacking	0.695	0.047	0.218

The quantitative results from this approach are compelling. Prior to augmentation, the Stacking ensemble model demonstrated the highest predictive performance, achieving an R^2 of 0.570. However, after supplementing the training data with a modest 20% of synthetic samples, a universal and significant improvement was observed across all evaluated models. The most pronounced enhancement occurred with the Random Forest model. Its performance escalated from an R^2 of 0.529 to 0.730, an increase of nearly 38% in explained variance. This was complemented by a substantial reduction in prediction error, as its Root Mean Square Error (RMSE) decreased from 0.280 to 0.206%. These findings illustrate that generative data augmentation provides more than an incremental benefit; it can fundamentally elevate a model's predictive power, enabling the development of highly accurate SOC maps from datasets that would otherwise be considered limited. A textual diagram outlining the process of data augmentation using Generative AI for improved SOC prediction is shown in Fig. 2.12.

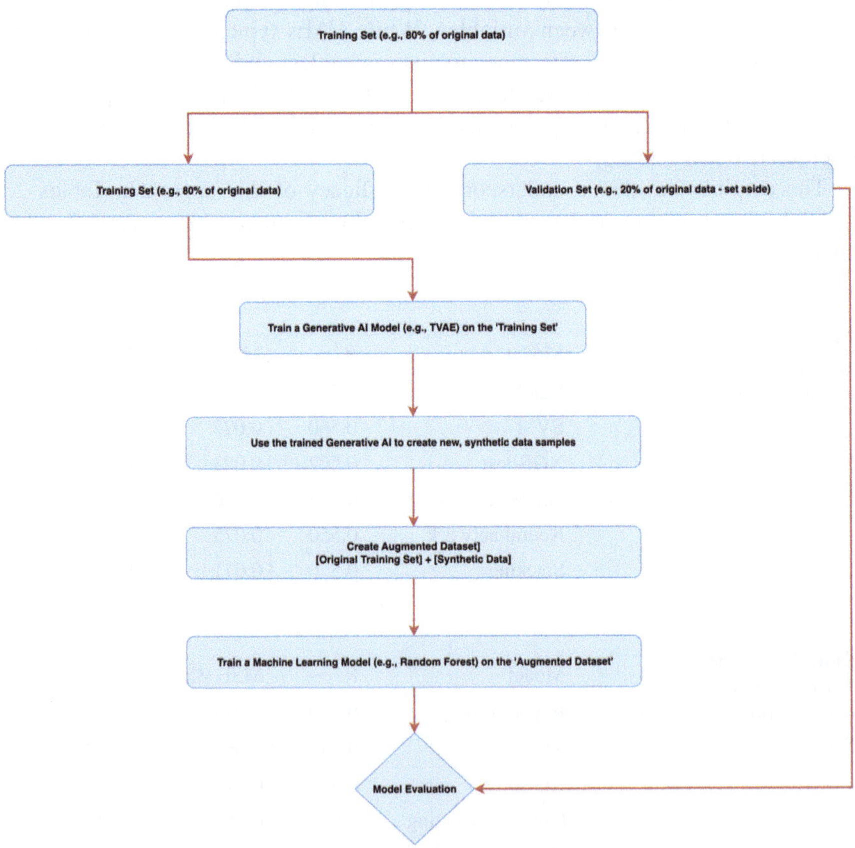

Fig. 2.12 Data augmentation using Generative AI for improved SOC prediction

2.4 Digital Soil Mapping

From a science-driven technology in the early 1990s, DSM has developed into a fully functional and operational procedure for measuring and assessing soil in space today. In contemporary soil science, DSM has become a game-changer, providing a methodical and effective way to forecast soil characteristics over time and space. By combining field measurements, lab data, and sophisticated computational modelling to produce spatially explicit maps of SOC, DSM offers a useful substitute. Monitoring SOC is one of its most important uses. The DSM approach for predictive modelling (and mapping) of soil was presented in the McBratney et al. (2003) paper. This method has its roots in earlier research by Jenny (1941) and a Russian soil scientist, Dokuchaev. SCORPAN is a mnemonic for the following factors that are used to predict soil attributes: soil, climate, organisms, relief, parent materials, age, and spatial position. The following formula forms the SCORPAN approach Eq. (2.4):

$$S = f(s, c, o, r, p, a, n) + \varepsilon \tag{2.4}$$

Remote sensing is a cornerstone of DSM because it provides wide-area coverage and repeated measurements of the land surface. Multispectral and hyperspectral imagery capture reflectance patterns that are closely linked to soil organic matter, mineral composition, and vegetation status, making them valuable for SOC estimation. Microwave radar data provide additional insights into soil moisture, texture, and surface roughness, which are critical variables for characterising soil heterogeneity. Light Detection and Ranging (LiDAR) provides additional detail by generating high-resolution elevation models that reveal terrain features, such as slope, drainage, and microtopography, that influence SOC dynamics. When combined, these remote sensing products complement each other by capturing both spectral and structural properties of soils and landscapes. Their integration within DSM frameworks substantially improves the accuracy and scalability of SOC prediction across diverse agro-ecosystems.

The accuracy of DSM predictions has been greatly improved through the integration of machine learning and deep learning methods, including Random Forest, Gradient Boosting, Support Vector Regression, Cubists, and Neural Networks. These approaches are effective in capturing nonlinear interactions and complex dependencies between SOC and its controlling factors. Moreover, the inclusion of auxiliary variables, such as terrain, climate, vegetation indices, and land-use information, provides additional explanatory power, thereby enhancing model reliability at broader spatial scales. Additionally, ancillary data such as topography, climate, vegetation indices, and land use markedly enhance SOC prediction accuracy by capturing landscape-scale variability.

Recent progress in proximal soil sensing technologies, particularly FTIR, Vis–NIR spectroscopy, and pXRF spectrometry, has considerably strengthened DSM applications by enabling fast, cost-effective, and on-site assessment of soil properties. In parallel, innovative techniques such as data augmentation and generative artificial intelligence (AI) are increasingly adopted to address the limitations of datasets that are limited, scattered, and uneven, thereby improving model stability and predictive power. DSM now plays a crucial role in generating accurate SOC maps across different spatial scales, which are essential for carbon pooling, soil quality monitoring, and sustainable farming practices (Dasgupta et al. 2023; Das et al. 2025). Its utility is especially evident in precision agriculture, where site-specific soil information can guide input optimisation, improve resource-use efficiency, and enhance yield. Moreover, DSM contributes to global initiatives in carbon sequestration assessment and climate change mitigation, highlighting its value as a modern, transformative approach in soil science (McBratney et al. 2003; Grunwald 2009; Rossel et al. 2016).

2.5 Future Trends and Technological Innovations

Future trends in SOC mapping emphasise the integration of real-time monitoring, mobile applications, and cloud-based platforms. The miniaturisation and cost reduction of hyperspectral sensors are making them increasingly accessible for UAV-based and even handheld applications. Coupled with advances in cloud computing and geospatial data platforms such as Google Earth Engine, SOC monitoring is becoming more scalable and user-friendly (Hengl et al. 2017; Pizarro et al. 2023).

Emerging directions also include the use of deep learning for spectral analysis, which has shown promise in capturing complex nonlinear relationships and feature extraction in high-dimensional data (Padarian et al. 2019); Nguyen et al. 2020). The fusion of hyperspectral sensing with other modalities such as LiDAR, SAR, and thermal imagery may further enhance SOC predictions by incorporating structural and moisture-related information (Castaldi et al. 2016; Forkuor et al. 2017).

Finally, policy-driven frameworks encourage harmonisation of global soil spectral libraries, such as the International Soil Spectroscopy Network (ISSN), which aims to standardise data sharing and model development (Rossel et al. 2016; Arrouays et al. 2020). Such collaborative initiatives will be essential for achieving global-scale SOC monitoring in an increasingly climate-sensitive world.

2.6 Conclusion

DSM offers a scalable, adaptable, and innovative solution for SOC monitoring by integrating hyperspectral sensing, a proximal spectroradiometer, multimodal sensor fusion, and advanced machine learning models. With the increasing accessibility of hyperspectral technologies and the power of data-driven algorithms, DSM provides detailed and accurate SOC maps across multiple spatial and temporal scales. Beyond scientific advancement, this approach directly contributes to climate change mitigation, sustainable land management, and global carbon sequestration initiatives. As technology continues to evolve, integrating mobile and real-time applications will make SOC monitoring more accessible and impactful, strengthening its role in both environmental planning and policy-making. Ultimately, DSM-based SOC mapping stands as a cornerstone of environmental sustainability in the twenty-first century.

References

Alam SS, Chakraborty S, Jain FC, Deb S, Singh R, Weindorf DC (2025) Proximal sensor integration for land use classification and soil analysis in a coastal environment. Case Stud Chem Environ Eng 11:101079

Arrouays D, Minasny B, Hengl T, McBratney AB (2020) Soil data and information systems. Adv Agron 161:179–229

Baumgardner MF, Silva LF, Biehl LL, Stoner ER (1986) Reflectance properties of soils. Adv Agron 38:1–44

Ben-Dor E, Chabrillat S, Demattê JAM, Taylor GR, Hill J, Whiting ML, Sommer S (2009) Using imaging spectroscopy to study soil properties. Remote Sens Environ 113:S38–S55

Carvalho GS, Weindorf DC, Mona-liza CS, Ribeiro BT, Chakraborty S, Li B, Weindorf WC, Acree A, Guilherme LRG (2024) Prediction of compost organic matter via color sensor. Waste Manag 185:55–63

Castaldi F, Hueni A, Chabrillat S, Ward K, Buttafuoco G, Bomans B, van Wesemael B (2016) Evaluating the capability of satellite hyperspectral missions to estimate soil organic carbon in croplands. Remote Sens Environ 179:54–65

Cean R, Mancini M, Silva SHG, Bakr N, Chakraborty S, Li B, Acree A, dos Anjos LHC, Curi N, Weindorf DC (2024) Proximal sensor characterization of Haitian agricultural soils: a case study in the central plateau. J S Am Earth Sci 143:105007

Curcio D, Ciraolo G, D'asaro F, Minacapilli M (2013) Prediction of soil texture distributions using VNIR-SWIR reflectance spectroscopy. Procedia Environ Sci 19:494–503

Das A, Mishra MK, Chakraborty S, Bhattacharya BK, Dave R, Kumar D, Patel K, Setia R, Weindorf DC (2025) Deep carbon: a multiscale feature-time fusion approach for field level digital soil organic carbon mapping. Eur J Soil Sci 76(4):e70161

Da Costa MV, de Souza Costa ET, de Oliveira JPD, de Oliveira Lima GJ, Guilherme LRG, Carvalho GS, Duarte MH, Chivale JJ, Weindorf DC, Chakraborty S, Ribeiro BT (2024) Assessment of coffee leaves nutritive value via portable X-ray fluorescence spectrometry and machine learning algorithms. Spectrochim Acta B At Spectrosc 219:106996

Dasgupta S, Debnath S, Das A, Biswas A, Weindorf DC, Li B, Shukla AK, Das S, Saha S, Chakraborty S (2023) Developing regional soil micronutrient management strategies through ensemble learning based digital soil mapping. Geoderma 433:116457

Dasgupta S, Pate S, Rathore D, Divyanth LG, Das A, Nayak A, Dey S, Weindorf DC, Li B, Silva SHG, Ribeiro BT, Srivastava S, Chakraborty S (2024) Soil fertility prediction using combined USB-microscope based soil image, auxiliary variables, and portable X-ray fluorescence spectrometry. Soil Adv 2:100016

FAO & ITPS (2015) Status of the world's soil resources (SWSR): Main report. FAO, Rome

Forkuor G, Hounkpatin OK, Welp G, Thiel M (2017) High resolution mapping of soil properties using remote sensing variables in South-Western Burkina Faso: a comparison of machine learning and multiple linear regression models. PLoS One 12(1):e0170478

Gain H, Patil RN, Malik K, Das A, Chakraborty S, Banerjee J (2024) Image processing and impact analyses of terminal heat stress on yield of lentil. 3 Biotech 14(188)

Gorthi S, Swetha RK, Chakraborty S, Li B, Weindorf DC, Dutta S, Banerjee H, Das K, Majumdar K (2021) Soil organic matter prediction using smartphone-captured digital images: use of reflectance image and image perturbation. Biosyst Eng 209:154–169

Grunwald S (2009) Multi-criteria characterization of recent digital soil mapping and modeling approaches. Geoderma 152(3–4):195–207

Han P, Dong D, Zhao X, Jiao L, Lang Y (2016) A smartphone-based soil color sensor: for soil type classification. Comput Electron Agric 123:232–241

Hank TB, Berger K, Bach H, Clevers JG, Gitelson A, Zarco-Tejada P, Mauser W (2019) Spaceborne imaging spectroscopy for sustainable agriculture: contributions and challenges. Surv Geophys 40(3):515–551

Hbirkou C, Pätzold S, Mahlein AK, Welp G (2012) Airborne hyperspectral imaging of spatial soil organic carbon heterogeneity at the field-scale. Geoderma 175:21–28

Hengl T, Mendes de Jesus J, Heuvelink GB, Ruiperez Gonzalez M, Kilibarda M, Blagotić A, Shangguan W, Wright MN, Geng X, Bauer-Marschallinger B, Guevara MA, Vargas R, MacMillan RA, Batjes NH, Leenaars JGB, Ribeiro E, Wheeler I, Mantel S, Kempen B (2017) SoilGrids250m: global gridded soil information based on machine learning. PLoS One 12(2):e0169748

Hively WD, McCarty GW, Reeves JB III, Lang MW, Oesterling RA, Delwiche SR (2011) Use of airborne hyperspectral imagery to map soil properties in tilled agricultural fields. Appl Environ Soil Sci 2011(1):358193

Hu W, Huang B, Weindorf DC, Chen Y (2014) Metals analysis of agricultural soils via portable X-ray fluorescence spectrometry. Bull Environ Contam Toxicol 92(4):420–426

Jenny H (1941) Factors of soil formation. McGrawHill Book Co, New York 281

Jha G, Sihi D, Dari B, Kaur H, Nocco MA, Ulery A, Lombard K (2021) Rapid and inexpensive assessment of soil total iron using nix pro color sensor. Agric Environ Lett 6(3):e20050

Jones AA (1982) X-ray fluorescence spectrometry. Methods Soil Anal: Part 2 Chem Microbiol Prop 9:85–121

Kalopesa E, Tziolas N, Tsakiridis NL, Safanelli JL, Hengl T, Sanderman J (2025) Large-scale soil organic carbon estimation via a multisource data fusion approach. Remote Sens 17(5)

Lavanya V, Nayak A, Dasgupta S, Urkude S, Dey S, Biswas A, Li B, Weindorf DC, Chakraborty S (2023) A smartphone-integrated imaging device for measuring nitrate and phosphate in soil and water samples. Microchem J 193:109042

Lavanya V, Das A, Nayak A, Chawla Y, Dasgupta S, Weindorf DC, Li B, Chakraborty S (2024) Digital soil mapping of available phosphorus using a smartphone-integrated RGB imaging device and ascorbic acid extraction method. Smart Agric Technol 9:100591

Lobell DB, Asner GP (2002) Moisture effects on soil reflectance. Soil Sci Soc Am J 66(3):722–727

Mahmood HS, Hoogmoed WB, van Henten EJ (2012) Sensor data fusion to predict multiple soil properties. Precis Agric 13(6):628–645

Malmir M, Tahmasbian I, Xu Z, Farrar MB, Bai SH (2019) Prediction of soil macro-and microelements in sieved and ground air-dried soils using laboratory-based hyperspectral imaging technique. Geoderma 340:70–80

Mancini M, Andrade R, Silva SHG, Rafael RBA, Mukhopadhyay S, Li B, Chakraborty S, Guilherme LRG, Acree A, Weindorf DC, Curi N (2024) Multinational prediction of soil organic carbon and texture via proximal sensors. Soil Sci Soc Am J 88(1):8–26

Margenot AJ, Parikh SJ, Calderón FJ (2023) Fourier-transform infrared spectroscopy for soil organic matter analysis. Soil Sci Soc Am J 87(6):1503–1528

McBratney AB, Santos MM, Minasny B (2003) On digital soil mapping. Geoderma 117(1–2):3–52

Mukhopadhyay S, Chakraborty S, Bhadoria PBS, Li B, Weindorf DC (2020) Assessment of heavy metal and soil organic carbon by portable X-ray fluorescence spectrometry and NixPro™ sensor in landfill soils of India. Geoderma Reg 20:e00249

Mulder VL, De Bruin S, Schaepman ME, Mayr TR (2011) The use of remote sensing in soil and terrain mapping—a review. Geoderma 162(1–2):1–19

Mulla DJ (2021) Satellite remote sensing for precision agriculture. In: Sensing approaches for precision agriculture. Springer International Publishing, Cham, pp 19–57

Mundada S, Jain P (2025) Predicting soil organic carbon with ensemble learning techniques by using satellite images for precision farming. Sci Rep 15(1):28760

Nayak A, Chakraborty S, Swain DK (2023) Application of smartphone-image processing and transfer learning for rice disease and nutrient deficiency detection. Smart Agric Technol 4:100195

Nguyen QH, Le TH, Bui DT (2020) Deep learning models for predicting soil properties from spectral data. Remote Sens 12(17):2748

Nocita M, Stevens A, van Wesemael B, Aitkenhead M, Bachmann M, Barthès B et al (2015) Soil spectroscopy: an alternative to wet chemistry for soil monitoring. Adv Agron 132:139–159

Padarian J, Minasny B, McBratney AB (2019) Using deep learning for digital soil mapping. Soil 5(1):79–89

Parsons C, Grabulosa EM, Pili E, Floor GH, Roman-Ross G, Charlet L (2013) Quantification of trace arsenic in soils by field-portable X-ray fluorescence spectrometry: considerations for sample preparation and measurement conditions. J Hazard Mater 262:1213–1222

Pate S, Donah K, Chakraborty S, Weindorf DC, Carvalho GS, Deb S, Paramanik B, Sirbescu MLC, Ray DP, Li B (2025) Toward sustainable compost use: prediction of organic matter via smartphone image analysis. Agron J 117(4):e70121

Paulette L, Man T, Weindorf DC, Person T (2015) Rapid assessment of soil and contaminant variability via portable x-ray fluorescence spectroscopy: Copşa Mică, Romania. Geoderma 243:130–140

Pizarro S, Pricope NG, Figueroa D, Carbajal C, Quispe M, Vera J, Alejandro L, Achallma L, Gonzalez I, Salazar W, Loayza H, Cruz J, Arbizu CI (2023) Implementing cloud computing for the digital mapping of agricultural soil properties from high resolution UAV multispectral imagery. Remote Sens 15(12):3203

Rizzo R, Demattê JA, Lepsch IF, Gallo BC, Fongaro CT (2016) Digital soil mapping at local scale using a multi-depth Vis–NIR spectral library and terrain attributes. Geoderma 274:18–27

Rossel RV, Behrens T, Ben-Dor E, Brown DJ, Demattê JAM, Shepherd KD, Shi Z, Stenberg B, Stevens A, Adamchuk V, Aichi H, Barthes BG, Bartholomeus HM, Bayer AD, Bernoux M, Bottcher K, Brodsky L, Du CW, Chappel A, Fouad Y, Genot V, Gomez C, Grunwald S, Gubler A, Guerrero C, Hedley CB, Knadel M, Morras HJM, Nocita M, Ramirez-Lopez L, Roudier P, Rufasto CEM, Sanborn P, Sellitto VM, Sudduth KA, Rawlins BG, Walter C, Winowiecki LA, Hong SY, Ji W (2016) A global spectral library to characterize the world's soil. Earth Sci Rev 155:198–230

Sharma A, Weindorf DC, Man T, Aldabaa AAA, Chakraborty S (2014a) Characterizing soils via portable X-ray fluorescence spectrometer: 3. Soil reaction (pH). Geoderma 232:141–147

Sharma A, Weindorf DC, Wang D, Chakraborty S (2014b) Characterizing soils via portable X-ray fluorescence spectrometer: 4. Cation exchange capacity (CEC). Geoderma 239:130–134

Stiglitz R, Mikhailova E, Post C, Schlautman M, Sharp J (2016) Evaluation of an inexpensive sensor to measure soil color. Comput Electron Agric 121:141–148

Swanhart S, Weindorf DC, Chakraborty S, Bakr N, Zhu Y, Nelson C et al (2014) Soil salinity measurement via portable X-ray fluorescence spectrometry. Soil Sci 179(9):417–423

Swetha RK, Bende P, Singh K, Gorthi S, Biswas A, Li B, Weindorf DC, Chakraborty S (2020) Predicting soil texture from smartphone-captured digital images and an application. Geoderma 376:114562

Swetha RK, Chakraborty S (2021) Combination of soil texture with nix color sensor can improve soil organic carbon prediction. Geoderma 382:114775

Swetha RK, Dasgupta S, Chakraborty S, Li B, Weindorf DC, Mancini M, Silva SHG, Ribeiro BT, Curi N, Ray DP (2022) Using nix color sensor and Munsell soil color variables to classify contrasting soil types and predict soil organic carbon in eastern India. Comput Electron Agric 199:107192

Tahmasbian I, Xu Z, Boyd S, Zhou J, Esmaeilani R, Che R, Bai SH (2018) Laboratory-based hyperspectral image analysis for predicting soil carbon, nitrogen and their isotopic compositions. Geoderma 330:254–263

Tan K, Wang H, Chen L, Du Q, Du P, Pan C (2020) Estimation of the spatial distribution of heavy metal in agricultural soils using airborne hyperspectral imaging and random forest. J Hazard Mater 382:120987

Ulmanu M, Anger I, Gamenţ E, Mihalache M, Plopeanu G, Ilie L (2011) Rapid determination of some heavy metals in soil using an X-ray fluorescence portable instrument. Research. J Agric Sci 43(3)

Veum KS, Sudduth KA, Kremer RJ, Kitchen NR (2017) Sensor data fusion for soil health assessment. Geoderma 305:53–61

Walkley A, Black IA (1934) An examination of the Degtjareff method for determining soil organic matter, and a proposed modification of the chromic acid titration method. Soil Sci 37(1):29–38

Weindorf DC, Acosta Martinez V (2025) Fundamentals of soil characterization. In: Weindorf DC, Chakraborty S, Li B (eds) Unlocking the secrets of soil: applying AI and sensor Technologies for Sustainable Land use. Elsevier

Weindorf DC, Zhu Y, Chakraborty S, Bakr N, Huang B (2012a) Use of portable X-ray fluorescence spectrometry for environmental quality assessment of peri-urban agriculture. Environ Monit Assess 184(1):217–227

Weindorf DC, Zhu Y, Haggard B, Lofton J, Chakraborty S, Bakr N, Zhang W, Weindorf WC, Legoria M (2012b) Enhanced pedon horizonation using portable X-ray fluorescence spectrometry. Soil Sci Soc Am J 76(2):522–531

Weindorf DC, Zhu Y, McDaniel P, Valerio M, Lynn L, Michaelson G, Clark M, Ping CL (2012c) Characterizing soils via portable x-ray fluorescence spectrometer: 2 Spodic and albic horizons. Geoderma 189:268–277

Weindorf DC, Herrero J, Castañeda C, Bakr N, Swanhart S (2013a) Direct soil gypsum quantification via portable x-ray fluorescence spectrometry. Soil Sci Soc Am J 77(6):2071–2077

Weindorf DC, Paulette L, Man T (2013b) In-situ assessment of metal contamination via portable X-ray fluorescence spectroscopy: Zlatna, Romania. Environ Pollut 182:92–100

Weindorf DC, Bakr N, Zhu Y (2014) Advances in portable X-ray fluorescence (PXRF) for environmental, pedological, and agronomic applications. Adv Agron 128:1–45

Wetterlind J, Stenberg B, Söderström M (2008) The use of near infrared (NIR) spectroscopy to improve soil mapping at the farm scale. Precis Agric 9(1):57–69

Xu S, Wang M, Shi X, Yu Q, Zhang Z (2021) Integrating hyperspectral imaging with machine learning techniques for the high-resolution mapping of soil nitrogen fractions in soil profiles. Sci Total Environ 754:142135

Zhu Y, Weindorf DC (2009) Determination of soil calcium using field portable x-ray fluorescence. Soil Sci 174(3):151–155

Zhu Y, Weindorf DC, Zhang W (2011) Characterizing soils using a portable X-ray fluorescence spectrometer: 1 soil texture. Geoderma 167:167–177

Chapter 3
Agromet Advisory Services in India for Climate Risk Management in Agriculture

Kripan Ghosh, Ashutosh Kumar Misra, and Asha Latwal

Abstract With 51% of India's net sown area still rainfed, Indian agriculture remains highly dependent on weather, especially rainfall quantity and distribution, which shape the rainy season's characteristics and limit the crop growing period. This dependency impacts agricultural techniques, crop productivity, and livestock management, as irregularities in monsoon patterns, such as delayed onset, interruptions, and early withdrawal, as well as other adverse weather events result in droughts, floods, and other extreme events like hailstorms, cold waves, frost, heatwaves, and high winds, all of which reduce productivity. Increasing weather uncertainties emphasize the need for informed agricultural risk management. Although weather patterns are uncontrollable, adverse effects can be mitigated by timely, interpreted weather information and advisories. To address this, India Meteorological Department (IMD), Ministry of Earth Sciences (MoES), along with Indian Council of Agricultural Research (ICAR), State Agricultural Universities (SAUs), Indian Institute of Technology (IIT), and other institutions, operates the Gramin Krishi Mausam Sewa (GKMS) scheme to extend Agromet Advisory Services (AAS). Under GKMS, agromet advisories are being prepared every Tuesday and Friday, based on 5-day forecasts and weekly outlook, covering all major agricultural districts through 130 Agromet Field Units (AMFUs). Impact-Based Forecasts (IBF) for extreme weather events are issued as needed. Supported by a robust observational network and advanced technology, these advisories help farmers adapt to climate variability. Major organizations like IMD, ICAR, Indian Space Research Organisation (ISRO), IITs, and SAUs contribute through research and operational support, strengthening these services for enhanced farm resilience.

K. Ghosh (✉) · A. K. Misra · A. Latwal
Agricultural Meteorology Division, Climate Research & Services, India Meteorological Department, Pune, India
e-mail: kirpan.ghosh@imd.gov.in

A. K. Misra
e-mail: dr.ashutosh.misra@imd.gov.in

A. Latwal
e-mail: asha.latwal@imd.gov.in

D. Kumar Swain et al. (eds.), *Climate-Smart Technologies for Sustainable Agricultural Production*, https://doi.org/10.1007/978-3-032-25666-9_3

Keywords Climate risk management · Agromet advisory services · Climate resilience · Indian agriculture · Weather forecasting · Crop management

3.1 Introduction

Indian farmers face a multitude of weather and climate-related challenges that significantly threaten their livelihoods and food security. The country's agricultural landscape is increasingly impacted by erratic rainfall patterns, rising temperatures and an uptick in extreme weather events, all of which disrupt traditional farming practices and yield stability. The agriculture sector alone contributes approximately 16% of the country's GDP and provides a livelihood for about 46.1% of the population (Ministry of Finance, 2025). These climatic challenges are not only notable for their immediate effects but also for their long-term implications on socio-economic conditions and food systems across the country.

Despite improvements in coverage of irrigation from 49.3 to 55% of the gross cropped area, there remains a pronounced imbalance across states. While regions like Punjab, Haryana, Uttar Pradesh and Telangana have largely irrigation access, other states, such as Jharkhand and Assam continue to face acute shortfalls, with coverage below 20% (Ministry of Finance, 2025). A major concern regarding this issue is the increasing pattern of intra-seasonal rainfall variability, observed as less number of rainy days but with more intense rainfall events (Mukherjee et al., 2018; Samantray and Gouda, 2024) that ultimately results in excess soil erosion and prolonged dry spells. It disrupts the hydrological cycle causing frequent floods and droughts and adversely impacting crop production (Dash and Maity, 2023; Yan et al., 2024).

Regional differences make this scenario even more complex: e.g., northern states like Punjab and Haryana are struggling with groundwater scarcity and irregular monsoons (Panda et al., 2021; Mishra et al., 2024), southern states face high pest infestations which increase under humid conditions; in eastern and central India monsoon trends are weakening (Sandeep et al., 2018; Bajrang et al., 2023) and Himalayan states experience high climate variability, limiting crop resilience. It is noteworthy, as indicated by research studies, that in recent years, extreme weather events have occurred on nearly 90% of days from January to September (CSE, 2022), causing significant damage to crops and livestock.

Rising temperatures pose another serious challenge, especially for tropical countries like India, which are already operating near the highest limits of temperature tolerance. Average temperatures in India are projected to rise by 1.2–3.5 °C by 2050 (Ravindra et al., 2024), significantly impacting agricultural productivity. These high temperatures also influence various other climatic factors which have a role in crop production; e.g., excessive heat reduces the fruit-setting ability of citrus plants and increases water loss through transpiration in vegetables. The frequency of extreme weather events has escalated in recent decades, complicating the agricultural landscape. Recently in 2022, the country has experienced an exceptionally high temperatures during March and April 2022 with temperatures higher by 8 to

10.8 °C above normal and rainfall plummeting by 60 to 99% in 10 out of 36 meteorological subdivisions. The heatwave coincided with the crucial grain-filling phase of wheat, causing issues like yellowing, shrivelling and forced maturity, leading to yield reductions of 15–25%. Livestock, particularly dairy animals, also suffered from heat stress, resulting in a 15% decline in milk production. Additionally, egg production dropped, and broiler chickens experienced higher mortality rates (Bal et al., 2022).

3.2 Climate Risks in Different Regions of the Country

The impacts of climate variability are highly heterogeneous due to various regional factors such as local climate regimes, soil properties, topography and water availability. The Indo-Gangetic plains and north-eastern states face different challenges compared to arid zones like Rajasthan, which is more susceptible to prolonged dry spells. Local forecasts and adaptive management strategies play a very significant role for addressing these inequalities and require initiatives and planning focused on promoting drought- and flood-resistant crops in some of the most vulnerable areas.

3.2.1 North India

North Indian agriculture includes the fertile Indo-Gangetic Plains (IGP), which contribute to about 50% of India's grain production and support nearly 40% of the country's population (Jat et al., 2009). This region which is spread across Punjab, Haryana, western Uttar Pradesh and parts of Rajasthan consists one of the densest agricultural systems in the world, which can be attributed by high productivity and substantial climate sensitivity (Ladha et al., 2003). Farmers of this region are prone to a number of weather and climate-related risks threatening their livelihoods and food security. The most significant challenge is rising temperatures (Singh et al., 2001; Panda, et al., 2017); Jha et al., 2024; Norgate et al., 2024) which cause heat stress in crops during critical developmental stages. For example, heatwaves adversely affect the flowering period of crops like wheat, resulting in reduced grain quality and yield. Projections indicate that a temperature increase of 2.5–4.9 °C could lead to a 41–52% reduction in wheat yields and a 32–40% reduction in rice yields (Gupta and Pathak, 2016). Higher temperature raises evapotranspiration, which in turn increases irrigation demand, while groundwater levels are continuously declining. Additionally, the reliability of monsoon rainfall has decreased and farmers are facing late-onset of monsoons, untimely rains and extended dry spells. Changing rainfall patterns not only affect crop growth but also exacerbate water scarcity in regions like Punjab and Haryana, where farmers are caught between increasing water demand and decreasing groundwater sources. Furthermore, longer and more frequent droughts are posing serious threats to food production. Drought conditions cause wilting of plants, hinder growth, and reduce yields, severely impacting the food supply chain.

The monsoon season is vital for North Indian agriculture, yet it has become increasingly unpredictable. The first two months of the monsoon are particularly crucial when sowing of *kharif* crops takes place and water for irrigation is needed. Farmers have reported significant variations in rainfall patterns, oscillating between deficit and surplus conditions during the monsoon months. The challenges are compounded by the impacts of climate change, which have led to rising temperatures and increased heat stress, thereby generating more extreme weather events such as unseasonal rains and altered storm cycles. These shifting patterns disrupt traditional agricultural practices and threaten crop yields, particularly for staple crops like rice, wheat, and maize etc. (Kumar and Khanna, 2023; Das et al., 2025). Under the RCP 4.5 climate scenario, projections for the Indo-Gangetic plains indicate a decline in yields by 24.23% for paddy, 10.09% for maize and 6.06% for wheat by 2050. The decline is even more pronounced by 2080, with paddy yields falling by 30.48%, maize by 12.51% and wheat by 7.43% (Kumar et al., 2024).

High-intensity precipitation causes flooding, waterlogging, soil erosion and mechanical damage to the standing crops with more fungal and bacterial diseases. On the other hand, dry spells, particularly during flowering and grain filling stages, significantly impact yields. Rice, the predominant *Kharif* crop, shows high sensitivity to water stress during reproductive phases, with yield reductions of 20–50% reported during drought years (O'Toole and Chang, 1978).

Several scientific studies conducted recently highlight increasing concerns over rising winter temperature that adversely affects *Rabi* crop production (Dhankher and Foyer, 2018) and an urgent need for climate-resilient crops and varieties to mitigate these challenges. The unusually higher temperatures accelerate phenological development of several temperature sensitive crops and shorten the grain-filling period. Another worrisome concern is occurrence of frost which results in cellular breakdown and halting the growth, phenological mismatches that disrupt flowering and pollination, and deteriorated grain quality due to temperature stress.

The increasing frequency of hailstorms and extreme weather events near the harvesting period of *rabi* crops poses serious risks to the farmers. Recent reports emphasize the devastating impacts, including total crop damage and yield loss, which severely undermines farmer incomes. Quality of crops that survive these events often deteriorates significantly, leading to a sharp drop in market prices. Furthermore, inadequate insurance coverage for weather-related losses further complicates the situation, as limited financial support hinders farmers' recovery from these extreme events.

The *Zaid* season also faces severe challenges due to water scarcity, making effective water management crucial for maintaining agricultural practices. The region is experiencing depletion of groundwater at an alarming rate due to overexploitation for irrigation, resulting in a significant decline in water levels. This has increased competition for water, with different regions vying for the same limited resource. Compounding this tension, rising energy costs and increasing pumping costs associated with aquifer depth are placing greater financial pressure on farmers. These challenges raise significant concerns about the long-term sustainability of intensive irrigation systems, especially in the context of climate change.

Severe storms are becoming more frequent during critical stages of crop growth, often causing significant losses. Hailstorms are another growing threat, as pre-monsoon storms destroy crops before they mature. Strong winds have also damaged crops, causing them to fall, and agricultural infrastructure is becoming increasingly vulnerable to these extreme weather events. Furthermore, the early arrival of monsoon rains is disrupting harvesting plans, affecting both the quality and timeliness of crops. All of these challenges highlight the urgent need for adaptive approaches to address the growing risks associated with climate change.

3.2.2 *South Peninsular India*

South India's agriculture is shaped by two principal rainfall systems i.e., the Southwest (SW) monsoon (June–September, critical for *kharif* crops and reservoir recharge) and the Northeast (NE) or retreating/post-monsoon (October-December, highly important for Tamil Nadu and parts of Andhra Pradesh and Kerala). Large parts of cultivation are rain-fed; soils and cropping calendars are tightly coupled to monsoon timing and intra-seasonal distribution. Climate change is altering mean temperatures and the statistics of extremes (more frequent/intense heatwaves and heavy precipitation events, shifts in drought patterns), increasing exposure and reducing predictability of seasonal water availability.

The pre-monsoon period is characterised by rising temperatures and sporadic rainfall. Heatwaves have become more frequent and intense, causing significant yield losses in maize, groundnut, and pulses during flowering and reproductive stages (Harris et al., 2022). High evapotranspiration during this period accelerates soil moisture depletion, particularly in rain-fed tracts of Karnataka and Telangana. The major agriculture-related challenges in this season are crop wilting, reduced grain filling, livestock stress, and increased pest infestations.

The SW monsoon supports major crops such as rice, maize, sorghum, and pulses. The primary challenges faced by the farmers during this season are delayed or erratic onset of rains, which disrupts sowing, and heavy rainfall and flooding, which damage standing crops and cause soil erosion. On the other hand, the NE monsoon is crucial for Tamil Nadu and parts of Andhra Pradesh and Kerala. However, rainfall variability and tropical cyclones pose major risks. Cyclone-induced flooding and saltwater intrusion further degrade coastal farmlands. Winter cropping relies heavily on residual soil moisture and irrigation. While cold stress is less common in South India, rising minimum temperatures threaten *rabi* crops such as pulses and oilseeds by altering flowering and pest cycles (Geethalakshmi et al., 2023). Warmer nights reduce grain quality and productivity in wheat grown in elevated areas of Karnataka and Andhra Pradesh (Murari et al., 2018; Huggi, 2021).

3.2.3 West India

West India (principally Maharashtra, Gujarat, Rajasthan and Goa) supports highly diverse cropping systems that are tightly coupled to seasonal weather. Recent analyses show pervasive warming, shifts in seasonal precipitation and growing extremes that combine with local vulnerabilities (groundwater depletion, coastal salinization, smallholder rain-dependence) to amplify agricultural risk. West India's agricultural mosaic includes rainfed cotton, pulses and oilseeds on the Deccan plateau, irrigated rice, sugarcane and horticulture in coastal plains, and arid/semi-arid mixed farming in Saurashtra and Rajasthan. Farmers face four operational seasons for planning and risk management: pre-monsoon (MAM), southwest monsoon (JJAS), post-monsoon (ON) and winter (DJF).

Monsoon variability (delays/advances) in onset, longer intra-seasonal dry spells, and local intense rainfall are key agricultural hazards. Erratic monsoon behaviour causes uneven germination of *kharif* crops (cotton, sorghum, pulses etc.) and complicates pest/disease cycles. Heavy convective events result in the flooding, crop lodging, and severe erosion on sloped fields. According to model studies, extreme rainfall intensity and variability are likely to increase, increasing the risk of both flood and drought during the same season, even though mean monsoon precipitation may not change uniformly (Salunke et al., 2023).

The post-monsoon (ON) season is critical for harvesting of many late-sown crops. Western coastal districts are increasingly exposed to Arabian Sea cyclones and intense post-monsoon convective storms (e.g., Cyclone Tauktae, May 2021, which produced strong winds, coastal storm surge and heavy rain in Maharashtra and Gujarat). Such events damage standing crops, infrastructure, and can cause salinization of low-lying irrigated lands and wells in coastal areas (Kannaujiya et al., 2024).

Pre-monsoon months see rising maxima and more frequent heatwaves; early-season heat reduces transplanting windows, accelerates crop phenology (reducing grain filling), and increases irrigation demand where irrigation is available. Heat events also stress livestock and reduce labour productivity for planting. The observed trend of increasing pre-monsoon warming has already shortened planting windows in several peninsular districts (Kumar et al., 2023).

Warmer winters may shorten chilling periods required for some horticultural crops and alter pest/disease dynamics. Winter rains remain important for soil moisture for *rabi* crops in rainfed districts; variability or reduction in winter precipitation reduces sowing windows for wheat and pulses. Where irrigation is required, competition for depleted groundwater becomes acute in late winter/spring.

3.2.4 East India

Farmers in Eastern India (primarily Bihar, Jharkhand, Odisha, and West Bengal) face significant weather and climate-related challenges due to the regions' vulnerability to monsoons, tropical cyclones, and shifting climatic patterns. These areas rely heavily on rainfed agriculture, making them susceptible to variability in rainfall, rise in temperature, and extreme events. Climate change aggravates these issues, leading to reduced crop yields, soil degradation, and threats to food security. Below, the challenges are outlined for the region, drawing from observed trends and projections.

East India's agriculture is dominated by crops like rice, wheat, maize, and pulses. Farmers experience unpredictable monsoon patterns, with overall reduced rainfall in some areas, leading to water shortages and lower soil moisture. This hinders crop growth, particularly for rainfed farming, and contributes to drought conditions in states like Jharkhand and Bihar (Kumar et al., 2011; Halder et al., 2024). Heavy, concentrated rainfall causes frequent flooding, especially in Bihar (along the Ganges) and West Bengal, resulting in crop submergence, soil erosion, and loss of arable lands. This disrupts planting cycles and reduces yields. On the other hand, warmer temperatures increase pest and disease incidence, altering crop phenology and reducing quality and yields of staples like rice and wheat. Projections indicate shifts to higher-temperature regimes, further stressing agriculture (Das et al., 2025; Shukla et al., 2025). Coastal states like Odisha and West Bengal are highly prone to cyclones, which bring storm surges, heavy rains, and winds that destroy crops, infrastructure, and soil fertility (Halder et al., 2024).

3.2.5 North Eastern India

Northeastern India encompasses a diverse range of geographical features and demographics. The North-East region includes states such as Assam, Meghalaya, Arunachal Pradesh, Nagaland, Manipur, Mizoram, Tripura and Sikkim, collectively occupying 7.97 percent of India's geographical area and home to 3.78 percent of its population (MHA, 2025). The region is predominantly situated within the Brahmaputra River Basin, which accounts for a significant portion of India's total water resources, aiding in the sustenance of agriculture and ecosystems (Singh et al., 2004).

Recent trends in climate patterns in northeast India are characterized by increasingly unpredictable rainfall patterns. Over the last few centuries, the northeast region of India has experienced substantial shifts in rainfall patterns leading to increasingly unpredictable flood-events and drought-cycles. Rainfed farming systems and smallholder peasant cultivation are important components of agricultural systems in eastern and northeast India. The consequences of seasonal variability, frequent flooding and increasing climate uncertainty, threaten crop production (and quality), household food security, and livelihoods.

The weather and climate risks that farmers in northeast India must confront have serious ramifications for their livelihoods and agricultural production. Increasing temperature, changing weather patterns, and seasonal water shortages–all compound weather and climate risks and hold substantial economic consequences for food security and the rural economy. Extreme weather, including heat extremes, droughts, and heavy rainfall events, continues to become more severe, and the frequency of these events is an alarming risk in northeast India. For example, irregular rainfall can disrupt sowing and harvesting cycles for crops during the vital growing stage, leading to crop loss and economic loss for farmers (Chakraborty et al., 2025). Additionally, significant and severe flooding during the monsoon season results in the widespread destruction of crops and agricultural land, while also degrading soil quality, now and for the future.

3.3 Agromet Advisory Services for Minimizing the Climate Risks in India

Efforts have been made by the mankind to predict the weather from the ancient times. Although IMD was established in 1875, Royal Commission on Agriculture recommended a collaboration between the Departments of Agriculture and Meteorology to understand the relationship between weather and crops, resulting in the establishment of the Agricultural Meteorology Division in Pune in 1932 with the financial support of Imperial Council of Agricultural Research, presently known as Indian Council of Agricultural Research (ICAR) (IMD, 1976). Under the leadership of Dr L. A. Ramdas who did pioneer work in this field (Narsimha, 1994), the Agricultural Meteorology became a permanent division of IMD in 1940 (IMD, 2001).

In order to cater to the weather information needs for the farming community, IMD started issuing Farmers Weather Bulletin in 1945 which were broadcasted through radio. With the advancement of science and forecasting capabilities, IMD started providing weather forecast based agrometeorological advisories at state level since 1976 based on short-range weather predictions (Venkatasubramanian et al., 2014). As these advisories were provided only one day in advance, the farmers found them inadequate to plan their agricultural activities in such a short lead time. During 1983, consequent upon the weather aberrations resulting in significant wheat loss, an apex committee was formed which recommended setting up of a specialized Centre for Medium Range Weather Prediction several days in advance, which led to the establishment of National Centre for Medium Range Weather Forecasting (NCMRWF) in 1988. Thereafter NCMRWF, in collaboration with IMD, ICAR and various States, started an AAS to provide quantitative medium range weather forecast for three days with five units in 1991 on an experimental basis (Rao et al., 2007). Due to the significant diversity in climate and crop patterns within a state, ICAR has adopted the concept of agro-climatic zoning through the National Agricultural Research Project (Raman and Balaguru, 1988) which was further used for setting up AAS units. This service was later gradually expanded to all 127 agro-climatic zones of India so that AAS bulletins could be prepared/issued/disseminated based on weather forecasts and user feedback could also be provided (Maini and Rathore, 2008) and this service has demonstrated itself as an example of successful multi-institutional and interdisciplinary operation, providing valuable services to the agricultural community in the country (Venkatasubramanian et al., 2014). In 2007, the AAS was brought under IMD and renamed as Integrated Agrometeorological Advisory Services (IAAS). Subsequently, in June 2008, the district-level agrometeorological advisory service was launched across the country through a network of 130 Agrometeorological Field Units (AMFUs) in collaboration with ICAR and state agricultural universities, aiming at providing relevant weather information and management advices at the district level. Later, these services were brought under the GKMS scheme, with the objectives of improving the existing district-level AAS and extending them to the block level to increase outreach to all farming households. With the introduction of new high-resolution models, this service has been further expanded to the block level since 2018. The network of AMFUs under GKMS is presented (Fig. 3.1).

3.4 IMD's Operational Framework for Agromet Advisory Services

IMD follows a diverse range of forecast model products to cater requirements of various sectors to ensure that forecasts are relevant for both general public safety and sector-specific applications like agriculture using a seamless forecasting strategy wherein forecasts are generated in a continuous, integrated manner across different time horizons and spatial scales.

3.4.1 Different Types of Forecast Products and Their Application in AAS

At the shortest end of the spectrum, nowcasting provides predictions up to three hours lead time, with a major focus on types of severe weather such as thunderstorms, lightning, hail, and squalls. These nowcasts are issued for all the districts with more than 1,200 observation stations. Short- to medium-range forecasts are produced for periods extending up to 10 days which include rainfall and other key weather parameters at multiple spatial resolutions, including city-level, block-level, district-level, and meteorological subdivision scales. Such forecasts are especially crucial for planning agricultural operations, transport logistics, and water management activities. For a broader planning horizon, IMD also issues extended range forecasts that provide guidance up to four weeks ahead. These are prepared for the

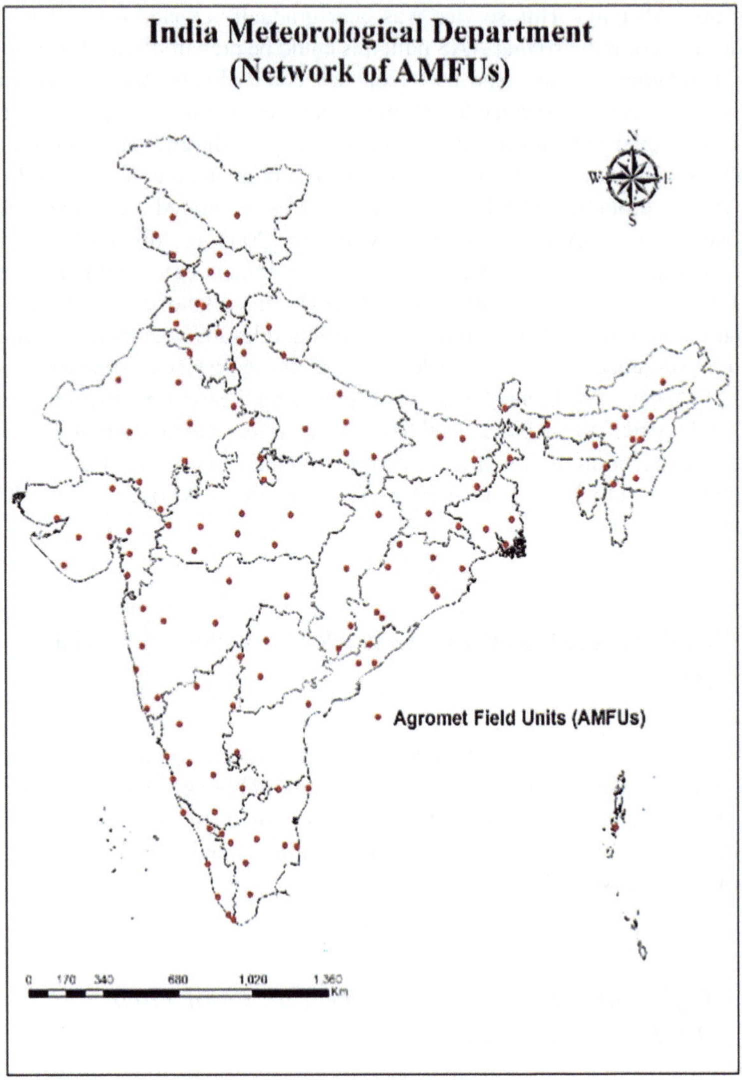

Fig. 3.1 Network of AMFUs and DAMUs under GKMS

entire country, for homogeneous meteorological regions, and for individual meteorological subdivisions, offering valuable insights into potential wet or dry spells well before they occur. At the seasonal scale, long-range forecasts are generated one season in advance, covering All-India, homogeneous regions, and the Monsoon Core Zone. These seasonal predictions play a key role in agricultural planning, reservoir management, and national food security strategies.

In the context of AAS, short- and medium-range forecasts form the backbone of actionable advisories provided to farmers. For this purpose, IMD utilizes a suite of advanced Atmospheric General Circulation Models (AGCMs). The GFS-T1534 global model, with a horizontal resolution of approximately 12.5 km and a forecast range of up to 10 days, is updated twice daily to ensure timeliness and accuracy. Complementing the global model, the WRF-ARW regional model, operating at resolutions of 27–9 km, delivers forecasts up to three days in advance, providing finer spatial details for regional-scale weather systems. For even more localized predictions, particularly in regions, prone to rapid weather changes, the WRF-ARW subregional model runs at an ultra-high resolution of 3 km, offering one-day forecasts that capture small-scale features such as local convective storms and wind patterns.

3.4.2 Agromet DSS

To improve coordination and operational efficiency in the provision of weather-based agricultural advice, an innovative digital tool, called the Agromet–Decision Support System (Agromet DSS) has been launched. The site is being developed as a single platform for the seamless exchange of data and information between all parties involved in the preparation and/or communication of agrometeorological services. The site's design seeks to provide objectivity, timeliness, transparency, and effectiveness, while improving service quality. Agromet DSS provides scientists and partner institutions with analytical tools to integrate real-time weather data with crop data. This enables the development of location-specific advice for farmers. The system efficiently processes multiple data sets, from weather forecasts to crop growth stages, saving time and improving the relevance and accuracy of the advice.

More importantly, the platform allows advisory units to operate on a smaller scale, issuing recommendations at the micro level (e.g., for individual blocks or village groups). This helps in considering a wide range of agroclimatic conditions across India and provides farmers with information to make informed decisions about sowing, irrigation, nutrient application, pest control, and other farming operations. Agromet DSS, through its close collaboration between meteorologists, agricultural scientists, and extension agencies, has become a vital tool for connecting cutting-edge weather forecasting with its applications to the farming community.

3.4.3 Translating Forecasts into Crop Advisories

An expert panel consisting of agricultural experts from various fields (agrometeorologists, agronomists, horticulturists, entomologists, plant pathologists, veterinarians, and district agricultural officers from the state agriculture department) is constituted at each AMFU to meet the requirements of the farming community by interpreting district and block-level weather forecasts for the next five days and meteorological subdivision level weather forecasts for the next week into crop and livestock management advisories specific to the growth stages of the crops and livestock conditions.

Agrometeorological advisories are issued every Tuesday and Friday for districts and blocks, based on medium-range forecasts, as well as extended-range forecasts for meteorological subdivisions. These advisories aim to help farmers minimize potential crop damage and losses and ultimately maximize agricultural yields. These advisories also assist farmers in planning critical activities such as chemical use, irrigation, pest and disease management, and other weather-sensitive agricultural operations. Additionally, these advisories provide guidance on crop variety selection, sowing dates, intercropping practices, harvesting times, and post-harvest management. In a nutshell, Agromet Advisory empowers farmers by providing them with timely and actionable weather information, enhancing their ability to make informed decisions that increase productivity and profitability. These advisories are generated at different spatial scales viz. block and district level along with state and national composites. The block and district level AAS bulletins mainly cater the need of farmers while state and national composite bulletins are designed to support decision-making by planners and administrators. District and Block Level Agromet Advisory Service System is presented in Fig. 3.2 and few representative AAS bulletins have been shown in Fig. 3.3.

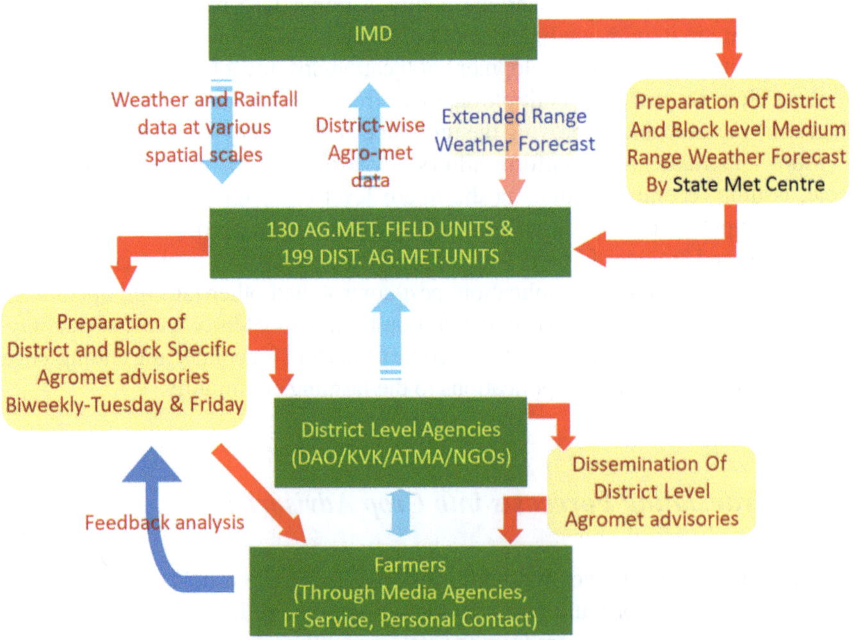

Fig. 3.2 District and block level agromet advisory service system

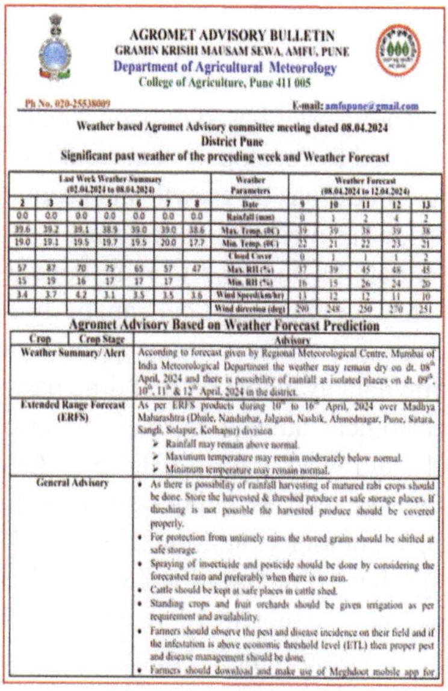

AGROMET ADVISORY BULLETIN
GRAMIN KRISHI MAUSAM SEWA, AMFU, PUNE
Department of Agricultural Meteorology
College of Agriculture, Pune 411 005

Ph No. 020-25538009 E-mail: amfupune@gmail.com

Weather based Agromet Advisory committee meeting dated 08.04.2024
District Pune
Significant past weather of the preceding week and Weather Forecast

Last Week Weather Summary (02.04.2024 to 08.04.2024)							Weather Parameters	Weather Forecast (08.04.2024 to 12.04.2024)				
2	3	4	5	6	7	8	Date	9	10	11	12	13
0.0	0.0	0.0	0.0	0.0	0.0	0.0	Rainfall (mm)	0	1	2	4	2
39.6	39.2	39.1	38.9	39.0	39.0	38.6	Max. Temp. (0C)	39	39	38	39	38
19.0	19.1	19.5	19.7	19.5	20.0	17.7	Min. Temp. (0C)	22	21	22	23	21
							Cloud Cover	0	1	1	1	2
57	87	70	75	65	57	47	Max. RH (%)	37	39	45	48	45
15	19	16	17	17	17		Min. RH (%)	16	15	26	24	20
3.4	3.7	4.2	3.1	3.5	3.5	3.6	Wind Speed(km/hr)	13	12	12	11	10
							Wind direction (deg)	290	248	250	270	251

Agromet Advisory Based on Weather Forecast Prediction

Crop	Crop Stage	Advisory
Weather Summary/ Alert		According to forecast given by Regional Meteorological Centre, Mumbai of India Meteorological Department the weather may remain dry on dt. 08th April, 2024 and there is possibility of rainfall at isolated places on dt. 09th, 10th, 11th & 12th April, 2024 in the district.
Extended Range Forecast (ERFS)		As per ERFS products during 10th to 16th April, 2024 over Madhya Maharashtra (Dhule, Nandurbar, Jalgaon, Nashik, Ahmednagar, Pune, Satara, Sangli, Solapur, Kolhapur) division ➢ Rainfall may remain above normal. ➢ Maximum temperature may remain moderately below normal. ➢ Minimum temperature may remain normal.
General Advisory		• As there is possibility of rainfall harvesting of matured rabi crops should be done. Store the harvested & threshed produce at safe storage places. If threshing is not possible the harvested produce should be covered properly. • For protection from untimely rains the stored grains should be shifted at safe storage. • Spraying of insecticide and pesticide should be done by considering the forecasted rain and preferably when there is no rain. • Cattle should be kept at safe places in cattle shed. • Standing crops and fruit orchards should be given irrigation as per requirement and availability. • Farmers should observe the pest and disease incidence on their field and if the infestation is above economic threshold level (ETL) then proper pest and disease management should be done. • Farmers should download and make use of Meghdoot mobile app for

(a) District AAS Bulletin

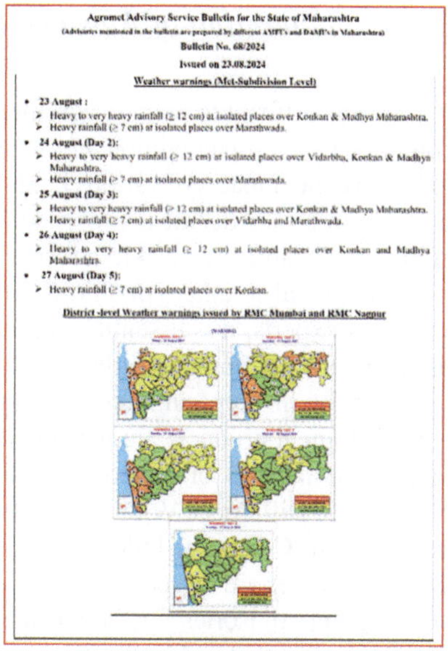

Agromet Advisory Service Bulletin for the State of Maharashtra
(Advisories mentioned in the bulletin are prepared by different AMFU's and DAMU's in Maharashtra)
Bulletin No. 68/2024
Issued on 23.08.2024

Weather warnings (Met-Subdivision Level)

- 23 August :
  - ➢ Heavy to very heavy rainfall (≥ 12 cm) at isolated places over Konkan & Madhya Maharashtra.
  - ➢ Heavy rainfall (≥ 7 cm) at isolated places over Marathwada.
- 24 August (Day 2):
  - ➢ Heavy to very heavy rainfall (≥ 12 cm) at isolated places over Vidarbha, Konkan & Madhya Maharashtra.
  - ➢ Heavy rainfall (≥ 7 cm) at isolated places over Marathwada.
- 25 August (Day 3):
  - ➢ Heavy to very heavy rainfall (≥ 12 cm) at isolated places over Konkan & Madhya Maharashtra.
  - ➢ Heavy rainfall (≥ 7 cm) at isolated places over Vidarbha and Marathwada.
- 26 August (Day 4):
  - ➢ Heavy to very heavy rainfall (≥ 12 cm) at isolated places over Konkan and Madhya Maharashtra.
- 27 August (Day 5):
  - ➢ Heavy rainfall (≥ 7 cm) at isolated places over Konkan.

District -level Weather warnings issued by RMC Mumbai and RMC Nagpur

(b) State Composite AAS Bulletin

(c) National AAS Bulletin

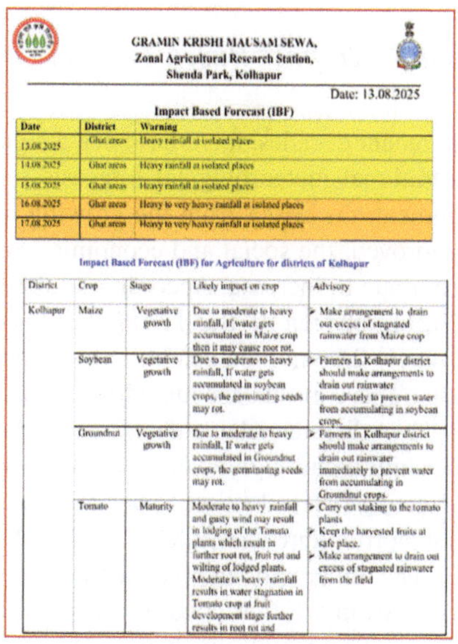

GRAMIN KRISHI MAUSAM SEWA,
Zonal Agricultural Research Station,
Shenda Park, Kolhapur

Date: 13.08.2025

Impact Based Forecast (IBF)

Date	District	Warning
13.08.2025	Ghat areas	Heavy rainfall at isolated places
14.08.2025	Ghat areas	Heavy rainfall at isolated places
15.08.2025	Ghat areas	Heavy rainfall at isolated places
16.08.2025	Ghat areas	Heavy to very heavy rainfall at isolated places
17.08.2025	Ghat areas	Heavy to very heavy rainfall at isolated places

Impact Based Forecast (IBF) for Agriculture for districts of Kolhapur

District	Crop	Stage	Likely impact on crop	Advisory
Kolhapur	Maize	Vegetative growth	Due to moderate to heavy rainfall, If water gets accumulated in Maize crop then it may cause root rot.	➢ Make arrangement to drain out excess of stagnated rainwater from Maize crop
	Soybean	Vegetative growth	Due to moderate to heavy rainfall, If water gets accumulated in soybean crops, the germinating seeds may rot.	➢ Farmers in Kolhapur district should make arrangements to drain out rainwater immediately to prevent water from accumulating in soybean crops.
	Groundnut	Vegetative growth	Due to moderate to heavy rainfall, If water gets accumulated in Groundnut crops, the germinating seeds may rot.	➢ Farmers in Kolhapur district should make arrangements to drain out rainwater immediately to prevent water from accumulating in Groundnut crops.
	Tomato	Maturity	Moderate to heavy rainfall and gusty wind may result in lodging of the Tomato plants which result in further root rot, fruit rot and wilting of lodged plants. Moderate to heavy rainfall results in water stagnation in Tomato crop at fruit development stage further results in root rot and	➢ Carry out staking to the tomato plants ➢ Keep the harvested fruits at safe place. ➢ Make arrangement to drain out excess of stagnated rainwater from the field

(d) IBF for Agriculture

Fig. 3.3 Sample AAS bulletins for different states **a** District AAS Bulletin for Pune **b** State Composite AAS Bulletin for Maharashtra **c** National AAS Bulletin and **d** IBF for Agriculture for Kolhapur

3.4.4 Generation of Agromet Products

It is always difficult to prepare agrometeorological advisories based only on weather conditions and crop condition and stage. There is need to develop systems and tools to analyse the impacts of various kinds of weather conditions on growth and productivity of crops and decide appropriate farm operations as measures for framing more effective advisories to tackle adverse weather situations and also to reap benefits from congenial weather conditions. In order to enhance the analytical capability of scientists/experts, it is essential to develop/produce and utilize various agricultural products using different weather, crop and satellite data available on real time basis.

To assist in preparing accurate agro-meteorological advisories, various agro-meteorological products are generated on a daily and weekly basis. These include maps and statistics of meteorological parameters, as well as indices such as the Normalized Difference Vegetation Index (NDVI), Vegetation Condition Index (VCI), Temperature Condition Index (TCI), and Standardized Precipitation Index (SPI). These resources are shared with AMFUs, enabling them to prepare more customized and relevant advisories for the farming community. Some sample gromet Products have been depicted in Fig. 3.4.

3.4.5 Impact Based Forecast in Agriculture

In recent decades, there has been a significant leap in the weather forecasting from traditional meteorological forecasts to more advanced, IBF systems. This development has arisen from the growing recognition that while forecast accuracy has improved, the social and economic losses resulting due to extreme weather events continue to increase. According to the World Meteorological Organization (WMO, 2021), between 1970 and 2019, weather- and climate-related disasters caused approximately 2.06 million deaths and economic losses exceeding USD 3.64 trillion globally. These events accounted for around 62% of all recorded disasters, resulting in 80% of disaster-related deaths and 99% of associated economic losses. Such statistics highlight a significant discrepancy between improved forecasting capability and effective risk reduction, especially in climate-sensitive sectors such as agriculture and horticulture (ARRCC et al., 2020).

The inherently unpredictable nature of the atmosphere often causes sudden changes in weather conditions, which can have serious consequences for standing crops. These sudden changes when associated with extreme weather, pose an increasing challenge to agricultural stability and productivity. In response to this challenge, IMD initiated issuing IBFs specifically for agriculture from December 2021, initially focusing on high-impact weather events such as heavy rainfall, hail, cold waves, frost, heat waves, and strong surface winds. These forecasts are disseminated at the district level, targeting areas where such weather events are likely to cause severe damage to crops. District-specific weather alerts are prepared by the

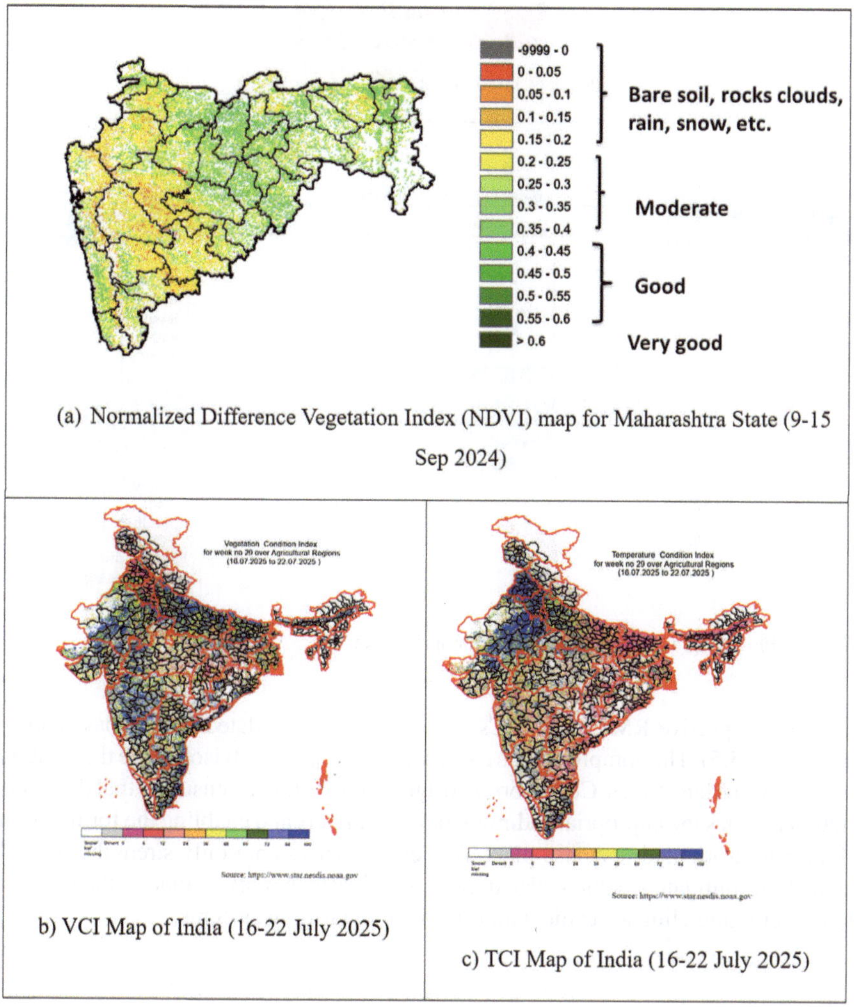

(a) Normalized Difference Vegetation Index (NDVI) map for Maharashtra State (9-15 Sep 2024)

b) VCI Map of India (16-22 July 2025)

c) TCI Map of India (16-22 July 2025)

Fig. 3.4 A few Agromet products to assess the crop condition: **a** NDVI for Maharashtra, **b** VCI and **c** TCI maps

respective Regional Meteorological Centers (RMCs) and Meteorological Centers (MCs). Based on these warnings, AMFUs, which operate under the GKMS framework, develop customized field advisories with the collaboration of experts from various agricultural disciplines, such as plant science, horticulture, entomology, soil science, plant pathology, animal husbandry, and fisheries. A sample IBF for Kolhapur district of Maharashtra has been depicted in Fig. 3.3d.

These advisories are prepared to help farmers make timely and informed decisions to protect their crops and livestock from the challenges of the upcoming weather. Once prepared, district-level IBFs are compiled at the state level by meteorologists

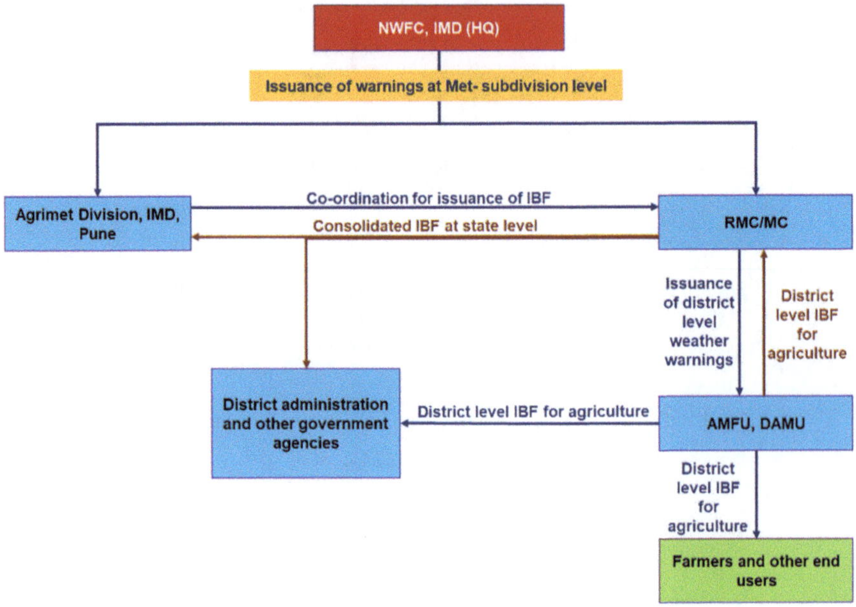

Fig. 3.5 Flow diagram of issuance of IBF for agriculture

from the respective RMCs and MCs, resulting in consolidated impact-based advisories (Fig. 3.5). The compiled IBFs along with appropriate advisories are then shared with District Magistrates, Collectors, and agricultural officers, ensuring that the information reaches the appropriate administrative channels and enabling the formulation of an action-oriented response plan. This entire process not only strengthens local capacity to mitigate weather-related agricultural risks but also supports the broader goal of creating climate-resilient and sustainable farming systems.

3.5 Dissemination Mechanisms in AAS

Effective dissemination is a very significant factor between forecast generation and farmers' actions. The success of any service including AAS doesn't solely rely on forecast accuracy but also on timely, accessible communication across heterogeneous agro-social systems in a user-friendly format in their local languages. Some of the most common communication channels currently used have been described below.

3.5.1 Traditional Media: All India Radio, Doordarshan

All India Radio (AIR) and Doordarshan (DD) have long been used to broadcast weather bulletins, advisory messages, and public awareness campaigns. Even in the digital age, they hold a significant outreach in every part of the country which especially valuable in remote areas with limited connectivity or among farmers who do not own smartphones. Advisory messages through radio and television are typically broadcast in local languages.

3.5.2 IMD Websites and Mobile Apps

With the expansion of internet and mobile penetration, digital platforms now play a central role in AAS dissemination. The Agrometeorological information and AAS bulletins are hosted at the websites of Agricultural Meteorology Division (https://imdagrimet.gov.in/) and main website of IMD (https://mausam.imd.gov.in/). These are also linked to several mobile apps like Meghdoot, Mausam, Kisan Suvidha, Kisan Sanchar etc.

3.5.3 Mobile SMS

Short Message Services (SMS) are a very good way to connect with farmers, and they work just as well for farmers without smartphones (Dheebakaran et al., 2020; Sharma et al., 2021). The Ministry of Agriculture's Kisan portal, which is currently in use during severe weather warnings like cyclones, was used to send brief and actionable SMS advisories. A summary of SMS sent during last 5 years have been given in Table 3.1.

3.5.4 Interactive Voice Response System (IVRS)

These are often operated through public private partnerships in which recorded advisories in their regional languages are delivered to farmers through telephone calls who have registered their numbers. This channel is particularly effective for reaching marginal farmers with limited literacy or smartphone access.

Table 3.1 Number of SMS sent during recent cyclones/deep depressions

Name of cyclone/deep depression	Origin	Period	No. of SMS sent	States covered
Deep depression	Bay of Bengal	September 2021	8,11,106	Madhya Pradesh, Odisha, Chhattisgarh
Gulab	Bay of Bengal	September 2021	84,53,702	Andhra Pradesh, Telangana, Odisha, West Bengal, Chhattisgarh, Maharashtra, Gujarat
Jawad	Bay of Bengal	December 2021	46,51,236	Andhra Pradesh, Odisha, West Bengal
Deep depression	Bay of Bengal	March 2022	7,56,408	Andhra Pradesh, Tamil Nadu, Puducherry
sitrang	Bay of Bengal	October 2022	31,73,651	Arunachal Pradesh, Assam, Manipur, Meghalaya, Nagaland, Odisha, Tripura, West Bengal
Mocha	Bay of Bengal	May 2023	17,15,493	Assam, Nagaland, Manipur, Mizoram, Tripura
Biparjoy	Arabian Sea	June 2023	2,08,36,313	Gujarat, Rajasthan
Michaung	Bay of Bengal	December 2023	60,00,884	Andhra Pradesh, Tamil Nadu, Odisha
Remal	Bay of Bengal	May 2024	1,44,86,563	Arunachal Pradesh, Manipur, Meghalaya, Mizoram, Nagaland, Odisha, Tripura, West Bengal

3.5.5 *Messaging Platforms like WhatsApp*

Farmers are getting weather information, alerts and AAS bulletins through various messaging platforms like WhatsApp groups created by AMFUs and various IMD Offices. It is a very effective two-way communication channel for AAS dissemination and getting the feedback. IMD has also partnered with a private entity viz., Green Alerts to deliver critical weather information to farmers across India through Panchayat Mausam Sewa programme.

3.5.6 *Social Media*

Social media platforms (e.g., Facebook pages, Twitter/X accounts, YouTube channels) are increasingly used by IMD, state agriculture departments, and extension services for enhancing the outreach among the farmers. These channels are used to disseminate weather alerts, infographics, explanatory videos, and links of detailed advisories. The interactive features allow farmers and stakeholders to ask questions and share local feedback. Yet, social media reach tends to skew toward more connected, literate farmers; rural areas with low bandwidth or low digital literacy may not benefit equally.

3.6 Role of AAS for Climate Risk Management in Agriculture

Variability in weather factors (especially rainfall and temperature) is one of the major factors contributing to inter-annual variability in crop production. It is well known that Indian agriculture is primarily dependent on the southwest monsoon. Therefore, the amount and distribution of rainfall during this period, along with the monsoon's interruptions, delays, and early withdrawals, can significantly impact crop production across the country. Furthermore, weather extremes such as prolonged droughts, floods, high winds, heat waves, and hailstorms adversely affect crop yields and are a major source of risk for farmers. Therefore, timely information on predicted weather helps farmers optimizing their resources, ultimately improving agricultural production, both qualitatively and quantitatively.

Climate risk management (CRM) refers to the various approaches used to minimize the adverse impacts of climate through the multi-dimensional efforts including climate adaptation, disaster management and sustainable development. The strategies used in CRM in agriculture are intended to reduce the adverse impacts of climate hazards and variability on agricultural production. It involves actions like risk assessment, insurance and financial planning, crop and water management, soil conservation along with response to the disaster. Thus, the aim of CRM is to make agriculture profitable and maintaining food security despite climate variability and changes in weather patterns and catastrophic weather events. CRM options for various kinds of hazards have been provided in Table 3.2.

In view of above, weather forecast based AAS helps farmers to minimize crop risks by providing advance information on impending weather conditions such as temperature, precipitation, wind, atmospheric pressure etc. With this information, farmers can make informed decisions on when to sow/plant, irrigate, harvest and how to protect their crops. For example, if a forecast predicts heavy rain or strong winds, farmers can adopt suitable measures to protect their crops well in advance for minimizing loss/damage. Similarly, if a forecast predicts a prolonged period of dry spell, farmers can adjust irrigation plans to conserve water and protect their crops from moisture stress. In such manner, AAS helps farmers to optimize their farm operations, avoid wastage of inputs, increase crop yields and minimize the risk of crop loss.

Table 3.2 Climate risk management option for various types of risks

Weather/climate hazard	Likely impacts on agriculture	Risk management Options suggested
Intense rainfall and floods	a. Waterlogging leading to root rot b. Lodging of mature crops c. Soil erosion, loss of topsoil and nutrients d. Loss of plant vigour and growth disruption to critical agricultural activities e. Attack of pests and diseases on the crops f. Anaerobic soil conditions g. Reduction in nitrate and the formation of nitrogen gas	a. Improved drainage in the fields b. Use of Raised beds for planting c. Postponement of fertilizer application and intercultural operations during excess rainfall d. Early harvesting of crops, if 80–85% matured and shifting to safer places e. Adoption of crops like rice that can function effectively in saturated and even submerged conditions for locations that are flooded regularly f. Effective Water storage systems (rivers, lakes, reservoirs etc.) in reducing flood damage g. Vegetation that uses more water and that can act as a barrier to moving water (horizontally and vertically) can reduce flood severity and impacts
Hailstorm	a. Physical damage to leaves, fruits, and stems b. Shattering of leaves, break or damage to shoots c. Wounds on scaffold branches etc d. Split and broken stems e. Complete damage to seedlings f. Fruit drop and fruit damage resulting reduction in harvest produce g. Enhance the chances of pest and disease infestation and rotting of fruits	a. Use hail net for orchards to protect from mechanical damage b. Cleaning up the debris and trimming off broken stems and leaves c. Pruning of the most affected branches d. Thinning off the damaged fruits e. Removal of damaged fruits, which may attract insects f. Application of fertilizer to the impacted plants for helping them to re-grow and develop new foliage g. Some plants are too heavily affected and fixing hail damage is not possible. These plants should be removed and replaced h. Crop insurance

(continued)

Table 3.2 (continued)

Weather/climate hazard	Likely impacts on agriculture	Risk management Options suggested
Heat waves/high temperatures	a. Adverse impact on mineral nutrition and pollen development resulting in poor yield b. Reduction in absorption and subsequent assimilation of nutrients c. Dehydration and water stress in plants d. Abnormal rates of photosynthesis and respiration e. Scorching and sunburn injuries on the exposed area of the horticultural plants known as 'Sun scald'	a. Provide life-saving irrigation to standing crops at critical growth stages b. Cover the small fruit plants and nurseries to prevent terminal heat stress c. Water storage with the help of the rainwater harvesting structures d. Keep the orchards weed free to reduce competition of water and undertake mulching in vegetables and fruit tree basins e. Provide adequate food and clean drinking water to livestock and keep them under the shed
Cold wave/frost/low temperature	a. Delayed germination b. Flower/fruit drop in sensitive crops c. Damage of tissues in seedlings d. Chilling and freezing injury to plants e. Suffocation due to formation of snow cover in hilly regions f. Heaving (mechanical lifting of plants along with soil)	a. Managing warm conditions around the fields and covering the crop by green net b. Heating the air throughout the agricultural farm c. Provide light Irrigation prior to frost events d. Sprinkling with water e. Putting a protective cover of craft paper over plants f. Use of shelterbelts and windbreaks

(continued)

Table 3.2 (continued)

Weather/climate hazard	Likely impacts on agriculture	Risk management Options suggested
Cyclonic Storms	a. Uprooting of trees b. Lodging of plants c. Water stagnation d. Mechanical damage due to strong wind e. loss of fertility due to salt deposition f. Salinity intrusion in coastal farms	*Pre cyclone CRM options:* a. Due to expected heavy rains harvesting of the matured crops and keeping the harvested produces in a safe place b. Arrangement of drainage facilities c. Staking of vegetables d. Propping of sugarcane e. Early harvesting of crops if crop maturity is more than 80–85% *Post cyclone CRM options:* a. Removal of excess water from the standing crop fields b. Propping of sugarcane, banana plants (to keep them erect) and thus, to minimize the damage c. Use of Salt-tolerant crop varieties d. Pest and disease management
Drought/Dry Spell	a. Stunted growth due to moisture stress b. Poor seed germination c. Flower and fruit abortion	a. Rainwater harvesting b. Use of Drip or sprinkler irrigation c. Sowing of drought-tolerant varieties d. Application of mulch to reduce evaporation

3.7 Challenges and Opportunities

3.7.1 Last-Mile Connectivity Issues

In many remote or hilly regions, mobile/internet coverage is patchy or unreliable, hindering delivery of advisories. Electricity outages or lack of smartphone literacy further constrain uptake. Ensuring offline fallback (radio, IVRS) and strengthening rural telecom infrastructure remain essential.

3.7.2 *Forecast Uncertainty and Interpretation Challenges for Farmers*

Even with high-quality models, forecast uncertainty is inherent, especially at micro-scales. Farmers may misinterpret probabilistic forecasts or receive conflicting messages from multiple sources. There is a need for training extension agents, improving clarity in advisory language (e.g., using "likely" vs "certain"), and emphasising impact-based thresholds rather than raw probabilities.

3.7.3 *Language and Local Context Adaptation*

India's linguistic diversity demands advisories in local languages and dialects. More than mere translation, advisories must be culturally adapted, using familiar analogies or practices. Local agronomic suitability (soil types, cropping systems) must also be incorporated–an advisory standard for Punjab may not suit Odisha or Meghalaya.

3.7.4 *Enhancing Spatial and Temporal Resolution of Forecasts*

Demand is growing for hyper-local (village-level) forecasts at hourly timescales. Advances in High-Resolution Rapid Refresh (HRRR) models, convection-permitting models, and data assimilation from IoT sensors (soil moisture, microclimate sensors) could bridge the "last few kilometres" gap.

3.7.5 *Integrating Climate Change Projections into Operational Advisories*

While short- and medium-range forecasts dominate, integrating climate change projections (10–30-year horizons) into routine advisories can support long-term adaptation planning (e.g., shifting crop calendars, introducing resilient varieties). This blending of climate and weather services remains a frontier.

3.7.6 *Strengthening Collaboration with Private Sector and Agri-Tech Startups*

Public advisories can be enriched by partnerships with private-sector and startups working in precision farming, IoT sensors, mobile apps, and AI-driven decision tools. These collaborations can help tailor content, scale delivery, crowdsource feedback, and co-develop farmer-facing decision aids.

3.7.7 *Use of AI/ML for Customized Micro-level Advisories*

Machine learning and artificial intelligence offer potential to predict localized anomalies, crop stress, pest/disease outbreaks, or farmer-level advisory suggestions. AI models can ingest satellite data, weather forecasts, soil and yield histories, plus farmers' feedback, and generate hyper-local recommendations. But caution is needed–models must be interpretable, validated, and integrated with local domain expertise to earn farmers' trust.

3.8 Conclusion

One of the most comprehensive attempts in the world to incorporate meteorological science into agricultural decision-making is India's Agromet Advisory Services (AAS), in collaboration with ICAR, SAUs, and extension agencies, which is based on IMD's forecasting infrastructure. AAS now provides customized advisories to millions of farmers to reduce the risks of drought, floods, heat, frost, and other climate hazards through a network of AMFUs and digital platforms.

Closing the last-mile communication gap, lowering forecast uncertainty, localizing messages to local languages and customs, improving spatial and temporal resolution of forecasts, and integrating climate change projections into routine advisory generation are still obstacles, though. Promising approaches to tackling these issues are provided by the developing fields of artificial intelligence, machine learning, agri-tech innovations, and private–public partnerships.

For further strengthening this service, efforts are required to make the advisories more farmer friendly with less technical content. Although a lot has been done to strengthening the outreach of AAS to farming communities but there is still a wide gap that needs to be filled by incorporating a greater number of farmers, NGOs, Panchayats, farmers self-help groups etc. Continued investment in observational networks (soil, microclimate sensors) and more robust evaluation of these services are required to quantify benefits, adoption rates, and cost-effectiveness. To ensure inclusivity for last-mile connectivity, it is necessary to expand multi-platform outreach, integrating traditional media with mobile, app-based, and community networks.

References

ARRCC, UKMO, climate centre, & UK Aid (2020) The future of forecasts : impact-based forecasting for early action. https://doi.org/10.13140/RG.2.2.12366.89920

Bajrang C, Attada R, Goswami BN (2023) Possible factors for the recent changes in frequency of central Indian summer monsoon precipitation extremes during 2005–2020. NPJ Climate Atmosp Sci 6(1): 1–9. https://doi.org/10.1038/s41612-023-00450-y

Bal SK, Prasad JVNS, Singh VK (2022) Heat wave 2022 causes, impacts and way forward for Indian Agriculture. Technical Bulletin No. ICAR/CRIDA/TB/01/2022, ICAR—Central research institute for dryland agriculture Hyderabad, Telangana, India. http://www.icar-crida.res.in

Chakraborty D, Roy A, Singh NU, Saha S, Das SK, Mridha N, Yumnam A, Paul P, Gowda C, Biam KP, Patra S, Amrutha T, Singh BP, Mishra VK (2025) Assessing climate change impact on rainfall patterns in Northeastern India and its consequences on water resources and rainfed agriculture. Earth 6(1):2. https://doi.org/10.3390/EARTH6010002

CSE (2022) Down to earth climate India 2022–An assessment of extreme weather events January–September. Centre Sci Environ, New Delhi

Das A, Kumar S, Kasala K, Nedumaran S, Paithankar P, Kumar A, Jain A, Avinandan V (2025) Effects of climate change on food security and nutrition in India: a systematic review. Current Res Environ Sustain 9:100286. https://doi.org/10.1016/J.CRSUST.2025.100286

Dash SS, Maity R (2023) Effect of climate change on soil erosion indicates a dominance of rainfall over LULC changes. J Hydrol Reg Stud 47:101373. https://doi.org/10.1016/J.EJRH.2023.101373

Dhankher OP, Foyer CH (2018) Climate resilient crops for improving global food security and safety. Plant Cell Environ 41(5): 877-884. https://doi.org/10.1111/pce.13207

Dheebakaran G, Panneerselvam S, Geethalakshmi V, Kokilavani S (2020) Weather based automated agro advisories: an option to improve sustainability in farming under climate and weather vagaries. Global Climate Change Environ Policy: Agricult Perspect, 329–349. https://doi.org/10.1007/978-981-13-9570-3_11

Geethalakshmi V, Gowtham R, Gopinath R, Priyanka S, Rajavel M, Senthilraja K, Dhasarathan M, Rengalakshmi R, Bhuvaneswari K (2023) Potential impacts of future climate changes on crop productivity of cereals and legumes in Tamil Nadu, India: a mid-century time slice approach. Adv Meteorol 2023(1):4540454. https://doi.org/10.1155/2023/4540454

Gupta A, Pathak H (eds) (2016) Climate change and agriculture in India. Department of Science and Technology, Govt of India

Halder B, Mundher Yaseen Z, Malaysia K (2024) Long-term climate change analysis in northeast and eastern India Long-term climate change analysis in northeast and eastern India. Geo Nat Hazards Risk 15(1): 2409206. https://doi.org/10.1080/19475705.2024.2409206

Harris F, Amarnath G, Joy EJ, Dangour AD, Green RF (2022) Climate-related hazards and Indian food supply: assessing the risk using recent historical data. Global Food Security 33: 100625. https://doi.org/10.1016/j.gfs.2022.100625

Huggi L (2021) Studies on agro-climatic onset of pigeonpea cropping season, effect of date of sowing, spacing on growth, yield and modelling in pigeonpea. Ph.D. Thesis, University of Agricultural Sciences, Bangalore

IMD (1976) 100 years of weather service (1875–1975). Director general of observatory (C&G) Poona

IMD (2001) 125 years of service to the nation–A retrospective and futuristic overview. The meteorological office press Pune

Jat ML, Gathala MK, Ladha JK, Saharawat YS, Jat AS, Kumar V, Sharma SK, Kumar V, Gupta R (2009) Evaluation of precision land leveling and double zero-till systems in the rice-wheat rotation: water use, productivity, profitability and soil physical properties. Soil Tillage Res 105(1): 112-121. https://doi.org/10.1016/j.still.2009.06.003

Jha R, Mondal A, Ghosh S, Murtugudde R (2024) Northward shift of pre-monsoon zonal winds exacerbating heatwaves Over India. Geophys Res Lett 51(19):e2024GL110486. https://doi.org/10.1029/2024GL110486

Kannaujiya VK, Rai AK, Malakar S (2024) Nature and impact of extremely severe cyclone Tauktae over India. Discover Oceans 1: 4. https://doi.org/10.1007/s44289-024-00004-x

Kumar A, Tripathi P, Singh KK, Mishra AN (2011) Impact of climate change on agriculture in eastern Uttar Pradesh and Bihar states (India). Mausam 62(2): 171-178. https://doi.org/10.54302/mausam.v62i2.280

Kumar R, Kuttippurath J, Gopikrishnan GS, Kumar P, Varikoden H (2023) Enhanced surface temperature over India during 1980–2020 and future projections: causal links of the drivers and trends. NPJ Climate Atmosph Sci 6:164. https://doi.org/10.1038/s41612-023-00494-0

Kumar NR, Abdulla KS, Singh NP, Jha GK, Srivastava SK (2024) Impact of climate change on agriculture in Indo-Gangetic Plains of India. Agricult Econo Res Rev 37(1):47–58

Kumar S, Khanna M (2023) Distributional heterogeneity in climate change impacts and adaptation: evidence from Indian agriculture. Agric Econ 54(2):147–160. https://doi.org/10.1111/AGEC.12765

Ladha JK, Dawe D, Pathak H, Padre AT, Yadav RL, Singh B, Singh Y, Singh Y, Singh P, Kundu AL, Sakal R, Ram N, Regmi AP, Gami SK, Bhandari AL, Amin R, Yadav CR, Bhattarai EM, Das S, Hobbs PR (2003) How extensive are yield declines in long-term rice-wheat experiments in Asia? Field Crops Res 81(2–3). https://doi.org/10.1016/S0378-4290(02)00219-8

Maini P, Rathore LS (2008) Economic impact assessment of agro- meteorological advisory service of NCMRWF. NMRF/PR/01/2008, National Centre for Medium Range Weather Forecasting, Ministry of Earth Sciences, Government of India, p 104

MHA (2025) North East division. Ministry of Home Affairs, Government of India. https://www.mha.gov.in/en/commoncontent/north-east-division

Ministry of finance (2025) Agriculture and food management: sector of the future. In: Economic survey 2024–25. Department of Economic Affairs, Government of India, New Delhi, pp 245–270

Mishra V, Dangar S, Tiwari VM, Lall U, Wada Y (2024) Summer monsoon drying accelerates India's groundwater depletion under climate change. Earth's Future 12(8):e2024EF004516. https://doi.org/10.1029/2024EF004516

Mukherjee S, Aadhar S, Stone D, Mishra V (2018) Increase in extreme precipitation events under anthropogenic warming in India. Weather Climate Extremes 20:45–53. https://doi.org/10.1016/J.WACE.2018.03.005

Murari KK, Mahato S, Jayaraman T, Swaminathan M (2018) Extreme temperatures and crop yields in Karnataka India. Review Agrarian Stud 8(2):92–113

Narsimha R (1994) The dynamics of the ramdas layer. Curr Sci 66(1):16–22

Norgate M, Tiwari PR, Das S, Kumar D (2024) On the heat waves over India and their future projections under different SSP scenarios from CMIP6 models. Int J Climatol 44(3):973–995. https://doi.org/10.1002/JOC.8367

O'Toole JC, Chang TT (1978) Drought resistance in cereals-rice: a case study. In: Mussell H, Staples RC (eds) Stress physiology in crop plants. Wiley

Panda DK, AghaKouchak A, Ambast SK (2017) Increasing heat waves and warm spells in India, observed from a multiaspect framework. J Geophys Res Atmos 122(7):3837–3858. https://doi.org/10.1002/2016JD026292

Panda DK, Ambast SK, Shamsudduha M (2021) Groundwater depletion in northern India: impacts of the sub-regional anthropogenic land-use, socio-politics and changing climate. Hydrol Process 35(2):e14003. https://doi.org/10.1002/HYP.14003

Raman KV, Balaguru T (1988) NARP-An innovative approach towards FSR in India. Agric Adm Ext 30(3):203–213. https://doi.org/10.1016/0269-7475(88)90055-4

Rao PS, Saseendran SA, Rathore LS, Bahadur J (2007) Medium range weather forecasts in India during Monsoon 1994. Meteorol Appl 3(4):317–324. https://doi.org/10.1002/met.5060030404

Ravindra K, Bhardwaj S, Ram C, Goyal A, Singh V, Venkataraman C, Bhan SC, Sokhi RS, Mor S (2024) Temperature projections and heatwave attribution scenarios over India: a systematic review. Heliyon 10(4):e26431. https://doi.org/10.1016/J.HELIYON.2024.E26431

Salunke P, Keshri NP, Mishra SK, Dash SK (2023) Future projections of seasonal temperature and precipitation for India. Front Climate 5: 1069994. https://doi.org/10.3389/fclim.2023.1069994

Samantray P, Gouda KC (2024) A review on the extreme rainfall studies in India. Nat Hazards Res 4(3):347–356. https://doi.org/10.1016/J.NHRES.2023.08.005

Sandeep S, Ajayamohan RS, Boos WR, Sabin TP, Praveen V (2018) Decline and poleward shift in Indian summer monsoon synoptic activity in a warming climate. Proc Natl Acad Sci USA 115(11):2681–2686. https://doi.org/10.1073/PNAS.1709031115/SUPPL_FILE/PNAS.1709031115.SAPP.PDF

Sharma U, Chetri P, Minocha S, Roy A, Holker T, Patt A, Joerin J (2021) Do phone-based short message services improve the uptake of agri-met advice by farmers? A case study in Haryana India. Climate Risk Manag 33:100321. https://doi.org/10.1016/J.CRM.2021.100321

Shukla S, Sijo A, Parekh S (2025) Advancing climate-resilient agriculture in India by strengthening institutional capacity–Lessons from Bihar and Odisha advancing climate-resilient agriculture in India by strengthening institutional capacity: lessons from Bihar and Odisha. Council on energy, environment and water (CEEW), New Delhi. www.creativecommons.org/licences/by-nc/4.0/legalcode

Singh S, Mall RK, Singh N (2001) Changing spatio-temporal trends of heat wave and severe heat wave events over India: an emerging health hazard. Int J Climatol 41:E1831–E1845. https://doi.org/10.1002/joc.6814

Singh VP, Sharma N, Ojha CSP (Eds.) (2004) The brahmaputra basin water resources (1st ed., Vol. 47). Springer Netherlands. https://doi.org/10.1007/978-94-017-0540-0

Venkatasubramanian K, Tall A, Hansen J, Aggarwal P (2014) Assessment of India's agro-meteorological advisory service from a farmer perspective. CCAFS working paper no. 54. In: CGIAR research program on climate change, agriculture and food security (CCAFS) (Issue 54)

WMO (2021) WMO atlas of mortality and economic losses from weather, climate and water extremes (1970–2019). In: WMO statement on the state of the global climate (WMO-No. 1267; vol 1267, Issue 1)

Yan Y, Hu Z, Wang L, Jiang J, Dai Q, Gan F, Almaliki AH, Hashim MA, Hussein EE, Ghoneim SSM (2024) Impact of extreme rainfall events on soil erosion on karst Slopes: a study of hydrodynamic mechanisms. J Hydrol 638:131532. https://doi.org/10.1016/J.JHYDROL.2024.131532

Chapter 4
Climate Smart Agricultural Practices in Southern India

T. Prathima

Abstract Agrometeorology, integrating agricultural practices with climate data, plays a crucial role in developing Climate-Smart Agriculture (CSA) to mitigate climate change impacts on agriculture. With India housing 17.6% of the global population and 15% of livestock on only 2.4% of the world's land, climate change challenges like unpredictable rainfall, extreme temperatures are significant threats. This study discusses how agrometeorology supports CSA through weather-based agro-advisories, real-time contingency planning, and climate-resilient technologies. Localized weather forecasting and agromet advisories aid farmers in timely decision-making, minimizing losses due to adverse weather. CSA strategies emphasize sustainable practices like in-situ soil moisture conservation, drought-resistant crops, and efficient water management systems, such as farm ponds and rainwater harvesting. Advanced weather information allows crop planning at a micro-level, increasing resilience and productivity by aligning crop cycles with climate patterns. Additionally, CSA promotes livestock management, crop rotation, and agroforestry to preserve biodiversity and enhance soil health. By adopting CSA practices supported by agrometeorology, farmers can improve productivity, reduce greenhouse gas emissions, and achieve sustainable development goals. This integrated approach highlights the importance of policy support, educational outreach, and research advancements in fostering resilient agricultural systems.

Keywords Agrometeorology · Climate-smart agriculture · Agro-advisory · Water management · Climate resilience

4.1 Introduction

Southern India, encompassing the states of Andhra Pradesh, Telangana, Tamil Nadu, Karnataka, and Kerala, is characterized by a rich diversity of agro-climatic zones ranging from coastal belts to semi-arid and hilly terrains Agriculture remains the

T. Prathima (✉)
Agronomy, Regional Agricultural Research Station, Tirupati, India
e-mail: t.prathima@angrau.ac.in

D. Kumar Swain et al. (eds.), *Climate-Smart Technologies for Sustainable Agricultural Production*, https://doi.org/10.1007/978-3-032-25666-9_4

backbone of rural livelihoods in this region, with a substantial proportion of the rural workforce dependent on agriculture and allied activities, consistent with national labour force data showing that agriculture accounted for 46.1% of total employment in 2023–24. However, climate change has emerged as a significant threat to agricultural sustainability. Rising temperatures, irregular monsoon patterns, droughts, floods and cyclones are affecting both rainfed and irrigated farming systems (IMD 2020a, b; MoEFCC 2020).

More intense heat waves of longer duration at a higher frequency are projected over India, with increase in surface air temperature by 0.5 °C during 1901–2007. In Southern India, observational records indicate a significant warming trend, with states such as Andhra Pradesh, Telangana, Tamil Nadu, and Karnataka experiencing increases in annual mean temperature of the order of 0.6–0.8 °C since the mid-twentieth century, along with a rising frequency of hot days and warm nights (IMD 2020a, b; Krishnan et al. 2020). Accelerated warming during 1971–2009, with a greater increase in minimum temperatures compared to maximum temperatures, further indicates changes in diurnal temperature patterns (IPCC 2023). Rainfall patterns have also shifted, showing an increase in precipitation over the west coast, north Andhra Pradesh, and northwest India, while a decreasing trend is observed in east Madhya Pradesh and adjoining regions. Furthermore, the incidence of one-day extreme rainfall events is on the rise, contributing to greater flood risks and soil erosion. Concurrently, the sea level in the Indian Ocean has been rising at a rate of 1.63 mm per year from 1993 to 2009, exacerbating coastal vulnerability.

Adding to these concerns, new evidence highlights the significant influence of local land use and land cover changes in intensifying heavy precipitation events, further amplifying regional climate variability and extreme weather risks. Frequency of cyclones during post monsoon seasons (2071–2100) is projected to much higher than baseline scenario (1961–1990) (IPCC 2023). All the above-mentioned points represent the physical evidence and effects of climate change. However, the biological changes include early blossoming of trees, appearance of grasses in Antarctica, changing cropping pattern etc. These combined factors pose serious challenges to agriculture, water resources, and overall socio-economic stability, underscoring the urgent need for adaptive climate-smart strategies.

Between 1901 and 2020, India's mean land surface air temperature increased by approximately 0.62 °C (IMD 2020a, b). This national trend is reflected in Southern India, where warming and climatic variability are increasingly impacting agricultural systems. Rainfed crops like millets and pulses, which dominate Karnataka and Telangana, and irrigated crops such as rice and sugarcane in Tamil Nadu and Andhra Pradesh, are under growing stress due to rising temperatures and erratic rainfall. In Kerala, coastal flooding and landslides are recurring threats, particularly affecting plantation crops such as coconut, banana, and rubber (Table 4.1). In Telangana, high temperatures during the *kharif* season, have led to crop losses of 1.65% for rice and 4.09% for groundnut, respectively (Moulkar and Dayakar 2023).

The agricultural sector in Southern India is also marked by fragmented landholdings, high dependency on monsoons, and limited access to advanced technologies-making it particularly vulnerable to climate extremes (ICAR-NICRA 2022). In this

Table 4.1 Climate vulnerabilities in southern India

State	Dominant crops	Climate risk factors	Vulnerability level
Andhra Pradesh	Rice, Groundnut, Cotton	Cyclones, salinity, drought	High
Telangana	Maize, Cotton, Pulses	Heat stress, dry spells	High
Tamil Nadu	Rice, Sugarcane, Banana	Erratic rainfall, groundwater stress	Medium-High
Karnataka	Millets, Pulses, Oilseeds	Frequent droughts, temperature rise	High
Kerala	Coconut, Rubber, Banana	Floods, landslides	Medium

Source ICAR-NICRA (2023); SAPCC Reports for Andhra Pradesh, Tamil Nadu, Telangana, and Kerala (2014–2015)

context, Climate Smart Agricultural Practices (CSAPs) such as drought-tolerant crop varieties, micro-irrigation systems, integrated nutrient and pest management, and agroforestry offer region-specific solutions that not only sustain productivity but also strengthen ecological resilience (FAO 2013; Dhivya et al. 2024).

To address these challenges, the concept of CSA, introduced by FAO (2013), provides a promising approach. CSA aims to achieve three key objectives: (1) Sustainably increase productivity, (2) Enhance resilience and adaptation to climate change, and (3) Reduce greenhouse gas (GHG) emissions. This chapter presents an in-depth exploration of CSAPs implemented in Southern India and evaluates their effectiveness in building resilience and improving sustainability.

4.2 Climate Smart Agricultural Practices (CSAPs)

4.2.1 Weather Smart Technologies

4.2.1.1 Real Time Contingency Planning (RTCP)

Any contingency measure, either technology related (land, soil, water, crop) or institutional and policy based, which is implemented based on real time weather pattern (including extreme events) in any crop growing season" is considered as Real Time Contingency Planning (RTCP). It involves preparing flexible and adaptive strategies to respond immediately to unexpected events or changing conditions as they occur. It is based on continuous monitoring of real-time data, such as weather updates, river levels, or crop status, allowing quick decisions to minimize risks and losses. If done timely and effectively, RTCP likely to contribute to household and village food and fodder security.

In the context of climate change, this approach helps manage sudden extreme events like heatwaves, floods, or unseasonal rainfall by implementing timely actions such as adjusting irrigation, changing sowing dates, or applying protective measures. The main objective is to enhance resilience, reduce vulnerability, and maintain stability in uncertain and rapidly changing environments.

Real Time Contingency
Early season drought—Dust Mulching

- Under early-season drought conditions, dust mulching is recommended as a moisture conservation practice to reduce soil evaporation and improve early crop establishment in rainfed systems, as outlined in national drought coping guidelines for dryland agriculture (ICAR-CRIDA 2012).
- Midseason drought—Dust mulching, straw mulching, Foliar spray-spraying of KNO_3

During mid-season drought, a combination of soil moisture conservation practices such as dust mulching or straw mulching, along with foliar nutrient sprays, is advised to alleviate moisture stress, maintain leaf turgor, and support crop recovery, as emphasized in crop contingency plans for southern India (Farooq et al. 2009; TNAU 2024).

- Terminal drought—Protective irrigation
- Under terminal drought situations, protective (life-saving) irrigation at critical growth stages is identified as an effective strategy to minimize yield losses in rainfed crops, particularly in dryland regions, and forms a key component of drought mitigationkl/.xstrategies recommended for rainfed production systems (ICAR-CRIDA 2012).

4.2.1.2 Crop and Genetic Interventions

(a) ***Stress-Tolerant Varieties***

Some of the stress-tolerant varieties for Southern India are mentioned in Table 4.2.

4.2.1.3 Crop Diversification, Intercropping, Crop Rotation and Green Cover

Crop diversification, especially intercropping systems like maize + cowpea or groundnut + red gram, improves resource use efficiency, stabilizes incomes, and enhances food security. Such systems also reduce pest incidence and improve soil cover.

The concept of 365 Green Cover is a key element of CSA, focusing on maintaining continuous vegetation cover on agricultural land throughout the year. This is achieved through practices such as cover cropping, mulching, agroforestry, and intercropping, which protect the soil from erosion, improve soil structure, enhance organic matter, and support beneficial organisms. By ensuring that the soil remains

Table 4.2 Some of the varieties suitable for Southern India's climate challenges have been promoted

Crop	Variety name	State(s)	Stress tolerance	Source	Adoption level
Rice	CR Dhan 801	Andhra Pradesh, Tamil Nadu, Telangana, Karnataka	Drought tolerance	ICAR-NRRI	Moderate
	Sahbhagi Dhan	Karnataka, Tamil Nadu	Drought tolerance		Moderate
	MTU 1010	Andhra Pradesh, Telangana	Heat tolerance	ANGRAU	High
	ADT 48	Tamil Nadu	Early maturity, partial drought tolerance	TNAU	Moderate
Maize	DHM 117/ DHM 121	Andhra Pradesh, Telangana	Heat and drought tolerance	ICAR-IIMR	High
	Bio 9637	Andhra Pradesh, Karnataka	Heat tolerance	AICRP-Maize	Low
Sorghum	SPV 2217	Karnataka	Drought tolerance (grain sorghum)	ICAR-Sorghum	Moderate
Pearl Millet	ICMV 221	Karnataka, Tamil Nadu	High temperature and drought resistance	ICRISAT	Low
Finger Millet	KMR 204	Karnataka	Drought tolerance	UAS Dharwad	Moderate
	GPU 28	Karnataka, Tamil Nadu	Drought and blast resistance	TNAU	High
Blackgram	LBG 752	Andhra Pradesh, Telangana	Drought tolerance	ANGRAU	Moderate
Greengram	VBN 4	Tamil Nadu	drought resilience	TNAU	High
Red Gram	BSMR 736	Andhra Pradesh, Telangana	Drought, sterility mosaic	IIPR	Moderate
Cowpea	VCP 09	Kerala, Tamil Nadu	Drought tolerance	KAU	Low
Groundnut	ICGV 91114	Andhra Pradesh, Telangana	Drought tolerance	ICRISAT	Moderate

(continued)

Table 4.2 (continued)

Crop	Variety name	State(s)	Stress tolerance	Source	Adoption level
	TG 37A	Karnataka, Tamil Nadu	Heat & drought tolerance		Low
	TMV 7	Tamil Nadu	Drought-resistant	TNAU	High
	Kadiri 6 (K-6)	Andhra Pradesh, Telangana	Drought tolerance	ICAR-DGR	High
Banana	BRS-1 (Udhayam)	Kerala, Tamil Nadu	Wind & salt tolerance	NRCB, Trichy	Low
Coconut	Kalpa Surya	Kerala	Drought tolerance	CPCRI	Moderate
Rubber	RRII 105	Kerala, Tamil Nadu	Drought & tapping panel dryness resistance	RRII	High
	RRII 430	Kerala, Karnataka	High temperature, wind & leaf disease tolerance		Low
	RRII 414	Kerala	Moderate drought tolerance, high yield		Low

covered, farmers can improve water retention, reduce surface runoff, regulate microclimates, and sequester atmospheric carbon dioxide, thereby contributing to climate change mitigation. In India, CSA initiatives like the National Mission for Sustainable Agriculture promote 365 Green Cover to improve crop productivity, enhance resilience to climate variability, and reduce greenhouse gas emissions. This approach not only supports the goal of increasing the country's carbon sink by 2030 but also aligns with global efforts to address food security and environmental sustainability.

Crop rotation is an important practice within CSA that helps improve agricultural sustainability, productivity, and resilience to climate change. In the context of CSA, crop rotation contributes to climate resilience by breaking pest and disease cycles naturally, reducing reliance on chemical inputs, and promoting soil structure and organic matter buildup, which helps retain moisture during droughts and prevents erosion during heavy rains. For example, rotating legumes with cereals not only helps fix nitrogen in the soil but also reduces the need for synthetic fertilizers, thereby lowering greenhouse gas emissions (Sahu et al. 2020). Moreover, by diversifying cropping patterns, farmers can spread economic risk and adapt better to shifting climate conditions (Table 4.3). National programs in India, such as the *National Mission on Sustainable Agriculture*, encourage crop rotation as a key CSA practice to enhance food security, promote soil conservation, and improve long-term farm productivity while mitigating climate impacts.

Table 4.3 Recommended intercropping systems for climate-smart agriculture in Southern India

State	Intercropping systems	Benefits	Source
Tamil Nadu	• Maize + Vegetable cowpea (2:2) • Maize + Blackgram	• High LER (1.53), B:C ratio (3.34) • Enhanced soil fertility • Reduced pest incidence	Parimaladevi et al. (2019)
Andhra Pradesh	• Maize + Groundnut • Redgram + Greengram • Sorghum + Chickpea	• Improved land use efficiency • Better drought resilience	ANGRAU & AICRP-based systems
Telangana	• Maize + Redgram • Castor + Cowpea • Pearl millet + Redgram	• Stable income in rainfed areas • Legume N-fixation benefits	ICRISAT & PJTSAU
Karnataka	• Finger Millet + Redgram • Groundnut + Sunflower • Maize + Dolichos	• Better soil cover • Moisture retention • Yield stability	UAS Dharwad & ICRISAT
Kerala	• Groundnut + Redgram (4:5) • Banana + Cowpea • Coconut + Pineapple	• Weed control • Soil moisture retention • Laterite soil suitability	Kerala Agricultural University

4.2.1.4 Micro Crop Planning

Micro Crop Planning is a localized and farmer centric approach to planning cropping systems. Unlike macro-level crop planning (at the state or national level), MCP focuses on village, block, or even farm-level conditions, considering Soil type and fertility.

- Rainfall distribution and irrigation potential.
- Local climate and micro-weather patterns.
- Market access and demand for crops.
- Farmer's resources such as labour, machinery, and capital.

Micro crop planning translates the weather data to useful information such as

- Mapping of rainfall variability and characterization of crop growing environment
- Evaluate suitable crop planning at micro level based on soil, rainfall and other weather elements

The goal of MCP is to "place the right crop in the right place at the right time", ensuring both ecological sustainability and economic profitability.

Table 4.4 Threshold limits of weekly rainfall for different agricultural operations

S.No.	Agricultural operation	Threshold limit of weekly rainfall (mm)
1	Land preparation	10
2	Sowing	25–30
3	Weeding and fertilizer application	30–40
4	Avoidance of fertilizer application	>80

Characterization of rainfall based on threshold limit
Rainfall < threshold limit—Dry
Rainfall > Wet

For instance, in a semi-arid region with shallow soils and erratic rainfall, micro crop planning may recommend short-duration millets and pulses instead of water-intensive crops like paddy. Conversely, in well-irrigated command areas, it may suggest rice–wheat rotation or high-value horticulture crops (Table 4.4).

A micro-level study on paddy-based cropping systems in the Nellore district (Andhra Pradesh) analysed the sustainability and profitability of different cropping combinations such as paddy-groundnut, paddy-paddy, and paddy-cotton using primary data from farmers during 2022–23. The results showed that the paddy-cotton system yielded the highest net income and sustainability index, demonstrating how local crop planning can influence both economic returns and long-term sustainability in rainfed agricultural environments (Sri et al. 2024).

4.2.1.5 Length of Growing Period (LGP) in Crop Planning

A critical scientific basis for micro crop planning is the Length of Growing Period (LGP). LGP refers to the number of days in a year when temperature and moisture conditions are suitable for crop growth. It is determined by:

- Rainfall availability and distribution.
- Soil moisture storage capacity.
- Evapotranspiration demand.

Broad categories of LGP are:

- Short LGP (<90 days): Areas suited for short-duration crops like pearl millet, cowpea, green gram.
- Medium LGP (90–180 days): Areas suited for cereals, pulses, oilseeds, and some double cropping.
- Long LGP (>180 days): Areas suited for multi-cropping, rice-based systems, and commercial crops.

By aligning crop choices with LGP, MCP minimizes risks of crop failure, optimizes water and nutrient use, and enhances system productivity.

ICT Based Agro-Advisories

Agriculture in South India is highly sensitive to fluctuations in weather and climate, primarily due to the region's heavy reliance on the southwest monsoon, which contributes over 70% of the annual rainfall. The growing occurrence of extreme weather events such as droughts, floods, unseasonal rains, and heatwaves largely driven by climate change, has further increased the vulnerability of farming communities. Studies have indicated that climate change poses a significant threat to overall food productivity. In this context, Weather-Based Agro-Advisory Services (AAS) have become an essential tool for promoting climate-resilient agriculture by delivering timely, location-specific, and actionable advice on weather and crop management to farmers. These advisories enable farmers to make informed decisions regarding sowing, irrigation, fertilization, pest and disease control, and harvesting, thus minimizing losses and improving productivity and income (Rathore 2013). Providing accurate and locally relevant climate and weather information is a crucial step in enhancing community resilience to climate change, diseases, and natural disasters (Srishailam et al. 2023; Manjunath et al. 2024).

4.2.2 Water-Smart Practices

4.2.2.1 Drip and Sprinkler Irrigation

Micro-irrigation technologies such as drip and sprinkler systems have been widely promoted in crops like sugarcane, groundnut, and vegetables in Tamil Nadu and Andhra Pradesh. These systems deliver water directly to the root zone, thereby minimizing evaporation losses, reducing weed growth, enhancing nutrient uptake, and increasing water-use efficiency. According to Kuppannan and Raman (2012), the adoption of drip irrigation across nine states-including Tamil Nadu and Andhra Pradesh has provided substantial benefits, resulting in water savings of up to 40–80%. In Coimbatore district of Tamil Nadu, adoption of micro-irrigation systems (MIS) in banana cultivation led to increases in cropped area (6.54%), irrigated area (6.55%), cropping intensity (69.19%), and irrigation intensity (70.95%). Financial analysis using discounted cash flow techniques showed that drip irrigation was economically viable, with a positive Net Profit Value of ₹ 43,851, a BCR of 1.42, and an Internal Rate of Returns (IRR) of 30.88%.

Government programs like the Per Drop More Crop (PDMC) component under the Pradhan Mantri Krishi Sinchayee Yojana (PMKSY) have significantly accelerated the adoption of micro-irrigation technologies such as drip and sprinkler systems in rain-scarce regions of Southern India. According to a 2023 Press Information Bureau release, approximately 83.06 lakh hectares have been covered under micro-irrigation through PDMC from 2015–16 to 2023–24, with ₹18,714.69 crore released as central assistance to states during this period. Since 2022–23, the scheme has been implemented under the Rashtriya Krishi Vikas Yojana (RKVY) framework, offering subsidized support, and is backed by a ₹5000 crore Micro Irrigation Fund (MIF) through NABARD. These initiatives have improved water-use efficiency, reduced input costs, and enhanced crop productivity and farm incomes (PIB 2023).

Studies show that water productivity (kg/m^3) improves significantly with micro-irrigation, particularly in high water-demanding crops such as sugarcane and banana (Gorain et al. 2020; Bhuvaneswari et al. 2020).

Water Use Efficiency (WUE) is a key metric in irrigation efficiency and is calculated as:

$$\text{WUE} = \frac{\text{Crop Yield (kg/ha)}}{\text{Water applied } \left(\text{m}^3/\text{ha}\right)}.$$

4.2.2.2 Alternate Wetting and Drying (AWD)

AWD is a water-saving practice in rice that involves periodic drying of the field instead of continuous submergence. This reduces percolation losses and allows oxygenation of the root zone. Lampayan et al. (2015) report that AWD can save 25–35% irrigation water and reduce methane emissions by 30–40% without compromising yield.

In Tamil Nadu and Karnataka, AWD is promoted through climate-resilient rice initiatives. It also lowers labour costs related to water management and reduces standing water-related pest incidence, particularly stem borers. Studies have shown that AWD can reduce insect pest infestation by up to 92%, with stem borer incidence reported to be as low as 5.6% under AWD, compared to 27.8% under conventional flooding (Katti and Padmavathi 2022).

4.2.2.3 Laser Land Levelling

In the arid and semi-arid regions of Asia, the challenges related to water availability in agriculture are expected to worsen with climate change. With every 1 °C rise in temperature, irrigation water demand is projected to increase by at least 10%, While water availability is likely to decline (Rajkumar et al. 2022). As a result, precision water management technologies will play a crucial role in ensuring future food security. Traditional land leveling methods are cumbersome, time-consuming, and costly, which is why modern techniques like Laser Land Levelling (LLL) with a zero gradient are gaining global popularity.

Laser land levelling involves smoothing the land surface to within ±2 cm of the average elevation using laser-equipped drag buckets. This method employs high-horsepower tractors and soil movers fitted with GPS and/or laser-guided systems to precisely cut or fill soil, achieving the desired level or slope. LLL is recognized as a key climate-smart intervention, particularly in intensively cultivated flood-irrigated systems in southern India, and has been applied in both irrigated and rainfed ecosystems.

LLL has emerged as an important precision land management practice in Southern India to improve irrigation efficiency and crop productivity. Unlike conventional leveling, which leaves uneven patches in fields, LLL uses laser-guided equipment to create a uniformly graded surface with minimal slope variation. This precision ensures more uniform water distribution, reduces waterlogging and runoff, and enhances fertilizer use efficiency. In semi-arid regions of Karnataka, field demonstrations have shown that LLL not only reduced the drudgery of land preparation but also resulted in 25–35% higher paddy yields, 25–30% irrigation water saving, and 30–40% reductions in labour and energy use, along with higher levelling index and more uniform soil moisture distribution (Kanannavar et al. 2020). Similarly, in Telangana, cultivation of rice under laser-leveled fields with conventional transplanting significantly enhanced growth and yield attributes of rice compared to conventional and unleveled fields (Ramya et al. 2021). Beyond water savings, farmers in these regions also report benefits such as improved timeliness of sowing, easier mechanization, and enhanced profitability due to reductions in labour and fuel use. Despite these advantages, adoption remains constrained by factors such as high service charges in some districts and the need for trained operators. Nevertheless, with increasing availability of LLL through custom hiring services and government support under schemes like Per Drop More Crop (PMKSY), the practice is gradually gaining acceptance as a climate-smart technology suited to the irrigated cropping systems of Southern India.

4.2.2.4 Rainwater Harvesting

Rainwater harvesting (RWH) is a vital climate-smart strategy for Southern India, where rainfall is highly seasonal and often erratic. By capturing, storing, and efficiently using rainwater, farmers can enhance water availability, reduce dependency on groundwater, and improve cropping intensity under variable climate conditions.

- **Farm Ponds**: Farm ponds are small, on-farm water storage structures that capture runoff during the monsoon and provide supplemental irrigation in dry spells. In Karnataka's Northern Dry Zone particularly in Vijayapura district where paddy, chilli, maize, sorghum, sunflower, cotton and chickpea are the major crops cultivated. In the district, farm ponds increased cropping intensity to 157% compared to 142% in farms without ponds, while farm incomes rose by 17–48% and rabi employment doubled (Dupdal et al. 2024). In Tamil Nadu, the Kalaignarin All Village Integrated Agriculture Development Programme (KAVIADP) has successfully constructed farm ponds free of cost, enabling farmers in Mayiladuthurai district to diversify their livelihoods. For example, beneficiaries practicing pisciculture reported potential incomes of up to ₹1 lakh annually, alongside improved groundwater recharge and opportunities for integrated farming (Charan

2023). However, challenges such as siltation, which can significantly reduce pond capacity over time, highlight the importance of regular maintenance and proper design for sustainability.

- **Watershed-Based Interventions**: Watershed development projects, combining check dams, contour bunds, and percolation tanks, enhance groundwater recharge and soil moisture across Southern India. Integrated watershed management programs have significantly improved agricultural productivity, water availability, and resilience to drought. In Tamil Nadu, such initiatives increased cropping intensity and improved soil fertility through soil and water conservation measures (Kuppannan and Devarajulu 2009). In Telangana, the Mission Kakatiya initiative, focusing on restoring minor irrigation tanks, resulted in a 78.5% increase in household agricultural income in the tank ayacut areas (Government of Telangana 2025). In Karnataka, the Upper Krishna Project facilitated irrigation in drought-prone regions covering an area of 1,536,000 acres of land supporting crop production and water availability (Government of Karnataka 2025). In the semi-arid region of Andhra Pradesh, watershed programs combining check dams, contour bunds, and percolation tanks significantly improved groundwater levels, crop yields, and incomes while reducing runoff losses (Wani 2002). Collectively, these programs contribute to improved water use efficiency, higher cropping intensity, and enhanced agricultural resilience across the region.

4.2.3 Carbon smart technologies

4.2.3.1 Conservation Agriculture (CA)

Conservation agriculture, characterized by minimum tillage, residue retention, and diverse crop rotations is emerging as a climate-smart strategy in Southern India, particularly in Andhra Pradesh, Telangana, and Karnataka, where dryland and rainfed systems dominate. Long-term field experiments in semi-arid Alfisols of Southern India reveal that minimum tillage paired with crop residue recycling significantly increases soil organic carbon (SOC) by 11% in the top 0–20 cm layer while enhancing labile carbon fractions like microbial biomass C and permanganate-oxidizable C by 16–47% compared to conventional tillage (Prasad et al. 2016).

In Southern Telangana experiments involving zero tillage with retained residues such as Sesbania rostrata and maize stubble demonstrated improved soil nutrient stratification, increased carbon sequestration rates, and stable maize yields across three years in continuous maize-cotton systems (Nthebere et al. 2025).

In rainfed millet systems, a study from Tamil Nadu showed that no-till combined with mulching enhanced yields of finger millet, little millet, and foxtail millet by 24–36%, improved water-use efficiency (RWUE from 5.16 to 7.03 kg/ha/mm), and significantly suppressed weed growth (Sathiya et al. 2024).

These outcomes clearly demonstrate that conservation agriculture not only enhances soil health and productivity in diverse agro-climatic zones of Southern

India but also serves as a resilient, low-emission strategy vital for climate-smart agricultural transformation in the region.

4.2.3.2 Agroforestry and Livelihood Diversification

Agroforestry systems that integrate multipurpose tree species such as *Gliricidia sepium*, *Leucaena leucocephala* (Subabul), neem, and fruit trees (mango, tamarind, amla) have emerged as vital components of climate-smart agriculture in Southern India. These systems offer significant ecological and economic benefits by improving soil structure, enhancing nutrient cycling, buffering microclimatic extremes, and sequestering atmospheric carbon.

Agroforestry contributes to:

- Carbon sequestration: By enhancing carbon storage through tree biomass and soil organic matter, agroforestry systems serve as effective carbon sinks and support climate change mitigation.
- Soil fertility and microclimate improvement: Tree litter improves organic matter and reduces evapotranspiration, particularly in dryland areas.
- Livelihood diversification: Sale of fodder, fruits, fuelwood, timber, and medicinal products provides supplementary income, reducing reliance on rainfed cropping.
- Climate resilience: Tree roots stabilize soils, mitigate wind erosion, and buffer crops from temperature extremes, reducing climate-induced risks.

In Karnataka, tree-based farming models implemented notably in dryland districts like Tumkur led by BAIF in collaboration with local communities, have successfully increased household income while providing consistent sources of fodder, fuelwood, and biomass for composting, enhancing both livelihoods and landscape resilience (Doreswamy 1999). In the rainfed regions of Tamil Nadu, agrosilvopastoral systems which integrate fodder trees, food crops (like finger millet), and legume fodders in the same unit of land have been shown to improve fodder availability throughout the year, enhance soil health, and provide multiple income streams for smallholder farmers. These tree-based, integrated systems not only meet livestock nutrition requirements during lean seasons but also boost food grain production and overall livelihood security, making them a sustainable and resilient agricultural option in dryland environments of South India (Venkatesh et al. 2024). Similarly, in Andhra Pradesh, models like *MPTS (Multipurpose Tree Species) on bunds* and *horti-pastoral systems* are being scaled under the National Agroforestry Policy (2014) and the Sub-Mission on Agroforestry (SMAF). These models enhance productivity and soil health in semi-arid regions.

Furthermore, Tamil Nadu and Telangana are promoting *horti-agricultural models* under state-supported watershed programs, integrating guava or amla with groundnut or millets. These models support pollinator activity, stabilize farm income, and improve water retention (Osman et al. 2019). In Tamil Nadu, the Salaiyur watershed project near Coimbatore introduced guava, tamarind, sapota, amla, coconut, and banana plantings across rainfed wastelands using improved micro-site techniques (Chand et al. 2015).

In Kerala, homestead agroforestry is a dominant land-use system, characterized by multi-tiered tree components. These systems sequester approximately 2.7 Mg C ha^{-1} $year^{-1}$ through tree biomass and maintain soil organic carbon levels around 2.5%, both higher than in adjacent monocropped plantation systems (Mina et al. 2023).

Agroforestry thus supports climate change mitigation through enhanced carbon sinks and contributes to adaptation by improving farm-level diversification, soil health, and ecological stability across Southern India.

4.2.4 Nutrient-Smart Practices

4.2.4.1 Integrated Nutrient Management (INM)

Integrated Nutrient Management (INM) emphasizes the judicious and synergistic use of chemical fertilizers, organic manures, green manures, crop residues, and biofertilizers to improve nutrient use efficiency, sustain soil fertility, and support long-term agricultural productivity. In the diverse agro-climatic conditions of Southern India, where soil types and rainfall patterns vary significantly, INM has emerged as a climate-smart approach that supports sustainable intensification and ecological resilience.

Long-term trials conducted under the All India Coordinated Research Project (AICRP) and documented by ICAR-Central Research Institute for Dry land Agriculture (CRIDA) have shown that integrated nutrient strategies significantly enhance soil organic carbon, improve nutrient balance, and increase crop yields, particularly under dryland farming conditions common to Andhra Pradesh, Tamil Nadu, and Karnataka (ICAR-CRIDA 2023). Similarly, the Nammazhvar Organic Farming Research Centre under Tamil Nadu Agricultural University (TNAU) reported that the use of INM practices has led to improved soil health and microbial activity, supporting higher returns and better nutrient synchronization over multiple cropping seasons (TNAU 2023a). TNAU's Agritech portal also highlights that combining chemical and organic inputs, along with the application of biofertilizers such as Rhizobium, Azospirillum, and phosphate-solubilizing bacteria, improves fertilizer use efficiency and reduces nutrient losses while enhancing overall soil fertility (TNAU 2023b). These practices are particularly beneficial in rainfed systems where nutrient availability is inconsistent, contributing both to climate change adaptation by increasing soil water retention and buffering capacity, and to mitigation by lowering nitrous oxide emissions through optimized nitrogen management.

4.2.4.2 Site-Specific Nutrient Management (SSNM)

SSNM involves customizing fertilizer recommendations to suit individual field conditions such as soil type, crop type, water availability, and nutrient availability rather than relying on generalized blanket guidelines. This approach ensures that nutrients are applied at the right rate, right time, right place, and right source (4R nutrient stewardship) (Richards et al. 2015)

Tools such as Nutrient Expert® and GreenSeeker® are practical decision-support systems built on SSNM principles. Nutrient Expert® computes fertilizer recommendations by integrating local soil nutrient supply, predicted yield potential, and historical farmer practices, while GreenSeeker® uses optical sensors (NDVI) to guide real-time nitrogen management (Sapkota et al. 2021; Pampolino et al. 2014)

Adoption of SSNM through these tools has consistently led to 15–25% reductions in fertilizer use without compromising yields, and in several rice-wheat systems, it boosted crop productivity by 10–15% while reducing greenhouse gas emissions by 2–20% compared to farmers' fertilizer practices (Sapkota et al. 2021). This makes SSNM a key climate-smart practice, enhancing nutrient use efficiency, reducing environmental pollution, and increasing farm income especially relevant for rainfed areas of Southern India with diverse agro-ecological variability.

4.2.5 Energy Smart Practices

4.2.5.1 Solar-based Technologies

These are increasingly vital for powering climate-smart farming. Solar irrigation pumps reduce dependence on diesel and unreliable electricity, lower GHG emissions, and provide timely irrigation for crops, especially in off-grid and drought-prone areas (Shinde and Wandre 2015). Solar-powered cold storage helps reduce post-harvest losses of fruits, vegetables, and milk, extending shelf life and improving farmers' market returns (Khemariya 2025). Solar dryers for spices, pulses, and fruits improve product quality and reduce spoilage (El-Sebaii and Shalaby 2012; Kumar et al. 2016). Integrating solar fencing reduces crop depredation by animals (Vibha et al. 2021) while solar-powered ICT kiosks enhance access to advisories and market information (Mittal et al. 2010). Collectively, solar solutions not only reduce energy costs but also enhance resilience, supporting sustainable intensification in Southern India.

4.2.5.2 Wind Based Technologies

Wind based technologies are emerging as a key strategy within CSA, offering a sustainable and climate-friendly solution to meet energy needs on farms while reducing greenhouse gas emissions. Integrating wind energy into agricultural systems helps farmers reduce dependence on fossil fuels for activities such as irrigation pumping, grain milling, cold storage, and electrification of farm operations. By harnessing clean, renewable wind power, farms can operate irrigation systems more efficiently, especially in regions prone to energy shortages or unreliable grid access, thereby improving water-use efficiency and ensuring timely irrigation under changing climate conditions. This approach is particularly relevant to Southern India, where states such as Tamil Nadu, Andhra Pradesh, and Karnataka possess high wind energy potential and extensive irrigated agriculture, making on-farm wind-based systems a viable option for enhancing energy security and climate resilience.

4.2.5.3 Residue Management

Crop residue burning is a major concern in India and, although it is most widespread in the Indo-Gangetic Plains, localized and increasingly reported instances have also been documented in parts of Southern India, particularly in Andhra Pradesh, contributing to air pollution, loss of soil organic matter, and greenhouse gas emissions. Recent reports indicate that paddy residue burning following the Kharif harvest has increased across several districts of Andhra Pradesh, where farmers burn rice straw and stubble to rapidly clear fields for the next cropping cycle, driven by mechanised harvesting and short turnaround periods between crops (Lanka 2025a). This practice has been associated with deterioration in local air quality, with elevated $PM_{2.5}$ concentrations recorded in and around Vijayawada and Guntur during peak burning periods, raising environmental and public health concerns (Lanka 2025b). While peer-reviewed research on residue burning in India largely focuses on northern regions, state-level reports and government advisories confirm that residue burning also affects soil health, air quality, and soil biological activity in southern states. In response, the Government of Andhra Pradesh has initiated awareness campaigns and advisories to discourage open burning and promote in-situ residue management practices (Government of Andhra Pradesh 2025).

Residue management technologies, such as the Happy Seeder, Super Straw Management System (Super-SMS), paddy straw chopper-cum-spreader, and mulching, allow farmers to incorporate or mulch residues directly into the soil, avoiding burning and saving fuel otherwise spent on repeated tillage. Residue recycling through these in-situ methods enhances soil fertility, moisture retention, and energy-use efficiency, thereby improving resilience under variable climate conditions. In-situ residue management has also been shown to improve the soil's physical, chemical, and biological properties, making it a sustainable and environmentally friendly alternative to residue removal. Ex-situ management options, including biofuel, biochar, electricity generation, or baling, further reduce pollutant emissions while providing additional economic benefits to farmers (Kaur et al. 2022).

4.2.5.4 Fuel-Efficient Engines

The adoption of fuel-efficient tractors, pumps, and harvesters can significantly reduce diesel consumption while maintaining or improving productivity. New-generation engines with better combustion efficiency, improved transmission, and smart power management have been shown to reduce fuel use compared to conventional engines (Classic Tractor News 2025; Jensen et al. 2025). In addition, the integration of precision land leveling, zero-till seed drills, and energy-optimized irrigation pumps further improves overall resource-use efficiency, thereby lowering both costs and greenhouse gas emissions. Precision land leveling has been reported to enhance water-use efficiency, reduce irrigation time, and improve crop establishment and productivity in rice wheat systems of South Asia (Jat et al. 2009). Similarly, zero-till seed drills not only reduce fuel consumption by minimizing the number of field operations but also mitigate soil erosion and enhance soil organic matter over time. Moreover, the use of energy-optimized irrigation pumps such as those designed with higher efficiency motors and variable-frequency drives has been shown to achieve significant energy savings and lower greenhouse gas emissions, aligning with the principles of precision agriculture. In recent years, the development of electric and hybrid tractors has emerged as a promising solution for improving engine efficiency and reducing emissions in the agricultural sector. These sustainable power sources provide environmentally friendly alternatives to traditional diesel tractors and have the potential to lower operational costs in the long run (Tractor Taishan 2025). However, challenges such as limited operating range, high initial investment, and inadequate charging infrastructure still need to be addressed for their widespread adoption. With continued advancements in battery technology and greater integration of renewable energy, electric and hybrid tractors are likely to play a significant role in the future of sustainable farming systems. Collectively, these technologies promote sustainable agricultural intensification by conserving energy, improving soil and water resources, and contributing to environmental protection.

4.2.5.5 Biofuels

In the past few decades, the world has seen an uncontrolled use of fossil fuels. It is undeniably true that fossil fuels hold the credit of shaping our world, but at the cost of environmental and related hazards. The primary sources of fossil fuels in India today are coal, petroleum, and natural gas, followed by renewable energy sources including bioenergy, wind, solar, and hydropower. In order to maintain the environment and lessen our long-term reliance on fossil fuels, it is necessary to diversify the present energy supply with more renewable energy sources as non-renewable energy sources diminish over time. One of the finest options is usage of renewable energy. Among the renewable sources, bioenergy has emerged as one of the most promising renewable energy sources globally. Renewable energy sources such as energy from biomass (bioenergy) are a key to address concerns for energy security and climate change.

Bioenergy is defined as renewable energy that is manufactured from biomass (biological commodity). It is equal to the total of municipal solid waste, industrial waste, primary solid biofuels, biogas, bioethanol, and other liquid biofuels, as well as charcoal and other fuels. The biodegradable portion of agricultural products, trash, and leftovers are referred to as biomass. Any plant material utilized to create energy is referred to as a bioenergy crop. The raw materials used to create another product are known as feedstock. Biomass feedstocks can be obtained from a variety of sources, including waste fuels, energy or agricultural products, etc. Bio-ethanol, biodiesel, and biogas are the main end products of bioenergy crops.

4.2.5.6 Characteristics of Bioenergy Crops

As potential bioenergy crops, researchers are looking at plant species with quick growth, tolerance to biotic and abiotic stresses (Somerville et al. 2010), and little requirement for biological, chemical, or physical pretreatments. Several crops such as sugarcane, sorghum, maize, switchgrass, and miscanthus already possess improved phenotypic, architectural, biochemical, and physiological traits that are all beneficial for the production of biofuels. Collectively, these attributes make bioenergy crops well suited for sustainable and large-scale bioenergy generation.

4.2.6 Knowledge smart technologies

4.2.6.1 Role of Climate Information and ICT Tools

The integration of climate information and ICT tools has revolutionized how farmers manage climate risks and make informed decisions, especially in regions vulnerable to climate variability. Early warning systems and mobile agro-advisories now bring timely, location-specific insights directly to farmers. For instance, Meghdoot, developed by the India Meteorological Department (IMD) and the Indian Institute of Tropical Meteorology (IITM), Indian Council of Agricultural Research (ICAR), and ICRISAT, delivers weather forecasts and crop advisories in local languages with support of State Agricultural Universities, supporting choices around sowing, irrigation, fertilization, and pest management (Dhulipala et al. 2021). This is complemented by advisories issued through Krishi Vigyan Kendras (KVKs), which translate scientific recommendations into actionable field-level guidance. Beyond traditional tools, platforms like Kisan Suvidha, Pusa Krishi, and Farm-o-Pedia offer real-time market information, pest alerts, and climate-smart agronomic tips that supplement forecasting services (Taruna et al. 2025). Early warning systems such as DAMINI, which alerts farmers to lightning risks, within a 20–40 km radius, and area-specific flood warnings issued through platforms like NDMA's SACHET portal and state-capacity models, have significantly reduced casualties and crop losses (Press Information Bureau 2022)

Participatory approaches like PICSA (Participatory Integrated Climate Services for Agriculture) involve farmers in analyzing historical climate data and weather forecasts to co-design context-appropriate adaptation strategies. The approach has been shown to empower farmers across Africa and Asia. In Mali and Senegal, for instance, 97% and 76% of participating farmers, respectively, rated PICSA as "very useful," and made strategic changes in sowing dates, soil and water management, crop varieties, and fertilizer use (Dayamba et al. 2018). Similarly, evaluations in Rwanda found that more than 90% of trainees implemented meaningful innovations and reported improved food security, incomes, planning confidence, and climate resilience (Clarkson et al. 2020). Additionally, impact evaluations in Malawi showed that farmers involved in modernized early warning systems and PICSA recorded yield increases of around 60% compared to non-participants, alongside stronger uptake of seasonal forecasts and adaptive practices (Abidoye et al. 2025). Together, these ICT-enabled advisory tools, when thoughtfully designed, translated, and delivered through multiple channels (including voice, SMS, and community radio), build climate resilience and reinforce farmers' agency to proactively adapt in the face of shifting environmental conditions.

4.2.6.2 Policy Support and Institutional Initiatives

To mainstream CSA in Southern India, a range of national and state-level policies, missions, and institutional programs have been launched to promote adaptation, mitigation, and resilience-building among farmers (Table 4.5). Central institutions like ICAR, CRIDA, SAUs, and KVKs play a pivotal role in conducting on-farm demonstrations, capacity-building programs, and climate risk assessment, while state-specific schemes complement national efforts.

Key National Programs and Initiatives

- **National Mission on Sustainable Agriculture (NMSA)**: A central pillar of India's National Action Plan on Climate Change (NAPCC), NMSA promotes integrated farming, soil health management, water-use efficiency, and agroforestry.
- **Pradhan Mantri Kisan Urja Suraksha evam Utthaan Mahabhiyan (PM-KUSUM)**: Supports installation of solar-powered irrigation pumps and decentralized solar energy use in agriculture to reduce emissions and dependency on conventional power sources.
- **State Action Plans on Climate Change (SAPCCs)**: Customized for each state, SAPCCs integrate climate-smart priorities into regional agricultural development plans.
- **NICRA (National Innovations in Climate Resilient Agriculture)**: Launched by ICAR in 2011, NICRA is a flagship initiative implemented across climate-vulnerable districts in Southern India. It supports:
  - Development of Climate Resilient Villages (CRVs).

Table 4.5 State-level climate-smart programs in Southern India

Program/initiative	State(s)	CSA relevance
Rythu Bandhu & Rythu Bharosa/ Rythu Seva	Telangana, Andhra Pradesh	Income support schemes that promote timely input use and sustainable cropping
Tamil Nadu Climate Resilient Agriculture Project (TNRTP)	Tamil Nadu	Focuses on CSA practices, climate information services, and livelihood diversification
Watershed Development Programmes	Karnataka	Promote soil-moisture conservation, water harvesting, and agroforestry
Polambadi (Field School) Programs	Andhra Pradesh, Tamil Nadu	Farmer participatory training on pest management, soil health, and weather-based advisories
Krishi Bhagya Scheme	Karnataka	Promotes farm ponds, micro-irrigation, and integrated nutrient management
Mission Kakatiya	Telangana	Restores minor irrigation tanks for water conservation and drought resilience
Haritha Keralam Mission	Kerala	Promotes organic farming, watershed conservation, and ecosystem restoration

- Promotion of drought, flood, and heat-tolerant varieties.
- Use of Automatic Weather Stations (AWS) and ICT tools for localized agro-advisories.
- Climate risk assessment, contingency planning, and crop simulation modeling.
- Resilient livestock, fisheries, and agroforestry interventions.
- Capacity building and community empowerment through KVKs and SAUs.

4.3 Challenges in Scaling CSAPs

- Limited access to quality inputs and CSA technologies
- Poor credit and insurance penetration
- Lack of awareness and technical knowledge
- Inadequate market linkages for diversified crops
- Institutional bottlenecks in extension delivery

4.4 Way Forward

To accelerate the adoption of CSA in Southern India:

- Strengthen extension services and promote farmer field schools
- Encourage public-private partnerships for technology dissemination
- Ensure credit access and risk insurance for smallholders
- Promote convergence of CSA with existing programs (e.g., MGNREGA, PMKSY)
- Invest in climate-resilient infrastructure (e.g., storage, irrigation, ICT)

4.5 Conclusion

Climate Smart Agriculture provides a holistic framework to address the challenges of climate change and agricultural sustainability. Given its diverse agro-ecological systems and climate vulnerabilities, Southern India holds significant potential for the adoption of CSAP. While many of these technologies and approaches have shown encouraging results, their widespread implementation depends on robust policy support, strong institutional collaboration, and active participation by farmers. By integrating scientific innovation with traditional knowledge, the region can pave the way for climate-resilient, productive, and inclusive agricultural growth.

References

Abidoye B, Aladysheva A, Haitz N, Montresor G, Nyekanyeka T, Orlic E, Prowse M (2025) Supporting farmers dealing with climate change: The impact of Participatory Integrated Climate Services for Agriculture (PICSA) on smallholder lead farmers in Malawi. Dev Policy Rev 43(1):e12822

Bhuvaneswari J, Thiyagarajan G, Manikandan M, Natarajan SK, Thenmozhi S, Prabhakaran NK (2020) Performance evaluation of drip fertigation in sugarcane and banana. Perform Eval 21:22

Chand S, Sikka AK, Sharda VN, Madhu M, Singh DV, Ragupathy R, Meena MS (2015) Sustainable agriculture development in rainfed areas through watershed programme: A case study, Coimbatore, Tamilnadu

Charan NS (2023) Farm ponds help transform livelihood of farmers in Mayiladuthurai district. The Hindu. https://www.thehindu.com/news/cities/Tiruchirapalli/farm-ponds-help-transform-livelihood-of-farmers-in-mayiladuthurai-district/article67318393.ece

Clarkson G, Dorward P, Nsengiyumva G, Kagabo D (2020) Participatory integrated climate services for agriculture (PICSA) as part of Rwanda climate services for agriculture: findings from quantitative evaluation of 2016/17 PICSA implementation (CCAFS working paper No. 338). CGIAR research program on climate change, Agriculture and food security (CCAFS). https://hdl.handle.net/10568/111111

Classic Tractor News (2025) How has tractor engine efficiency changed over the decades?. https://classictractornews.com

Dayamba DS, Ky-Dembele C, Bayala J, Dorward P, Clarkson G, Sanogo D, Diop ML, Traore I, Diakite A, Nenkam A, Binam JN, Ouedraogo M, Zougmore RB (2018) Assessment of the use of Participatory Integrated Climate Services for Agriculture (PICSA) approach by farmers to manage climate risk in Mali and Senegal. Clim Serv 12:27–35. https://oar.icrisat.org

Dhivya C, Murugan PP, Senthilkumar M, Asokhan M, Dheebakaran G, Gangai SR, Arunkumar R (2024) Determinants of climate-smart agricultural technology adoption: A comprehensive systematic review. Plant Sci Today 11:5156

Dhulipala R, Gogumalla P, Rao KPC, Palanisamy R, Smith A, Nagaraji S, Suryachandra Rao A, Vishnoi L, Singh KK, Bhan SC, Whitbread AM (2021) Meghdoot: a mobile app to access location-specific weather based agro-advisories pan India (CCAFS working paper no. 370). CGIAR research program on climate change, Agriculture and food security (CCAFS)

Doreswamy C (1999) Tree-based farming systems for poverty alleviation in semi-arid tropics. In: Proceedings of the regional workshop on forests for poverty reduction: can community forestry make money? BAIF Institute for Rural Development, Tiptur, Karnataka, India. FAO

Dupdal R, Naik BS, Ramesha MN, Ravi KN, Rao BK (2024) Water harvesting structure-farm pond for increasing crop yields and farm income in rainfed semi-arid ecosystem of Karnataka. Indian J Soil Conserv 52:68–73

El-Sebaii AA, Shalaby SM (2012) Solar drying of agricultural products: A review. Renew Sust Energ Rev 16(1):37–43

Farooq M, Wahid A, Kobayashi NSMA, Fujita DBSMA, Basra SM (2009) Plant drought stress: effects, mechanisms and management. In: Sustainable agriculture. Springer Netherlands, Dordrecht, pp 153–188. https://doi.org/10.1051/agro:2008021

Food and Agriculture Organization (FAO) (2013) Climate-smart agriculture sourcebook. FAO. https://www.fao.org/climate-smart-agriculture-sourcebook

Gorain S, Singh DR, Kumar P, Venkatesh P, Jha GK (2020) Economics of sugarcane and banana cultivation under drip irrigation system: a case study of northern Maharashtra. Econ Aff 65(2):151–159

Government of Andhra Pradesh (2025) AP govt urges farmers not to burn crop residues, stresses soil health benefits. Department of Agriculture & Cooperation, Andhra Pradesh. https://www.deccanchronicle.com/southern-states/andhra-pradesh/ap-govt-urges-farmers-not-to-burn-crop-residues-stresses-soil-health-benefits-1922651

Government of Karnataka (2025) Upper Krishna project. https://en.wikipedia.org/wiki/Upper_Krishna_Project

Government of Telangana (2025) Mission Kakatiya: restoration of 46,000 tanks and lakes. https://en.wikipedia.org/wiki/Mission_Kakatiya

ICAR-Central Research Institute for Dryland Agriculture (CRIDA) (2012) Guidelines on drought coping plans for rainfed production systems. ICAR-CRIDA, Hyderabad, India. https://www.icar-crida.res.in/assets/img/Books/2003-04/Guidelines.pdf

ICAR-CRIDA (2023) Soil and nutrient management. https://icar-crida.res.in/Soil_and_nutrient_management.html

ICAR-NICRA (2022) National innovations on climate resilient agriculture: annual report 2021–22. Indian council of agricultural research. https://www.nicra-icar.in

ICAR-NICRA (2023) Vulnerable districts under national innovations on climate resilient agriculture. Press Information Bureau. https://pib.gov.in/PressReleaseIframePage.aspx?PRID=1982673

India Meteorological Department (IMD) (2020a) Assessment of climate change over the Indian region. Ministry of Earth sciences, Government of India

India Meteorological Department (IMD) (2020b) Climate of India during 2020. Ministry of Earth sciences, Government of India. https://mausam.imd.gov.in

Intergovernmental Panel on Climate Change (IPCC) (2023) Climate change 2023: synthesis report. Contribution of working groups I, II and III to the sixth assessment report of the intergovernmental panel on climate change, IPCC, pp. 35–115. https://doi.org/10.59327/IPCC/AR6-9789291691647

Jat ML, Gupta R, Ramasundaram P, Gathala M, Sidhu H, Singh S et al (2009) Laser assisted precision land leveling: A potential technology for resource conservation in irrigated intensive production systems of Indo-Gangetic plains, vol 223. International Rice Research Institute

Jensen TA, Antille DL, Tullberg JN (2025) Improving on-farm energy use efficiency by optimizing machinery operations and management: A review. Agric Res 14(1):15–33

Kanannavar PS, Premanand BD, Subhas B, Anuraja B, Basavaraj BP (2020) Laser land levelling—An engineering approach for scientific irrigation water management in irrigation command areas of Karnataka, India. Int J Curr Microbiol App Sci 9(5):2393–2398. https://doi.org/10.20546/ijcmas.2020.905.272

Katti G, Padmavathi C (2022) Impact of rice cultivation methods on insect pest incidence and their management. J Rice Res 15:84–89

Kaur M, Malik DP, Malhi GS, Sardana V, Bolan NS, Lal R, Siddique KH (2022) Rice residue management in the Indo-Gangetic Plains for climate and food security: A review. Agron Sustain Dev 42(5):92. https://doi.org/10.1007/s13593-022-00817-0

Khemariya S (2025) A review on role of solar cold storage system in India. Int J Res Appl Sci Eng Technol 13(6). https://doi.org/10.22214/ijraset.2025.72409

Krishnan R, Sanjay J, Gnanaseelan C, Mujumdar M, Kulkarni A, Chakraborty S (2020) Assessment of climate change over the Indian region: a report of the ministry of Earth sciences (MoES). Springer, Government of India. https://doi.org/10.1007/978-981-15-4327-2

Kumar M, Sansaniwal SK, Khatak P (2016) Progress in solar dryers for drying various commodities. Renew Sust Energ Rev 55:346–360

Kuppannan P, Devarajulu SK (2009) Impacts of watershed development programmes: Experiences and evidences from Tamil Nadu. 387–396

Kuppannan P, Raman S (2012) Potential and challenges in up-scaling micro-irrigation in India: Experiences from nine states

Lampayan RM, Rejesus RM, Singleton GR, Bouman BA (2015) Adoption and economics of alternate wetting and drying water management for irrigated lowland rice. Field Crop Res 170:95–108

Lanka V (2025a) Paddy residue burning on the rise across state. The Times of India, Vijayawada edition. https://timesofindia.indiatimes.com/city/vijayawada/paddy-residue-burning-on-the-rise-across-state/articleshow/125874421.cms

Lanka V (2025b) PM2.5 levels spike in Vijayawada as farmers burn crop waste. The Times of India, Vijayawada edition. https://timesofindia.indiatimes.com/city/vijayawada/pm2-5-levels-spike-in-vijayawada-as-farmers-burn-crop-waste/articleshow/126206639.cms

Manjunath KV, Maiti S, Garai S, Reddy DAK, Sahani S, Panja A, Jha SK (2024) Impact of climate services on the operational decision and economic outcome of wheat (Triticum aestivum) and rice (Oryza sativa) cultivation in Haryana. Indian J Agric Sci 94(3-S1):116–123

Mina U, Geetha G, Sharma R, Singh D (2023) Comparative assessment of tree carbon sequestration potential and soil carbon dynamics of major plantation crops and homestead agroforestry of Kerala. India Anthropocene Science 2(1):93–100

Ministry of Environment, Forest and Climate Change (MoEFCC). (2020). India's updated first biennial update report to UNFCCC. https://moef.gov.in

Mittal S, Gandhi S, Tripathi G (2010) Socio-economic impact of mobile phones on Indian agriculture (No. 246). Working paper

Moulkar R, Dayakar P (2023) Climate sensitivity of major crops yield in Telangana state, India. J Asia Pac Econ 29:2023–2040. https://doi.org/10.1080/13547860.2023.2230007

Nthebere K, Tata RP, Gudapati J, Bhimireddy P, Admala M, Chandran LP, Yadav MBN (2025) Assessment of conservation agriculture on soil nutrient's stratification ratio, carbon sequestration rate, management indices and crop productivity in Southern Telangana, India. Sci Rep 15(1):15038

Osman M, Rao GR, Govindarajan V (2019) Role of agroforestry in watershed management. In: Proceedings, pp 363–391

Pampolino MF, Witt C, Pasuquin JM, Johnston AM, Fisher MJ (2014) Development and evaluation of Nutrient Expert® decision support tool for cereal crops. Better Crops–South Asia 8(1):4–6

Parimaladevi C, Ramanathan SP, Senthil Kumar N, Suresh S (2019) Evaluation of maize based intercropping systems in Thamirabarani basin of Tamil Nadu. J Pharmacogn Phytochem 8(3):4051–4056

Prasad JVNS, Rao CS, Srinivas K, Jyothi CN, Venkateswarlu B, Ramachandrappa BK, Dhanapal GN, Ravichandra K, Mishra PK (2016) Effect of ten years of reduced tillage and recycling of organic matter on crop yields, soil organic carbon and its fractions in Alfisols of semi-arid tropics of southern India. Soil Tillage Res 156:131–139

Press Information Bureau (PIB) (2022) Death due to lightning strikes. In: Ministry of Earth Sciences. Government of India. https://pib.gov.in/PressReleasePage.aspx?PRID=1813993

Press Information Bureau (PIB) (2023) Per drop more crop. Government of India. https://pib.gov.in/PressReleaseIframePage.aspx?PRID=1985487

Rajkumar RH, Dandekar AT, Vishwanatha J, Karegoudar AV, Singh YK (2022) Laser guided land levelling: A precision technique for enhancing water productivity. J Farm Sci 34(Spl 5):497–501. https://doi.org/10.61475/jfm.v34iSpl5.285

Ramya SMS, Mahesh N, Revathi P, Raju B (2021) Effect of laser land levelling and establishment methods on growth and yield of Rice. Biological Forum—An. Int J 13(3a):73–78

Rathore LS (2013) Weather information for sustainable agriculture in India. J Agric Phys 13(2)

Richards MB, Butterbach-Bahl K, Jat ML, Ortíz Monasterio JI, Sapkota TB, Lipinski B (2015) Site-specific nutrient management: implementation guidance for policymakers and investors

Sahu G, Rout PP, Mohapatra S, Das SS, Pradhan PP (2020) Climate Smart Agriculture: A new approach for sustainable intensification. Curr J Appl Sci Technol 39(23):138–147

Sapkota TB, Jat ML, Rana DS, Khatri-Chhetri A, Jat HS, Bijarniya D, Sutaliya JM, Kumar M, Singh LK, Jat RK, Kalvaniya K (2021) Crop nutrient management using Nutrient Expert improves yield, increases farmers' income and reduces greenhouse gas emissions. Sci Rep 11:1564

Sathiya K, Nirmalakumari A, Sukanya T, Vanitha C, Karthikeyan R, Ayyadurai P (2024) Enhancing small millet farming resilience through conservation agriculture in rainfed regions. J Exp Agric Int 46(5):68–75

Shinde VB, Wandre SS (2015) Solar photovoltaic water pumping system for irrigation: A review. Afr J Agric Res 10(22):2267–2273

Somerville C, Youngs H, Taylor C, Davis SC, Long SP (2010) Feedstocks for lignocellulosic biofuels science. 329(5993):790–792

Sri K, Navya K, Sinha DK, Kumari T, Singh KM, Ahmad N (2024) Paddy-based cropping systems and their sustainability: A micro-level study of Nellore District of Andhra Pradesh, India. Asian J Agric Ext Econ Sociol 42(10):35–44. https://doi.org/10.9734/ajaees/2024/v42i102560

Srishailam BV, Sailaja P, Ganesh Kumar D, Subramanyam SNU (2023) Index to assess the impact of weather based agro-advisory services among the cotton growers of Telangana and Andhra Pradesh States. Biological Forum—An. Int J 15(4):740–746

Tamil Nadu Agricultural University (TNAU) (2023a) Directorate of crop management—Nammazhvar Organic Farming Research Centre. https://tnau.ac.in/site/directorate-cm/nammazhvar-organic-farming-research-centre-cropping-system

Tamil Nadu Agricultural University (TNAU) (2023b) Agritech portal—Integrated nutrient management in agriculture. https://agritech.tnau.ac.in/agriculture/agri_nutrientmgt.html

Tamil Nadu Agricultural University (TNAU) (2024) Contingency plan for Tamil Nadu: drought and moisture management practices. In: Directorate of Extension Education, TNAU, Coimbatore, India. https://agritech.tnau.ac.in/pdf/SWM25%20Contigencyplan.pdf

Taruna D, Sharma K, Sharma N, Sethi D, Devi N, Ankita K (2025) Role of ICT in bridging the knowledge gap in rural agriculture. Int J Agric Ext Soc Dev 8(1S):20–24. https://doi.org/10.33545/26180723.2025.v8.i1Sa.1503

Tractor Taishan (2025) Farm tractors 2025: What's new in engine efficiency?. https://tractortaishan.com

Venkatesh G, Gopinath KA, Ramana DBV, Kumari VV, Srinivas I, Shanker AK, Rao KV, Prasad JVNS, Reddy KS, Sridhar KB, Sarkar B, Raju BMK, Rajkumar B, Chary GR, Singh VK, Timsina J (2024) Agrosilvopastoral systems for improved crop and fodder productivity and soil health in the rainfed environments of South India. Agric Syst 214:103812. https://doi.org/10.1016/j.agsy.2023.103812

Vibha G, Lingaraju HG, Venktaramana GV (2021) Effectiveness of solar fence in reducing human–elephant conflicts in Manchahalli village, Mysuru, Karnataka, India. Curr Sci 120(4):707–711

Wani SP (2002) Integrated watershed management for minimizing land degradation and sustaining productivity in Asia. In: Integrated land management in the dry areas: Proceedings of Joint UNU-CAS International Workshop, Beijing, China, 2002, United Nations University

Chapter 5
Reforming Agricultural Price Support in India: Fiscal Costs, WTO Constraints, and Policy Trade-Offs

Arun Reddy Anugu and A. Amarender Reddy

Abstract For policymakers concerned with farm income stability and national food security, the Minimum Support Price (MSP) has long been a central policy instrument in Indian agriculture. However, demands by some farmer organisations to extend assured physical procurement beyond paddy and wheat to all 23 MSP-notified crops raise serious concerns regarding fiscal sustainability, operational feasibility, and compliance with the World Trade Organization (WTO) Agreement on Agriculture. Administered price support with physical procurement constitutes trade-distorting Amber Box support, whereas alternative instruments such as Price Deficiency Payments (PDP) can deliver income support with lower market distortions. This paper quantifies the fiscal implications of alternative MSP implementation pathways for the 2024–25 marketing year to inform evidence-based policy reform. We evaluate three policy scenarios: (i) partial physical procurement, (ii) a mixed regime combining procurement for selected crops with partial PDP for others, and (iii) a universal PDP framework. Using crop-wise data on production, MSP, and modal market prices from Agmarknet, we estimate farmer payouts, disposal revenues, and the resulting net fiscal burden under each scenario. The results show that limiting procurement to 40% of paddy and wheat output and 20% of other crops yields total farmer payouts of ₹3.70 lakh crore (around 1.12% of GDP), with a net fiscal burden of ₹3,731–₹52,171 crore (0.01–0.07% of GDP) after accounting for disposal revenues. A mixed MSP-PDP model, which retains physical procurement for paddy and wheat while extending PDP to 20% of output for other crops, results in broadly similar fiscal costs but delivers significant efficiency gains by sharply reducing storage, handling, and logistics requirements. A universal PDP scenario marginally increases gross expenditure but eliminates inventory-related costs entirely, simplifies administration, and reduces trade distortion. Based on these estimates, the paper recommends a hybrid price support framework: continued physical procurement for paddy

---

A. R. Anugu
Carnegie Mellon University, Pittsburgh, PA, USA

A. A. Reddy (✉)
Policy Support Research, ICAR-National Institute of Biotic Stress Management (NIBSM), Raipur, India
e-mail: amarender.reddy@icar.org.in

D. Kumar Swain et al. (eds.), *Climate-Smart Technologies for Sustainable Agricultural Production*, https://doi.org/10.1007/978-3-032-25666-9_5

and wheat to safeguard food security, combined with PDP for other crops. Such a transition balances fiscal prudence with administrative feasibility, reduces market distortions, and moves India's agricultural support system closer to WTO-consistent income support practices while maintaining effective protection for farmers.

Keywords Minimum support price · Price deficiency payments · WTO agreement on agriculture · Fiscal burden · Agricultural policy reform · India

5.1 Introduction

India's agricultural price policy is at a critical juncture. The Minimum Support Price (MSP) system, among the most interventionist price support mechanisms in the developing world has played a central role in stabilising farm incomes and ensuring national food security since its introduction during the Green Revolution of the mid-1960s. Conceived to incentivise production under conditions of yield and price uncertainty, MSP contributed decisively to foodgrain self-sufficiency and reduced vulnerability to chronic shortages. Over time, it has become a cornerstone of India's food security and rural welfare architecture.

However, the structure and operational logic of MSP have not evolved in line with the changing realities of Indian agriculture or the international trade environment. While MSPs are announced for 23 crops, including cereals, pulses, oilseeds, nutri-cereals, and selected commercial crops, their effective enforcement remains highly uneven across crops and regions. Prices are recommended by the Commission for Agricultural Costs and Prices (CACP) using a cost-plus framework, with an intended margin of at least 50% over Cost A2 + FL (All paid-out cost plus family labour) (Gupta et al. 2021). In practice, however, MSP functions less as a universal price floor and more as a selective, quantity-backed guarantee concentrated in a narrow set of crops and geographies.

This gap between policy intent on paper and implementation has become increasingly problematic. Through the Food Corporation of India (FCI) and state agencies, MSP enforcement is overwhelmingly centred on large-scale procurement of paddy and wheat to meet Public Distribution System (PDS) requirements and allied welfare programmes. For most other MSP crops, particularly pulses, oilseeds, coarse cereals, and crops such as copra, procurement remains sporadic, limited in scale, and contingent on market conditions (Fig. 5.1). As a result, MSP is binding primarily for farmers in surplus rice–wheat regions, while the majority of producers, especially in rainfed and diversified systems, remain exposed to price volatility despite formal MSP announcements.

The cereal-centric procurement model has generated mounting fiscal, efficiency, and distributional pressures. Procured grains are distributed through the PDS at highly subsidised issue prices of ₹2–₹3 per kg to nearly 80 crore beneficiaries, creating a persistent wedge between procurement costs and consumer prices. The fiscal burden associated with procurement, storage, transport, distribution, and wastage has

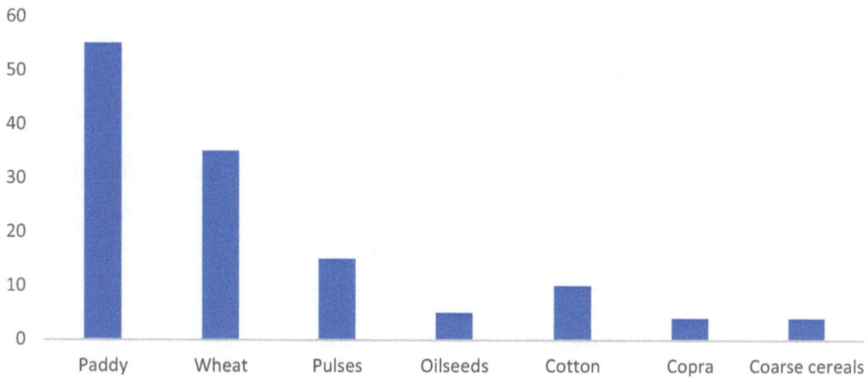

Fig. 5.1 Procurement share (%) in production from 2020–25. *Source* Food Corporation of India, 2024

increased steadily, with India's food subsidy exceeding ₹2.11 lakh crore in 2023–24 (Government of India, Ministry of Finance 2025). Rice and wheat alone account for more than 85% of procurement expenditure. In this context, extending assured physical procurement to all MSP-notified crops as frequently demanded by farmer organisations, would imply a sharp escalation in fiscal commitments and administrative complexity, raising concerns about long-term sustainability.

Beyond fiscal pressures, the limitations of procurement-led MSP have become more visible as Indian agriculture diversifies and confronts new challenges. Persistent price volatility in pulses and oilseeds, regional inequities in market access, environmental stress arising from cereal monocropping, and the systematic exclusion of rainfed and marginal regions from effective price support highlight the inadequacy of physical procurement as a universal income stabilisation instrument (Reddy 2009; Reddy et al. 2009; Reddy and Bantilan 2012). These structural concerns are compounded by international trade commitments, as administered price support with procurement constitutes trade-distorting Amber Box support under the WTO Agreement on Agriculture and which is having upper-limit.

Against this backdrop, alternative instruments such as Price Deficiency Payments (PDP) and Direct Benefit Transfers (DBT), which compensate farmers for the difference between MSP and realised market prices without physical procurement (Fig. 5.2), have gained prominence in policy debates. By delinking income support from physical grain acquisition, PDP-type mechanisms offer the potential for wider crop and regional coverage, reduced market distortions, lower storage and logistics costs, and improved compatibility with WTO disciplines, while preserving the core objective of protecting farm incomes.

This paper responds to the growing need for reform by analysing five alternative policy scenarios that progressively transition from the existing cereal-centric procurement system towards graded Price Deficiency Payment mechanisms. The scenarios range from the status quo of rice–wheat procurement, through partial procurement, to PDP coverage for 20%, 40%, and ultimately 100% of production of the remaining

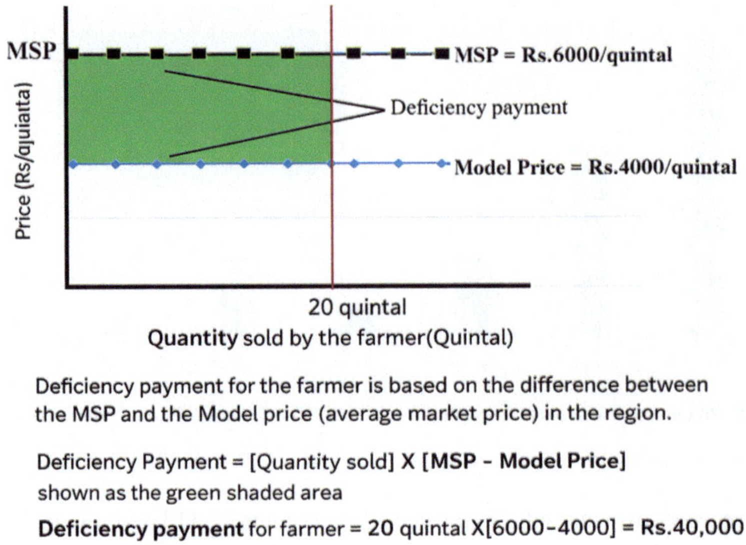

Fig. 5.2 Price deficiency payment mechanism

MSP-notified crops. Using crop-wise data on production, MSP, and market prices for 2024–25, the study estimates farmer payouts, disposal revenues, and net fiscal burdens under each scenario. The analysis evaluates fiscal costs, administrative feasibility, and efficiency implications, with particular attention to crops and regions historically excluded from effective price support. By doing so, the paper seeks to inform a pragmatic restructuring of India's agricultural price policy that balances farmer income protection, fiscal sustainability, market efficiency, and international trade compatibility.

5.2 Theoretical and Empirical Foundations

Economic theory establishes that minimum support price policies influence market outcomes only when the state credibly commits to purchasing output at the announced support price in effectively unlimited quantities (Timmer 1989; Gardner 2006; Chand 2022). In the absence of such commitment, support prices function primarily as reference signals with weak enforcement, offering limited income protection to producers. This distinction between announced and operationalised minimum prices is fundamental to understanding the growing disconnect in India's MSP regime.

The broader literature on agricultural price stabilisation in developing economies suggests that procurement-backed price floors are most effective during early stages of food security transitions, when markets are thin and supply risks are high (Timmer 1997; Dorward, et al. 2004; Dorward et al. 2006). Over time, however, such systems tend to generate significant market distortions, fiscal pressures, and spatial concentration of benefits, particularly when procurement capacity and budgetary resources are uneven. Under these conditions, price support increasingly favours surplus regions and larger producers, while marginal farmers, rainfed areas, and diversified cropping systems remain excluded (Negi et al. 2020; Korekallu Srinivasa et al. 2023).

India's MSP system exhibits these theoretical shortcomings in practice. Although MSPs are announced for 23 crops, effective procurement is overwhelmingly concentrated in paddy and wheat. For pulses, oilseeds, and coarse cereals, market prices frequently fall below MSP with limited compensatory intervention, rendering MSP largely notional for a majority of producers (World Bank 1999; Gulati et al. 2014; CACP 2024; Reddy et al. 2025). As a result, MSP has evolved into a selective and asymmetric policy that operates more as a procurement guarantee for specific crops than as a comprehensive income stabilisation mechanism (Government of India 2024, 2025; Reddy et al. 2025).

Fiscal and environmental analyses further reinforce the case for reorienting price support. A substantial share of food subsidy expenditure arises not from MSP per se, but from the downstream costs of physical procurement, including storage, handling, transport, distribution, and losses under the Public Distribution System (Jha and Ramaswami 2010; Gulati and Juneja 2022a, b). Procurement-led incentives have also contributed to environmental externalities, notably groundwater depletion, soil degradation, and cereal monocropping in the rice-wheat belt (OECD 2023a, b; OECD/FAO 2023a, b). From a trade perspective, administered price support linked to procurement constitutes Amber Box support under the WTO Agreement on Agriculture, further constraining the scope for indefinite expansion.

International experience provides evidence of viable alternatives that preserve income support objectives while reducing market distortions. In the United States, agricultural policy has progressively shifted away from procurement-based price support towards decoupled income support and counter-cyclical deficiency payments, substantially lowering trade distortions and stock accumulation (Gardner 2006). Similarly, China's transition from state procurement to target price and deficiency payment systems for maize and cotton demonstrates that farm income stabilisation can be achieved without excessive fiscal leakage or public stockpiling (Huang and Yang 2017).

Within India, policy research increasingly supports the use of Price Deficiency Payment mechanisms, particularly for crops such as pulses and oilseeds where public distribution is neither essential nor cost-effective (Gulati et al. 2014; NITI Aayog 2017). Evaluations of state-level initiatives, notably Madhya Pradesh's Bhavantar Bhugtan Yojana, indicate that despite implementation challenges related to price discovery and beneficiary targeting, PDP schemes can expand income support coverage while significantly reducing procurement-related distortions (Chand et al. 2015).

Anchored in this theoretical and empirical literature, the present study conceptualises MSP not as a single policy instrument but as a continuum of price support options, ranging from full physical procurement to partial and full price compensation through deficiency payments. By systematically comparing procurement-PDP hybrids, the paper contributes to the policy debate by quantifying fiscal costs, administrative feasibility, and efficiency trade-offs associated with alternative support regimes. In doing so, it explicitly addresses the persistent exclusion of non–paddy–wheat crops from effective price support and situates India's reform choices within internationally accepted, WTO-consistent policy frameworks.

5.3 Review of Literature

The literature on agricultural price support in India consistently highlights the inherent trade-offs between farm income stabilisation, fiscal sustainability, and market efficiency under MSP-based interventions. Early analytical work emphasised that minimum support price policies become fiscally and administratively burdensome when enforcement relies heavily on open-ended physical procurement rather than price signals alone (Timmer 1997; Gardner 2006). More recent assessments by the Commission for Agricultural Costs and Prices (CACP) and policy analyses by Gulati et al. (2021, 2024) and Chand (2022, 2025) underscore that extending assured MSP-backed procurement to all 23 notified crops would entail prohibitively high fiscal outlays, potentially running into several lakh crore rupees annually, alongside severe operational constraints. A key insight from this literature is that fiscal stress arises less from MSP announcements per se and more from the downstream costs of procurement, storage, transport, and distribution under the Public Distribution System.

Empirical studies document persistent inefficiencies and spatial concentration in India's procurement architecture. Jha and Ramaswami (2010) show that a substantial share of food subsidy expenditure is absorbed by handling, carrying, and distribution costs rather than direct income transfers to farmers. More recent evaluations highlight rising buffer stocks, escalating storage losses, and institutional rigidities within the Food Corporation of India, which together amplify subsidy requirements while narrowing the effective reach of MSP (Gulati and Juneja 2022a, b; Kumar et al. 2023). These dynamics disproportionately benefit surplus rice–wheat regions, while producers of pulses, oilseeds, and coarse cereals, often located in rainfed and ecologically vulnerable regions, remain weakly integrated into price support mechanisms (Negi et al. 2020; Korekallu Srinivasa et al. 2023).

A growing strand of the literature links procurement-centric MSP to adverse environmental outcomes. Studies by OECD (2023a, b) and OECD/FAO (2023a, b) document how price incentives concentrated on cereals have reinforced monocropping, groundwater over-extraction, and declining soil health in north-western India. Recent empirical work using farm-level and remote-sensing data further corroborates that procurement-linked price signals have constrained crop diversification despite rising demand for pulses and oilseeds (Birthal et al. 2021; Narayanan 2023). These findings strengthen the argument for decoupling income support from physical procurement.

International experience offers important lessons for reform. In the United States, the gradual shift from procurement-based price support to decoupled income payments and counter-cyclical deficiency mechanisms has significantly reduced market distortions and public stockholding, while maintaining income protection for farmers (Gardner 2006; OECD 2023a, b). China's transition from state procurement to target price and deficiency payment systems for maize, cotton, and soybeans provides further evidence that income stabilisation objectives can be achieved with lower fiscal leakage and greater market orientation when supported by reliable price discovery and digital payment systems (Huang and Yang 2017; Xu et al. 2025).

Within India, policy-oriented research increasingly supports Price Deficiency Payment (PDP) mechanisms, particularly for pulses and oilseeds where public distribution is neither essential nor efficient (Gulati et al. 2014; NITI Aayog 2017; Gulati et al. 2024). Evaluations of state-level initiatives, most notably Madhya Pradesh's Bhavantar Bhugtan Yojana, suggest that while early implementations faced challenges related to reference price determination, beneficiary identification, and moral hazard, PDP schemes have demonstrated potential to expand coverage, reduce procurement-related distortions, and improve fiscal transparency (Chand et al. 2015; Reddy et al. 2021).

Despite these advances, several gaps remain in the literature. First, most studies treat procurement-based MSP and PDP as binary alternatives, offering limited guidance on feasible transition pathways that combine food security objectives with income support. Second, scenario-based fiscal comparisons using recent crop-wise production and price data remain scarce. Third, the distributional implications of alternative instruments for crops historically excluded from effective price support have not been systematically analysed. This paper addresses these gaps by evaluating graded procurement-PDP scenarios using up-to-date data, thereby contributing evidence to the ongoing debate on reforming India's agricultural price support system in a fiscally sustainable and internationally consistent manner.

5.4 International Experience with Price and Income Support: WTO AoA Perspective

International experience with agricultural support policies reveals a clear long-term shift away from procurement-backed price support towards income-based and decoupled instruments. This transition has been driven by concerns over fiscal sustainability, market distortions, and compliance with the World Trade Organization (WTO) Agreement on Agriculture (AoA). A comparative overview of major agricultural economies shows that while many countries initially relied on administered prices and public procurement, most have progressively restructured support systems to reduce Amber Box exposure and expand Green Box measures.

This cross-country comparison highlights two stylised facts (Table 5.1). First, physical procurement and administered prices are increasingly confined to a narrow set of strategic objectives, such as food security or transitional support. Second, income stabilisation objectives are predominantly pursued through decoupled or deficiency-based mechanisms, which are less trade-distorting and more fiscally predictable.

India's current MSP regime remains an outlier in its continued reliance on large-scale physical procurement for income support, particularly for rice and wheat. While comparable countries such as the United States and China initially employed procurement-led systems, they have since transitioned towards deficiency payments, revenue insurance, and direct income transfers, thereby reducing stock accumulation and Amber Box exposure (Table 5.2).

From a WTO perspective, Price Deficiency Payments occupy an intermediate reform space. When designed as compensation for price gaps without mandatory production or procurement, PDP-type instruments can substantially reduce trade distortion relative to administered price support. International experience thus suggests that a hybrid model, retaining procurement for food security crops while shifting other MSP crops to PDP, represents a pragmatic and WTO-consistent reform pathway.

By situating India's price support debate within this international context, the analysis underscores that the proposed transition from MSP-backed procurement to PDP is not an institutional rupture but rather a convergence towards globally observed best practices in agricultural support design as country evolved in to a developed country by 2047.

5.5 Objectives of the Study

In light of growing fiscal pressures, operational constraints, and India's multilateral commitments under the WTO Agreement on Agriculture (AoA), this study pursues the following objectives:

(i) To compare the fiscal implications of alternative MSP implementation scenarios, distinguishing between procurement-based price support (Amber Box–relevant) and Price Deficiency Payments (PDP) and ultimately decoupled income support (Green box).
(ii) To assess administrative efficiency gains from a gradual transition away from cereal-centric physical procurement towards graded, DBT-enabled PDP mechanisms consistent with WTO-aligned income support practices.
(iii) To evaluate whether alternative policy scenarios improve the inclusiveness of price and income support, particularly for non-cereal crops and rainfed regions that remain weakly covered under the current MSP-procurement regime.

Table 5.1 International examples of farm support aligned with WTO AoA boxes

Country	Policy instrument	Nature of support	WTO AoA box	Why it falls in this box
India	Minimum support price (MSP) with procurement	Administered price support	Amber box	Market price support linked to current production; counts toward aggregate measure of support (AMS)
India	PM-KISAN (cash transfer)	Decoupled income support	Green box	Direct income transfer not linked to production
United States	Price loss coverage (PLC)	Price-contingent income support	Amber box	Payments triggered when prices fall below reference levels
United States	Agriculture risk coverage (ARC)	Revenue insurance	Amber box	Compensates revenue loss linked to market outcomes
United States	Conservation reserve program (CRP)	Environmental payment	Green box	Payments for land retirement and environmental services
European Union	CAP basic income support (pillar I)	Area-based income support	Green box	Decoupled from production; subject to cross-compliance
European Union	Coupled support (selected crops/ livestock)	Production-linked support	Amber box	Linked to output or animal numbers
China	Minimum purchase price (rice, wheat)	Price support	Amber box	Administered prices above market levels
China	Direct payments to grain farmers	Decoupled payments	Green box	Area-based payments not linked to output
Japan	Direct income support for farmers	Price-gap compensation	Amber box	Compensates difference between domestic and world prices
Japan	Multifunctionality payments	Environmental and rural support	Green box	Non-trade-distorting objectives
Canada	AgriStability	Revenue-based income support	Amber box	Triggered by income decline relative to reference
Canada	AgriInvest	Savings-based income stabilization	Green box	Government matching not linked to production
Brazil	Minimum guaranteed price (PGP)	Market intervention	Amber box	Price support when market prices fall
Mexico	PROCAMPO/ producción para el bienestar	Per-hectare income support	Green box	Decoupled payments based on historical area

(continued)

Table 5.1 (continued)

Country	Policy instrument	Nature of support	WTO AoA box	Why it falls in this box
Norway	Target prices and budget transfers	Price and income support	Amber box	Negotiated administered prices
Australia	Farm household allowance	Welfare-style income support	Green box	Social safety net, not production-related
New Zealand	No farm subsidies	Market-led	Not applicable	Post-reform elimination of trade-distorting support

Source World Trade Organization (WTO). (Various years). Notifications on domestic support under the Agreement on Agriculture. Geneva: WTO

Table 5.2 WTO AoA box classification—analytical snapshot

WTO AoA box	Core characteristics	Typical instruments
Amber box	Trade-distorting; linked to prices or production	MSP, administered prices, output-linked payments
Green box	Minimally trade-distorting; decoupled from production	Direct income support, environmental payments
Blue box	Production-limiting support	EU'S earlier CAP set-aside and acreage control payments
De minimis	Aggregate measure of support ($\leq$10% for developing countries; $\leq$5% for developed countries) for agriculture	Minor input subsidies

Source World Trade Organization (WTO). (Various years). Notifications on domestic support under the Agreement on Agriculture. Geneva: WTO

5.6 Data and Methodology

5.6.1 Data Sources

The analysis relies exclusively on official and routinely used government datasets to ensure transparency, policy credibility, and consistency with WTO reporting norms. MSPs for the 2024–25 marketing year are taken from notifications of the Directorate of Economics and Statistics (DES), Government of India. Crop-wise production data are drawn from the Third Advance Estimates for 2024–25 issued by the Ministry of Agriculture and Farmers Welfare.

Domestic market prices are proxied using national modal price averages from AGMARKNET (Agricultural Marketing Information Network) and CEDA (Centre for Electronic Data Analysis), which are widely used for market monitoring and for estimating market price support under the AoA. For rice and wheat, international reference prices are sourced from the World Bank Pink Sheet (2024) to assess export-linked disposal revenues and implicit subsidy components.

All fiscal estimates are computed at the national level and expressed in ₹ crore. Simulations assume full government participation for the specified coverage under each scenario, consistent with prevailing Food Corporation of India (FCI) operational practices. Reliance on official data ensures that results are directly usable for budgetary analysis as well as for assessing implications under India's Aggregate Measurement of Support (AMS) commitments.

5.6.2 Analytical Framework

From both a fiscal and WTO-compliance perspective, the cost of MSP implementation arises not only from price support to farmers but also from procurement-related operations and disposal outcomes. Accordingly, the net fiscal burden is estimated as:

$$\textbf{Net Fiscal Burden} = \textbf{Farmer Payout} + \textbf{Incidentals} - \textbf{Disposal Revenue}$$

where:

Farmer Payout includes payments through physical procurement at MSP (classified as market price support under the AoA) or through PDP delivered via DBT.

Incidentals (assumed at 10–15% of payouts) capture procurement, handling, storage, transport, and administrative costs incurred by FCI and state agencies, costs that largely explain the fiscal and trade-distorting nature of procurement-led MSP.

Disposal Revenue reflects receipts from domestic sales (assumed at 90% of MSP) or exports of rice and wheat at prevailing international prices.

For PDP-based support, where physical procurement is replaced by direct income transfers, the fiscal cost is computed as:

PDP Cost = Coverage Share x Production x MSP x Deficiency Rate

While PDP remains price-contingent and therefore not fully Green Box-compliant, the framework allows policymakers to explicitly compare procurement-based MSP (highly trade-distorting) with income-support-oriented alternatives that substantially reduce operational and market distortions.

5.6.3 Policy Scenarios

Five policy scenarios are examined, representing a graduated transition from procurement-intensive MSP (Amber Box-heavy) towards income support mechanisms that are administratively simpler and more closely aligned with WTO-consistent practices. Rather than presenting procurement and PDP as mutually exclusive, the scenarios reflect politically and economically feasible reform pathways.

- **Scenario 1**: **Full Procurement (Benchmark, High Amber Box Exposure)**
  100% procurement of all 23 MSP-notified crops at MSP.
  Incidentals of 10–15% applied uniformly.
  This scenario serves as a theoretical upper bound, illustrating the fiscal magnitude and WTO exposure associated with universal MSP enforcement through procurement.
- **Scenario 2**: **Operational Procurement (Baseline, Current Practice)**
  Procurement of 40% of paddy and wheat output and 20% of output of the remaining 21 crops.
  Paddy and wheat disposed through exports at international prices; other crops sold domestically at 90% of MSP. The 90% is arrived based on the model price of all 23 crops for the year 2024–25 based on AGMARKNET data.
  This scenario broadly reflects existing operational capacity, but covers all crops and current AMS-relevant support patterns.
- **Scenario 3**: **Partial PDP (Hybrid MSP–PDP Model)**
  Paddy and wheat procurement maintained at Scenario 2 levels to meet food security objectives.
  For the remaining 21 crops, 20% of output receives a PDP equivalent to 10% of MSP via DBT. The 10% is uniformly applied to arrive scenario-based results rather than actuals, actual vary widely, but on average they are near to 10–20%.
  This hybrid approach reduces procurement-related distortions while extending income support to crops and regions currently outside effective MSP coverage.
- **Scenario 4**: **Enhanced PDP (Expanded Hybrid Model)**
  Paddy and wheat procurement unchanged from Scenario 2.
  PDP coverage increased to 40% of output of the remaining 21 crops, with a deficiency payment of 20% of MSP.
  This scenario strengthens income stabilisation for pulses, oilseeds, and rainfed crops while further limiting the expansion of procurement-linked support.
- **Scenario 5**: **Universal PDP (Predominantly Income Support Model)**
  Paddy and wheat procurement retained at Scenario 2 levels to meet PDS and strategic buffer requirements.
  The remaining 21 crops receive PDP for 100% of production.
  Two variants are analysed:

- **Variant 5A**: PDP at 10 per cent of MSP
- **Variant 5B**: PDP at 20 per cent of MSP

This scenario represents the most administratively streamlined design, eliminating procurement, storage, and disposal costs for non-cereal crops and moving India's MSP framework closer to internationally prevalent income-support models with lower trade distortion.

5.7 Results and Discussion

5.7.1 Scenario 1: Full Procurement (Benchmark—Maximum Amber Box Exposure)

Procuring 100% of production of all 23 MSP-notified crops would require annual farmer payouts exceeding ₹12.37 lakh crore. After accounting for procurement, handling, storage, and distribution costs, total expenditure would rise to ₹13.6–14.23 lakh crore. Beyond the prohibitive fiscal burden, this scenario is operationally infeasible. Existing national storage capacity of approximately 150 million tonnes is grossly inadequate relative to the estimated 460 million tonnes of marketable surplus (Central Warehousing Corporation 2025) implying severe risks of congestion, spoilage, and inefficiency (Table 5.3).

From a WTO Agreement on Agriculture (AoA) perspective, this scenario represents the upper bound of market price support, with nearly the entire MSP expenditure contributing to India's Aggregate Measurement of Support (AMS). The results clearly demonstrate that a legally enforceable, universal procurement-based MSP would be neither fiscally sustainable nor compatible with long-term trade policy constraints. It is to be noted that, we have not considered disposal revenue in this, as it is practically impossible to adopt this strategy.

5.7.2 Scenario 2: Operational Procurement (Baseline—Current AMS-Intensive Regime)

Under a procurement structure broadly consistent with existing operational capacity, total farmer payouts are estimated at ₹3.70 lakh crore, rising to ₹4.07–4.25 lakh crore after incidentals. After accounting for disposal revenues from domestic sales and exports, the net fiscal burden ranges between ₹3,731 and ₹22,240 crore (Table 5.4). It should be noted that, food subsidy of government of India is not included, as we assume that the procured output is sold either at domestic on-going market prices or international prices.

While fiscally manageable, this model remains structurally cereal-centric, with paddy and wheat accounting for the overwhelming share of procurement and implicit price support. From an AoA standpoint, the bulk of support continues to accrue as market price support for a narrow set of crops, leaving producers of pulses, oilseeds, and coarse cereals largely exposed to price volatility despite formal MSP announcements.

Table 5.3 Full procurement scenario (Baseline: S-1)

S.No.	Crop	MSP (Rs/qtl)	Production (lakh tons)	Payout to farmers (Rs.crore)	Total payout with 10% incidentals (Rs.crore)	Total payout with 15% incidentals (Rs.crore)
1	Paddy	2,300	1,501.8	3,45,423	3,79,966	3,97,237
2	Wheat	2,275	1,179.5	2,68,325	2,95,157	3,08,574
3	Maize	2,225	434.1	96,585	1,06,244	1,11,073
4	Jowar	3,371	49.6	16,703	18,374	19,209
5	Bajra	2,625	112.1	29,413	32,354	33,825
6	Ragi	4,290	19.8	8,481	9,329	9,754
7	Barley	1,850	19.2	3,550	3,905	4,083
8	Gram	5,440	111.1	60,460	66,506	69,529
9	Tur	7,550	36.2	27,361	30,097	31,465
10	Moong	8,682	42.4	36,846	40,531	42,373
11	Urad	7,400	22.4	16,591	18,250	19,079
12	Lentil	6,425	16.5	10,627	11,690	12,221
13	Other pulses	7,099	28.0	19,906	21,896	22,891
14	Groundnut	6,783	119.4	81,003	89,103	93,153
15	Sunflower	7,280	2.3	1,645	1,810	1,892
16	Soyabean	4,892	152.7	74,691	82,160	85,895
17	Sesamum	9,267	8.9	8,275	9,103	9,517
18	Nigerseed	8,717	0.4	349	384	401
19	R&M	5,650	126.7	71,569	78,725	82,304
20	Safflower	5,800	0.4	249	274	287
21	Copra	11,582	12.9	14,987	16,486	17,235
22	Jute	5,335	15.3	8,163	8,979	9,387
23	Cotton	7,121	50.5	35,954	39,549	41,347
	Total		4,062.3	12,37,157	13,60,872	14,22,730

Note Authors calculations. Figures are rounded, total may not sum exactly

5.7.3 Scenario 3: Partial PDP (Hybrid Model—Reduced Market Distortion)

Introducing a Price Deficiency Payment equivalent to 10% of MSP for 20% of output of non-paddy and wheat crops significantly reduces reliance on physical procurement while extending income support to historically excluded crops. The DBT payout for these crops is estimated at ₹12,468 crore. After accounting for cereal procurement operations (including disposal at domestic or international prices) revenues, the net fiscal burden declines sharply to ₹-9,887–₹2,388 crore.

Table 5.4 Policy analysis matrix of different scenarios

Scenario	Procurement share	PDP coverage	Deficiency rate	Incidentals (%)	Estimated net fiscal burden (₹ crore)	Policy assessment
1. Full procurement	100% of all 23 crops	None	–	10–15	13,60,872–14,22,730	Fiscally infeasible and logistically unmanageable; ensures full price assurance but unrealistic for implementation
2. Operational procurement (baseline)	40% of paddy & wheat, 20% of others	None	–	10–15	3,731–22,240	Feasible within current fiscal space; cereal-centric, limited inclusivity
3. Partial PDP	40% paddy and wheat procured	20% of non-paddy and wheat crop output	10% of MSP	10–15	-9,887–2,388	Hybrid approach; modest fiscal impact, lower storage needs
4. Enhanced PDP	40% paddy and wheat procured	40% of non-paddy and wheat crop output	20% of MSP	10–15	27,518–39,793	Broader farmer coverage; efficient and administratively simpler
5A. Universal PDP (base)	40% paddy and wheat procured	100% of non-paddy and wheat crop output	10% of MSP	10–15	39,986–52,261	Most inclusive and scalable; optimal balance between fiscal control and welfare
5B. Universal PDP (enhanced)	40% paddy and wheat procured	100% of non-paddy and wheat crop output	20% of MSP	10–15	1,02,327–1,14,602	Strongest income protection; higher fiscal outlay but administratively efficient

Source Authors calculations

Although PDP remains price-contingent and therefore falls within the Amber Box, this scenario substantially reduces operational costs and lowers the intensity of market price support embedded in procurement. For policymakers, it represents a low-risk transition pathway that expands effective MSP coverage while limiting incremental AMS exposure.

5.7.4 Scenario 4: Enhanced PDP (Expanded Hybrid—Broader Coverage at Moderate AMS Cost)

Expanding PDP coverage to 40% of output with a deficiency payment of 20% of MSP raises DBT outlays to approximately ₹49,873 crore. The resulting net fiscal burden increases to ₹27,518–₹39,793 crore.

This scenario delivers significantly improved income stabilisation for farmers in rainfed and diversified cropping regions, particularly pulses and oilseeds. By eliminating procurement, storage, and post-harvest losses for these crops, it improves allocative efficiency and reduces non-price distortions. However, the results also highlight the sensitivity of fiscal outcomes and AMS implications to PDP coverage and payment rates, underscoring the importance of careful policy calibration.

5.7.5 Scenario 5: Universal PDP (Predominantly Income Support Model)

Under a Universal PDP at 10% of MSP, DBT payouts for non-paddy-wheat crops amount to ₹62,341 crore, yielding a net fiscal burden of ₹39,986–₹52,261 crore after accounting for paddy and wheat procurement adjustments. Increasing the deficiency rate to 20% raises the fiscal cost to approximately ₹1.02–₹1.14 lakh crore.

Despite higher budgetary requirements, both variants eliminate procurement inefficiencies for non-paddy-wheat crops and offer a predictable, transparent, and scalable form of income support through existing DBT infrastructure. From a WTO perspective, while these payments remain price-linked, the shift away from physical procurement substantially reduces trade and market distortions relative to conventional MSP enforcement.

5.7.6 Policy Implications

The scenario analysis highlights the limits of procurement-intensive MSP regimes, which impose substantial fiscal, logistical, and trade-distorting costs while delivering benefits that remain concentrated across crops, regions, and producer groups. Maintaining large-scale physical procurement across a wide crop basket represents the upper bound of AMS exposure and is increasingly misaligned with diversification and income-support objectives.

A more balanced reform pathway emerges through hybrid MSP frameworks that retain procurement for paddy and wheat, where food security and public distribution considerations remain paramount, while extending price deficiency payments (PDPs) to other crops. Within this framework, partial PDP designs with modest deficiency rates offer the most fiscally sustainable option, enabling broader crop and regional coverage with limited incremental AMS exposure and reduced market intervention relative to procurement-based support.

Enhanced and universal PDP variants, while effective in stabilising farm incomes during periods of sharp price decline, are better suited as counter-cyclical or crisis-response instruments rather than permanent entitlements, given their higher budgetary requirements and continued Amber Box classification under the WTO. As PDPs remain explicitly linked to administered prices and current production, they are treated as equivalent in economic effect to market price support and therefore contribute to AMS.

Implementation feasibility is a critical determinant of reform success. Scaling PDP-based support requires targeted strengthening of price discovery and information systems (AGMARKNET), market-level data capture, farmer registries, and direct benefit transfer (DBT) infrastructure—foundational systems that already exist but require upgrading for accuracy, timeliness, and coverage, particularly in rainfed and remote regions.

The counterfactual scenario (Table 5.5) of fully decoupled direct benefit transfers provides an important benchmark for interpreting the residual WTO exposure of all PDP-based designs. Unlike procurement and price-deficiency mechanisms, decoupled income or area-based transfers are independent of current prices, output, and crop choice, and therefore fall within the Green Box with no AMS liability. This comparison underscores that reducing physical market intervention alone is insufficient to eliminate trade distortion unless price contingency is also removed. While PDP reforms substantially lower market distortion relative to procurement-heavy regimes, a transition toward decoupled transfers represents the only structurally WTO-consistent long-run reform option.

Overall, the findings support a calibrated and phased transition from procurement-led MSP toward PDP-based income support as an intermediate reform, with a longer-term shift toward decoupled transfers. Such a sequencing can reduce market distortions, improve fiscal management, promote crop diversification, and extend income protection to neglected crops and regions, while remaining broadly aligned with India's WTO commitments.

Key Takeaways for WTO Compliance

- All MSP- and PDP-linked scenarios fall within the Amber Box, but their trade distortion intensity varies sharply.

Table 5.5 WTO AoA classification of alternative MSP–PDP policy scenarios

Scenario	Policy instrument mix	Dominant WTO AoA box	Nature of AMS exposure	Degree of trade distortion	Key WTO interpretation
Scenario 1: full procurement (benchmark)	100% physical procurement of all 23 MSP crops	Amber box	Very high—market price support across all crops	Severe	Represents maximum AMS liability; universal administered prices linked to current production
Scenario 2: operational procurement (baseline)	Procurement of paddy & wheat (40%); limited procurement of other crops (20%)	Amber box	High but concentrated—mainly rice & wheat	High (crop-specific)	Current de facto regime; AMS exposure remains cereal-centric and regionally concentrated
Scenario 3: partial PDP (hybrid)	Paddy–wheat procurement + PDP (10% MSP on 20% output of other crops)	Amber box (lower intensity)	Moderate—price-contingent DBT replaces procurement	Moderate	Reduces market intervention and stockholding; AMS persists but with lower distortion
Scenario 4: enhanced PDP	Paddy–wheat procurement + PDP (20% MSP on 40% output)	Amber box	Moderate–high (DBT-driven)	Moderate	Broader income protection with reduced market distortion relative to procurement
Scenario 5A: universal PDP (10%)	Paddy–wheat procurement + PDP for 100% of other crops	Amber box (minimal market distortion)	Moderate (price-linked)	Low	Eliminates procurement-related distortions; AMS arises only from deficiency payments
Scenario 5B: universal PDP (20%)	Same as 5A with higher deficiency rate	Amber box	High (budgetary)	Low–moderate	Best suited as counter-cyclical support; requires AMS space management
Counterfactual (not modelled): decoupled DBT	Area- or income-based transfers unrelated to prices	Green box	Nil	Minimal	WTO-compliant long-run reform option (e.g., PM-KISAN-type expansion)

Note Author's calculations

- Procurement-based MSP is the most AMS-intensive and trade-distorting instrument.
- PDP significantly lowers market distortion even when classified as Amber Box, because it avoids administered prices and stock accumulation.
- Hybrid and partial PDP scenarios offer the best balance between farmer income protection, fiscal prudence, and manageable AMS exposure.
- A future transition to fully decoupled income support would be required for Green Box compliance, but lies outside the scope of MSP reform per se.

References

Aayog NITI (2017) Changing structure of rural economy of India implications for employment and growth. National Institution for Transforming India, Government of India

Birthal PS, Negi DS, Roy D (2021) Crop diversification in India: determinants and policy implications. J Clean Prod 313:127906

Central Warehousing Corporation (2025) Ministry of Consumer Affairs, Food & Public Distribution, Government of India. https://www.cewacor.nic.in/Home/. Accessed 12 Aug 2025

Chand R (2022) Indian agriculture towards 2030—need for a transformative vision. In: Indian agriculture towards 2030: pathways for enhancing farmers' income, nutritional security and sustainable food and farm systems. Singapore, Springer Nature Singapore, pp 1–7

Chand R (2025) Reimagining agriculture for growth, poverty, nutrition, and sustainability. Agric Econ 56(3):485–492

Chand R, Raju SS, Reddy AA (2015) Assessing performance of pulses and competing crops based on market prices and natural resource valuation. J Food Legum 28(4):335–340

Commission for Agricultural Costs and Prices (CACP) (2024) Price policy for Kharif crops: marketing season 2024–05. Department of Agriculture and Farmers Welfare, Government of India. Retrieved from https://cacp.dacnet.nic.in

Dorward A, Kydd J, Morrison J, Urey I (2004) A policy agenda for pro-poor agricultural growth. World Dev 32(1):73–89. https://doi.org/10.1016/j.worlddev.2003.06.012

Dorward A, Sabates-Wheeler R, MacAuslan I, Buckley CP, Kydd J, Chirwa E (2006) Promoting agriculture for social protection or social protection for agriculture: strategic policy and research issues. Growth Soc Prot Discuss Pap 4

Gardner B (2006) American agriculture in the twentieth century: how it flourished and what it cost. Harvard University Press

Government of India (2024) Agricultural statistics at a glance 2024. Directorate of Economics and Statistics, Department of Agriculture and Farmers Welfare, New Delhi. Retrieved from https://eands.dacnet.nic.in

Government of India (2025) Third advance estimates of agricultural production 2024–25. Department of Agriculture and Farmers Welfare, New Delhi. Retrieved from https://agricoop.nic.in

Government of India, Ministry of Finance (2025) Union budget 2025–26: budget speech of the finance minister. https://www.indiabudget.gov.in

Gulati A, Juneja R (2022a) Reforming India's food subsidy regime. Indian J Agric Econ 77(1):1–20

Gulati A, Juneja R (2022b) Transforming Indian agriculture. In: Chand R, Joshi P, Khadka S (eds) Indian agriculture towards 2030. India studies in business and economics. Springer, Singapore. https://doi.org/10.1007/978-981-19-0763-0_2

Gulati A, Jain S, Satija N (2014) Rising farm wages in India—the 'pull' and 'push' factors. J Land Rural Stud 2(2):261–286

Gulati A, Roy R, Saini S (2021) Revitalizing Indian agriculture and boosting farmer incomes. Springer Nature

Gulati A, Das R, Winter-Nelson A (2024) Reducing post-harvest losses in India: farmer-level interventions and grain management strategies. In: Indian Council for Research on international economic relations (ICRIER), vol No. 24-r-03, New Delhi, India

Gupta P, Khera R, Narayanan S (2021) Minimum support prices in India: distilling the facts. Rev Agrar Stud 11(1):49–70

Huang J, Yang G (2017) Understanding recent challenges and new food policy in China. Glob Food Sec 12:119–126

Jha S, Ramaswami B (2010) How can food subsidies work better? Answers from India and The Philippines. Asian Dev Bank Econ Work Pap Ser 221

Korekallu Srinivasa A, Praveen KV, Surendran Padmaja S, Nithyashree ML, Jha GK (2023) Does a farmer's knowledge of minimum support price (MSP) affect the farm-gate price? Evidence from India. J Econ Dev 25(4):302–316

Kumar A, Dey K, Gupta KB (2023) Food Corporation of India: making public procurement more inclusive and efficient. Asian J Manag Cases 09728201231168246

Narayanan S (2023) Agricultural incentives, crop choice, and sustainability in India. World Dev 165:106206

Negi DS, Birthal PS, Roy D, Hazrana J (2020) Market access, price policy and diversification in Indian agriculture (no. 2020–009). Indira Gandhi Institute of Development Research, Mumbai, India

OECD (2023a) Agricultural policy monitoring and evaluation. OECD Publishing

OECD (2023b) Agricultural policy monitoring and evaluation 2023: adapting agriculture to climate change. OECD Publishing, Paris. https://doi.org/10.1787/b14de474-en

OECD/FAO (2023a) OECD–FAO Agricultural Outlook. OECD Publishing

OECD/FAO (2023b) OECD–FAO Agricultural Outlook 2023–2032. OECD Publishing, Paris. https://doi.org/10.1787/08801ab7-en

Reddy AA (2009) Pulses production technology: status and way forward. Econ Polit Wkly:73–80

Reddy AA, Bantilan MCS (2012) Competitiveness and technical efficiency: determinants in the groundnut oil sector of India. Food Policy 37(3):255–263

Reddy AA, Rani CR, Reddy GP (2009) Policy options for edible oil complex in India under WTO regime. J Rural Dev 30(1):11–24

Reddy AA, Bhattacharya A, Reddy SV, Ricart S (2021) Farmers' distress index: an approach for an action plan to reduce vulnerability in the drylands of India. Land 10:1236

Reddy AA, Anugu AR, Jirli B (2025) Market and technological reforms in Indian agriculture: needs for a Wholistic approach. IASSI Q: Contrib Indian Soc Sci 44(4):854–891

Timmer CP (1989) Food price policy: the rationale for government intervention. Food Policy 14(1):17–27

Timmer CP (1997) Farmers and markets: the political economy of new paradigms. Am J Agric Econ 79(2):621–627

World Bank (1999) India foodgrain marketing policies: reforming to meet food security needs (vol. 1 of 2): main report. World Bank, Washington, DC. http://documents.worldbank.org/curated/en/575711468750268959

Xu L, Zhu J, Wang X (2025) China's new subsidy policy for corn producers: does it alleviate market distortions? Agribusiness

Chapter 6
Waste-Energy-Agriculture Nexus and Resource Recycling for Smart Farming: An ICT-Based Circular Economy Model from Japan

Natsumi Matsui, Yoshitaka Taniguchi, Masashi Yamamoto, Norio Horie, Mohan Geetha, and Naoya Wada

Abstract Rapid economic growth driven by mass production and consumption has generated substantial quantities of waste globally. This development has intensified pressure on landfill capacities, increased greenhouse gas emissions, and contributed to broader environmental degradation. Addressing these issues requires implementing integrated waste management systems that not only ensure the safe treatment and disposal of waste but also facilitate the recovery of valuable resources from waste streams, reintegrating them into productive economic activities. Such strategies align with the principles of the circular economy and directly support the achievement of the Sustainable Development Goals (SDGs). This chapter provides a comprehensive case study of an integrated waste management enterprise situated in Toyama Prefecture, central Japan, operating under the guiding principle that "waste is a resource." The company oversees the entire waste lifecycle-from collection and transportation to incineration, recycling, and final disposal-within a centralized operational framework. Moreover, thermal energy and electricity generated during waste processing are repurposed for agricultural production through information and communication technology (ICT). This low-carbon, resource-circulating system sustains greenhouse cultivation of horticultural commodities, including flowers (lisianthus) and vegetables (tomatoes), enabling consistent, year-round production even under challenging climatic and geographical conditions. By integrating waste treatment with agricultural production, the system diminishes reliance on fossil fuels, enhances overall

N. Matsui · Y. Taniguchi (✉)
Toyama Kankyo Seibi Co. Ltd., Toyama, Japan
e-mail: y-taniguti@tks-co.jp

M. Yamamoto
Faculty of Economics, Kanagawa University, Yokohama, Japan

N. Horie · M. Geetha · N. Wada (✉)
Global Research Centre for Advanced Sustainability Science, University of Toyama, Toyama, Japan
e-mail: wada@sci.u-toyama.ac.jp

D. Kumar Swain et al. (eds.), *Climate-Smart Technologies for Sustainable Agricultural Production*, https://doi.org/10.1007/978-3-032-25666-9_6

energy efficiency, and promotes responsible consumption and production. At the regional level, it addresses structural challenges such as farmland abandonment, labor shortages, and demographic aging, while fostering cooperation and trust among waste management operators, agricultural producers, and local communities. Overall, this chapter illustrates how integrated waste–agriculture systems can serve as practical, place-based solutions to advance decarbonization and circular-economy principles, offering valuable insights for developing sustainable regional systems aligned with the SDGs.

Keywords Waste-to-energy · Circular economy · Smart agriculture · ICT-based agriculture · Resource recycling · Japan

6.1 Introduction

6.1.1 Global Trends in Waste Generation

The rapid expansion of industrial production, urbanization, and consumption-driven economic systems has led to a significant increase in global waste generation. In 2017, global waste output was estimated at approximately 20 billion tons annually, with projections indicating an increase to 45 billion tons by 2050 (Maalouf and Mavropoulos 2023). Municipal solid waste accounts for a substantial portion of this total, amounting to 2.7 billion tons in 2019, and it is anticipated to increase by 26–45% by the middle of the century. Despite technological advancements in waste management, nearly one-third of municipal waste worldwide remains uncollected, and a considerable proportion of collected waste is inadequately treated or disposed of in environmentally harmful manners (Maalouf and Mavropoulos 2023). These trends pose multiple environmental and social risks, including land and ecosystem degradation, contamination of soil and water resources, air pollution, and increased greenhouse gas emissions. Simultaneously, waste streams contain considerable quantities of potentially valuable materials and energy that remain underutilized. From a sustainability perspective, treating waste solely as a disposal issue rather than as a resource constitutes a missed opportunity to enhance resource efficiency and mitigate climate change.

UNEP (2024) summarized this concern about the improper treatment of waste worldwide. Figure 6.1 illustrates projected global trends in household (municipal solid) waste generation and treatment methods from 2020 to 2050. Total municipal solid waste generation is expected to increase substantially over the coming decades, reflecting population growth, urbanization, rising incomes, and changes in consumption patterns, particularly in low- and middle-income countries. The overall height of each bar shows that global waste generation rises steadily from just over 2 billion tonnes in 2020 to nearly 4 billion tonnes by 2050.

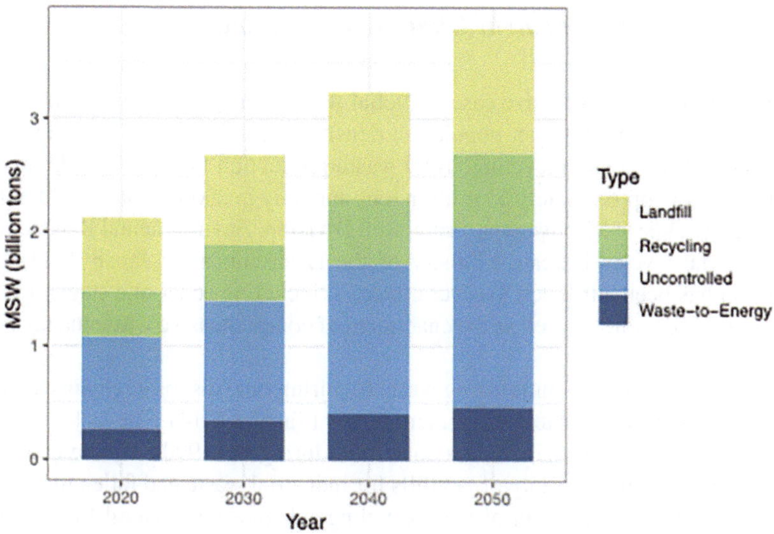

Fig. 6.1 Projected global municipal solid waste destinations in 2030, 2040 and 2050. *Source* UNEP (2024, p 22)

Figure 6.1 also highlights how waste management practices are projected to evolve over time. Uncontrolled disposal, such as open dumping or poorly managed landfills, remains a major component throughout the period and, in absolute terms, continues to grow. This indicates that, without significant policy intervention and investment, many regions will struggle to keep pace with rising waste volumes, potentially exacerbating environmental pollution, public health risks, and greenhouse gas emissions.

Concurrently, the proportion of waste processed via recycling and waste-to-energy technologies is anticipated to rise. Recycling demonstrates consistent growth, attributable to incremental enhancements in collection systems, sorting infrastructure, and policies promoting a circular economy. Nevertheless, by 2050, recycling will constitute only a modest portion of the total waste, indicating ongoing economic, institutional, and technological limitations. Waste-to-energy also increases progressively, especially in regions experiencing landfill shortages or robust climate policy incentives; however, it remains a supplementary rather than a primary method of waste treatment option.

Landfilling continues to play a central role in global waste management, increasing both in absolute volume and as a stable share of total waste. Overall, the figure underscores a critical policy challenge: while waste treatment technologies improve gradually, total waste generation continues to grow. This implies that achieving sustainable waste management will require not only downstream treatment solutions but also upstream measures, such as waste prevention, product redesign, and changes in consumption behavior.

6.1.2 Waste Management Challenges in Japan

Japan exemplifies a distinctive case in global waste management, attributable to its limited land area, high urban population density, and a longstanding policy focus on waste reduction and recycling. Total waste production in Japan has diminished from a peak of approximately 55 million tons annually in 2000 to roughly 40 million tons in 2021 (Waste Management Promotion Division, Environmental Regeneration and Material Cycles Bureau, Ministry of the Environment of Japan 2023). This reduction has been attributed to several factors, including economic stagnation and demographic decline, as well as the enactment of comprehensive waste management and recycling legislation.

It is worth noting that minimizing waste requiring final disposal remains a central objective in Japanese waste management policy (Tanaka 2014). The lack of enough final disposal sites was a significant constraint during the 1990s. By the end of the 1990s, the remaining capacity of landfills for industrial waste had fallen to less than three years. The introduction of new recycling schemes was central to preventing the exhaustion of Japan's final disposal sites. Figure 6.2 summarizes the recycling schemes established during this period. For example, packaging waste accounted for approximately 60 percent of household waste by volume, making it one of the most significant contributors to pressure on final disposal sites. Because packaging materials are generated ubiquitously through daily consumption, their accumulation poses a systemic challenge for municipal waste management. In response, dedicated recycling schemes for containers and packaging played a critical role in reducing the overall volume of waste requiring final disposal.

The Home Appliance Recycling Act made a substantial contribution to landfill reduction by targeting bulky, durable goods that were particularly difficult for municipalities to process. The Act covered televisions, refrigerators, washing machines, and air conditioners—items characterized by large size, complex material composition, and high treatment costs. Prior to the Act, municipalities faced significant technical and financial constraints in recycling these appliances. By shifting responsibility to producers and establishing a dedicated collection and recycling system, the Act significantly reduced the amount of large-scale waste sent to final disposal sites. As a result, Japan's remaining landfill capacity has increased significantly (Fig. 6.3). Nonetheless, landfill capacity remains limited, and acquiring new

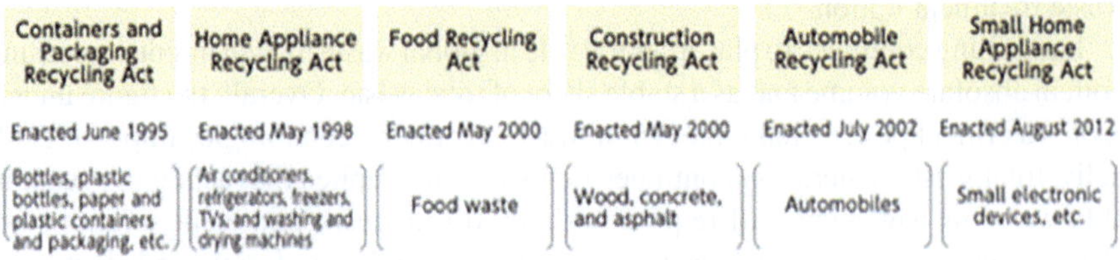

Fig. 6.2 Recycling-related policy framework in Japan. *Source* MOE (2014)

Fig. 6.3 Changes in remaining year of landfill capacity in Japan. *Source* annual report on environmental statistics by MOE (2024)

sites is progressively challenging due to environmental considerations and public opposition.

An additional challenge pertains to the spatial distribution of waste disposal facilities. Approximately 70% of final disposal sites in Japan are located in hilly and mountainous regions, whereas most waste is generated in densely populated lowland urban areas (Waste Management Promotion Division, Environmental Regeneration and Material Cycles Bureau, Ministry of the Environment of Japan 2023). This spatial disparity imposes an undue burden on rural communities, which are tasked with accepting waste generated elsewhere while safeguarding environmental quality and local livelihoods. Consequently, waste management operators in these rural areas are compelled to implement environmentally sound practices and foster strong relationships with local communities.

6.1.3 Objectives and Scope of the Chapter

In this context, this chapter presents a regional case study of Toyama Kankyo Seibi Co., Ltd., a waste management enterprise primarily operating in the hilly and mountainous regions of Toyama City, Toyama Prefecture, in central Japan. The corporation has established a centralized waste management system that encompasses waste collection, transportation, incineration, recycling, and final disposal. Beyond traditional waste treatment methods, the organization has expanded its activities to involve agricultural production, leveraging waste-derived thermal energy and electricity through ICT-based control systems.

The aim of this chapter is to critically examine how this integrated waste–agriculture model facilitates resource circulation, decarbonization, and regional sustainability. Special emphasis is placed on the role of ICT in enabling efficient energy utilization and ensuring stable agricultural output amidst challenging climatic and geographical conditions. By documenting and analyzing this case, the chapter seeks to offer practical insights into the application of circular economy principles at the regional level, aligning with the Sustainable Development Goals (SDGs).

6.2 Overview of Toyama Kankyo Seibi Co., Ltd.

Toyama Kankyo Seibi Co., Ltd., established in 1972, has evolved into a leading waste management provider within Toyama Prefecture. The organization employs approximately 350 personnel and primarily operates in the hilly and mountainous regions of Toyama City. An overview of the company's facilities and operational sites is presented in Fig. 6.4.

Since its inception, the company has prioritized environmental responsibility and harmonious coexistence with local communities. This focus reflects the geographical context of its operations, in which waste-disposal facilities are located near residential and agricultural zones. Consequently, maintaining trust and transparency has been vital for the company's sustained success.

A distinctive characteristic of Toyama Kankyo Seibi is its centralized waste-processing system, which integrates multiple waste-management functions into a

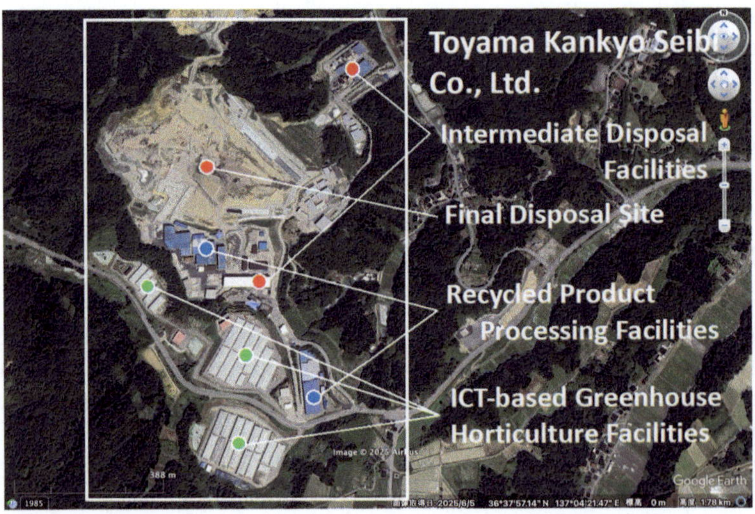

Fig. 6.4 Overview of the waste management company Toyama Kankyo Seibi Co., Ltd., in Toyama City, Toyama Prefecture, Japan (Maps data: Google, ©2025 Airbus)

single operational framework. This system encompasses waste collection, transportation, incineration, recycling, and landfill disposal. Such integration is relatively uncommon in Japan, where waste management responsibilities are often dispersed among various operators. The centralized approach enhances operational efficiency, improves environmental oversight, and improves resource recovery from waste streams. The company's activities are categorized into three principal business sectors: (1) waste management, (2) recycled products, and (3) agricultural production. The integration of these three sectors within a unified operational framework is depicted in Fig. 6.5.

These sectors are intricately interconnected through the company's guiding philosophy that "waste is a resource." Materials and energy recovered during waste treatment are reintroduced into productive processes, thereby establishing a resource-circulating system that transcends conventional waste management boundaries. This integrated approach underpins the linkage between waste management and agricultural production.

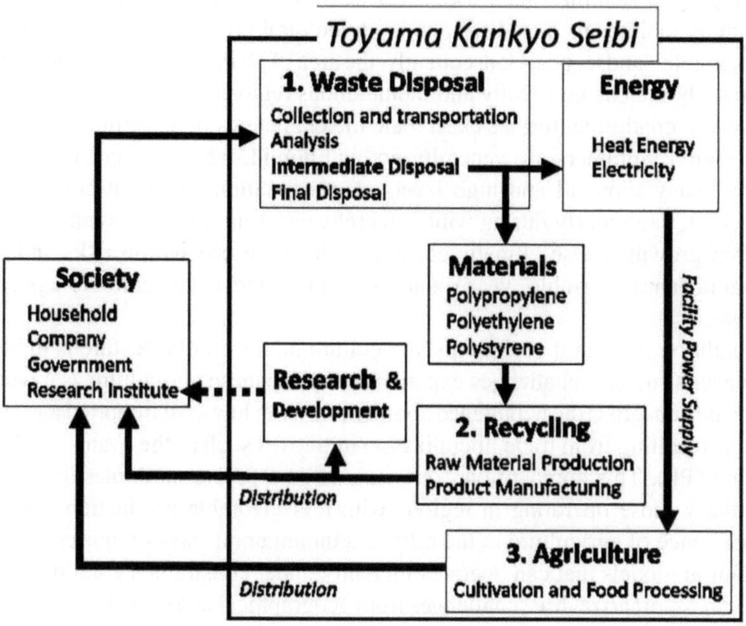

Fig. 6.5 Three main businesses of Toyama Kankyo Seibi Co., Ltd.

6.3 Agriculture in Hilly and Mountainous Areas of Toyama

6.3.1 Geographical and Climatic Conditions

Toyama Prefecture encompasses extensive mountainous terrain, commonly referred to in Japan as "hilly and mountainous areas", which are characterized by steep slopes, fragmented farmland, and limited accessibility, all of which constrain agricultural productivity. These regions occupy 74.1% of Japan's total land area (OECD 2025), 38.1% of the country's total agricultural land, and approximately 40% of gross agricultural output, according to the 2020 Census of Agriculture and Forestry in Japan (MAFF n.d.).

Hilly and mountainous areas are severely affected by depopulation and aging. The settlements are called "genkai shuraku" (marginal settlements), defined as communities in which individuals aged 65 years or older account for more than half of the total population (Feldhoff 2013). These settlements face significant challenges in managing agricultural land use, securing sufficient labor for both farming and community maintenance, and sustaining agricultural production across cultural and environmental landscapes. Consequently, the area of abandoned agricultural land has been rapidly increasing in hilly and mountainous regions.

Climatic conditions further exacerbate the challenges of agricultural production in the region. Summers are generally hot and humid, while winters are characterized by heavy snowfall and high levels of precipitation. Daylight hours are relatively short, particularly during winter, thereby constraining photosynthetic activity and crop growth. These climatic constraints increase production risks and hinder the maintenance of stable, year-round cultivation under conventional agricultural practices.

In addition to natural constraints, agriculture in Toyama Prefecture is influenced by broader structural challenges experienced by Japanese agriculture as a whole. A significant concern is the heightened competition from low-cost imported agricultural products resulting from trade liberalization initiatives such as the Trans-Pacific Partnership (TPP). This competition exerts economic pressure on domestic producers, particularly those operating in regions with less favorable production conditions. The resilience of agriculture in the hilly and mountainous areas requires innovative agricultural models that can operate efficiently under constrained conditions.

We can summarize three challenges from geographical and climate conditions: (1) Revitalizing agricultural land use, (2) Maintaining stable agricultural production, and (3) Innovating agricultural models that are feasible under the constrained conditions.

6.3.2 Waste Management Industrial Location and its Challenges

Waste management companies frequently face constraints due to their industrial locations, as waste management facilities and disposal sites are often considered *locally unwanted land uses* (LULUs). Previous research indicates that such facilities tend to be located in areas with relatively low levels of public opposition (Ishimura and Takeuchi 2019). Toyama Kankyo Seibi similarly demonstrates a strong sensitivity to local perceptions of its waste management activities and has therefore placed particular emphasis on environmental responsibility and harmonious coexistence with surrounding communities. The company owns and operates its own facilities and disposal sites in a hilly, mountainous region that has experienced significant depopulation and population aging. As a result of these demographic changes, the area has experienced a rapid increase in abandoned and underutilized agricultural land, leading to gradual transformations in the regional landscape. In response to these challenges, Toyama Kankyo Seibi has positioned agricultural engagement as a core component of its corporate social responsibility, focusing on three key objectives: revitalizing agricultural land use, stabilizing agricultural production, and developing innovative agricultural models suited to the regional context.

6.4 Next-Generation Greenhouse Horticulture and Circular Agriculture

6.4.1 Policy Background and System Development

In order to address the challenges facing Japanese agriculture, the Ministry of Agriculture, Forestry, and Fisheries (MAFF) has spearheaded the development of next-generation greenhouse horticulture systems. As part of this initiative, MAFF established "bases for next-generation greenhouse horticulture" across ten locations nationwide (MAFF 2018). These bases are designed to demonstrate advanced production technologies, enhance productivity, and bolster the competitiveness of domestic agriculture.

Toyama Kankyo Seibi was designated as one of these bases in 2014, reflecting its innovative integrated waste–agriculture approach and its potential contribution to regional revitalization.

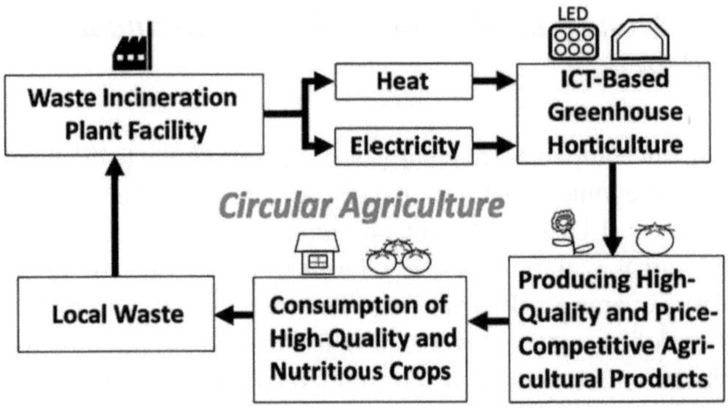

Fig. 6.6 Next generation greenhouse horticulture and circulation agriculture

6.4.2 *Waste-Derived Energy Use and Circular Agriculture*

The greenhouse system developed by Toyama Kankyo Seibi uses thermal energy and electricity generated during waste processing to provide heating, lighting, and environmental control. This approach reduces reliance on fossil fuels and diminishes greenhouse gas emissions associated with agricultural activities.

Agricultural products cultivated within these greenhouses are supplied to markets and consumed by households. The organic waste generated through consumption is subsequently collected and processed within the waste management system, where it is repurposed as an energy source. This cyclical exchange of materials and energy between waste treatment and agricultural production is schematically depicted in Fig. 6.6. This closed-loop system exemplifies the concept of circular agriculture, in which resources are continuously cycled between production, consumption, and waste management.

6.4.3 *Production Characteristics and Certification*

Lisianthus is cultivated on a 1.2-hectare plot of land, with an annual yield of approximately 204,000 stems. Tomatoes are grown on 2.8 ha using the Imec film-farming technique outlined below, resulting in an estimated annual production of approximately 320 metric tons. The tomato cultivation system is certified under GLOBALG.A.P. for a two-year validity period, and the production practices have continued to be maintained in line with the same standards for food safety, labor conditions, and environmental sustainability following the completion of the certification period.

6.5 Imec Film Farming Technology

Imec film farming (Mori 2013) is a root-zone management technology that uses a specialized film embedded with a hydrogel compound that retains substantial water. The configuration and operation of the Imec film farming system are illustrated in Fig. 6.7. This technology enables crop cultivation with minimal water input, particularly beneficial in mountainous regions where water resources may be scarce. As the system isolates plant roots from the soil, it can also be employed in areas contaminated with soil.

By positioning the Imec film between the plant roots and the water-supplying cloth, the uptake of water and nutrients by the roots is meticulously regulated. This controlled stress promotes enhanced nutrient absorption, resulting in tomatoes with elevated sugar content and improved nutritional quality. Although fruit size tends to be slightly smaller, the average sugar content is 8.99°Brix, with some varieties exceeding 10°Brix.

Given the frequent cloudy conditions and limited solar radiation in Toyama Prefecture, supplementary LED lighting is used (Fig. 6.8). Compared with traditional cultivation methods, the introduction of LED lighting has led to an approximate 30% increase in tomato yields.

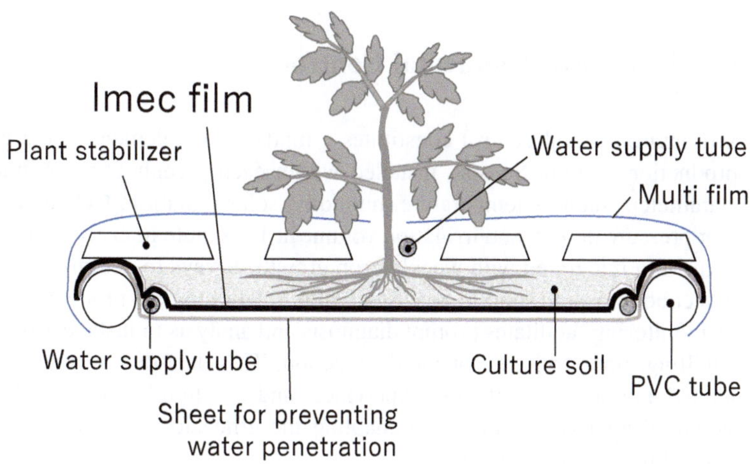

Fig. 6.7 Imec film farming

Fig. 6.8 Supplementary LED lighting for enhancing tomato yields in ICT-based greenhouse horticulture

6.6 ICT-Based Environmental Control

ICT-based environmental control constitutes a fundamental element of the greenhouse production system. Sensors installed within each greenhouse continuously record parameters such as temperature, humidity, solar radiation, CO_2 concentration, vapor pressure deficit, and irrigation volume at 15-minute intervals. These data are subsequently integrated with weekly crop growth surveys to determine optimal cultivation conditions and support consistent, high-quality tomato production. Real-time data monitoring facilitates prompt diagnosis and analysis in instances of issues such as wilting, poor fruit set, or yield reduction. This data-driven methodology has promoted a paradigm shift from experience- and intuition-based agriculture to a science-based approach, while also enabling the efficacious training of novice farmers with limited prior experience.

6.7 Additional Agricultural Initiatives and Value Creation

In response to issues such as farmland abandonment and labor shortages, Toyama Kankyo Seibi acquires or leases agricultural land that has become challenging for local residents to manage due to aging demographics. The organization engages in the cultivation of rice paddies, buckwheat, fast-growing trees, sweet potatoes, vegetables, and hazelnuts. Drone technology is employed for pest and disease management in rice fields, thereby improving labor efficiency. Additionally, the company operates its own processing facilities that enable the production of value-added products, including tomato juice and puree made from imperfect agricultural produce. This approach aims to reduce food loss, enhance value creation, and promote local production for local consumption.

6.8 Conclusion

The integrated waste–agriculture model outlined in this chapter exemplifies how circular economy principles can be effectively operationalized at the regional level through the combined utilization of waste-derived resources and information and communication technology (ICT)-based agricultural systems. By integrating waste management with agricultural production, the model advances decarbonization, responsible consumption and production, and sustainable food systems, while simultaneously addressing local challenges such as farmland abandonment and labor shortages. Beyond environmental and economic advantages, the model cultivates trust and cooperation among waste management operators, agricultural producers, local governments, and residents. Consequently, it offers a pragmatic and adaptable framework for establishing sustainable, low-carbon regional economies aligned with the Sustainable Development Goals (SDGs), especially in regions facing geographical and demographic constraints.

References

Feldhoff T (2013) Shrinking communities in Japan: community ownership of assets as a development potential for rural Japan? Urban Des Int 18:99–109. https://doi.org/10.1057/udi.2012.26

Ishimura Y, Takeuchi K (2019) The spatial concentration of waste landfill sites in Japan. Resour Energy Econ 58:101121. https://doi.org/10.1016/j.reseneeco.2019.101121

Maalouf A, Mavropoulos A (2023) Re-assessing global municipal solid waste generation. Waste Manag Res 41(4):936–947. https://doi.org/10.1177/0734242X221074116

MAFF (Ministry of Agriculture, Forestry and Fisheries) (2018) Situation of greenhouse horticulture. Available online: https://www.maff.go.jp/e/policies/agri/attach/pdf/index-15.pdf (Accessed on 31 Dec 2025)

MAFF (Ministry of Agriculture, Forestry and Fisheries) (n.d.) About hilly and mountainous areas (Chūsankan Chiiki): Cultivated land share in hilly and mountainous areas. Retrieved January 19, 2026, from https://www.maff.go.jp/j/nousin/tyusan/siharai_seido/s_about/cyusan/

MOE (Ministry of the Environment) (2014) History and current state of waste management in Japan. https://www.env.go.jp/content/900453392.pdf

MOE (Ministry of the Environment). (2024) Annual report on environmental statistics (Reiwa 6). https://www.env.go.jp/doc/toukei/tokeisyu.html

Mori Y (2013) New agro-technology (Imec) by hydrogel membrane. React Funct Polym 73(7):936–938. https://doi.org/10.1016/j.reactfunctpolym.2012.11.015

OECD (2025) Enhancing rural innovation in Japan: sustainable land-use management for rural innovation. OECD Publishing, Paris. https://doi.org/10.1787/8d3b682a-en

Tanaka M (2014) Municipal solid waste Management in Japan. In: Pariatamby A, Tanaka M (eds) Municipal solid waste Management in Asia and the Pacific Islands, environmental science and engineering. Springer, Singapore, pp 157–172. https://doi.org/10.1007/978-981-4451-73-4_9

UNEP (2024) Global waste management outlook 2024: beyond an age of waste—turning rubbish into a resource, Nairobi. https://wedocs.unep.org/20.500.11822/44939

Waste Management Promotion Division Environmental Regeneration and Material Cycles Bureau, Ministry of the Environment of Japan (2023) Waste disposal in Japan. Reiwa 3rd edn. p 82 (in Japanese). https://www.env.go.jp/recycle/waste_tech/ippan/r3/data/disposal.pdf

Chapter 7
Climate-Smart, Resource-Recycling Organic Farming with Managed Weed Coexistence in Toyama, Central Japan

Ariful Md Haque, Yuto Minemura, Mohan Geetha, and Naoya Wada

Abstract The extensive use of chemical fertilizers and pesticides has increased agricultural productivity; however, it has also led to significant environmental degradation, underscoring the need for sustainable, climate-resilient farming systems. Resource-recycling organic agriculture, which employs locally available inputs, presents a promising alternative; nonetheless, the ecological outcomes of integrated organic practices remain inadequately documented. This chapter assesses an integrated organic farming system implemented in Hakkayama, Toyama Prefecture, central Japan, which combines spent mushroom substrates, wood chips, biomass ash, and managed weed coexistence (SMS–CW). Soil carbon and nitrogen contents, soil microarthropod abundance, and seasonal vegetation dynamics were compared among SMS–CW, conventional, and other organic farming systems in soybean and vegetable fields. Although total soil carbon and organic nitrogen contents did not differ significantly among management practices, the SMS–CW system consistently demonstrated higher mean values. Soil microarthropod abundance was notably greater under the SMS–CW system, indicating enhanced soil biological activity. Multispectral imagery further revealed that SMS–CW fields maintained more stable vegetation cover throughout the growing season. These findings imply that the synergistic effects of organic amendments and weed coexistence enhance soil biological functioning and microclimatic stability. Under projected climate change scenarios characterized by increased summer temperatures and heavy rainfall, such integrated organic systems may enhance soil resilience by mitigating overheating and erosion risks.

Keywords Organic agriculture · Resource recycling · Soil quality · Climate resilience · Soil microarthropods

A. M. Haque · Y. Minemura
Graduate School of Science and Engineering for Education, University of Toyama, Toyama, Japan

M. Geetha · N. Wada (✉)
Global Research Centre for Advanced Sustainability Science, University of Toyama, Toyama, Japan
e-mail: wada@sci.u-toyama.ac.jp

D. Kumar Swain et al. (eds.), *Climate-Smart Technologies for Sustainable Agricultural Production*, https://doi.org/10.1007/978-3-032-25666-9_7

7.1 Introduction

Agrochemicals, including fertilizers and pesticides, have notably increased crop yields and contributed to global food security, while simultaneously reducing the need to expand agricultural land (Barretto et al. 2013; Stewart and Roberts 2012; Xiang et al. 2020). However, the widespread use of chemical inputs has led to significant environmental issues, particularly nitrogen pollution from fertilizer runoff, leaching, and gaseous emissions, which negatively affect terrestrial and aquatic ecosystems (Erisman et al. 2013; Gu et al. 2023; Wang et al. 2000). These issues have heightened interest in alternative agricultural systems designed to lessen reliance on synthetic inputs. Organic agriculture has emerged as a promising approach that emphasizes the use of locally available organic resources while minimizing chemical fertilizers and pesticides (Dhaliwal et al. 2021). Organic management commonly relies on inputs such as compost, animal manure, and mineral amendments to maintain soil fertility (Assefa et al. 2019; Bergstrand 2022; Hua et al. 2020), and has been demonstrated to enhance soil nutrient status, organic matter content, and soil biological activity (Bengtsson et al. 2005; Lori et al. 2017; Sahu et al. 2024). Furthermore, numerous organic farming systems permit partial weed coexistence rather than complete eradication, which can reduce nitrate leaching and foster more heterogeneous, vegetated field conditions that support soil biodiversity (Bond and Grundy 2001; Cussans 1992).

Agricultural management significantly influences soil quality, which is characterized by integrated physical, chemical, and biological properties that determine soil functionality (Bünemann et al. 2018; Karlen et al. 1997). Among biological indicators, soil microarthropods play a pivotal role because they occur across multiple trophic levels and contribute to organic matter decomposition, nutrient cycling, and soil structure formation through feeding and burrowing (Gergócs et al. 2022; Menta and Remelli 2020). It has been frequently reported that higher abundances of soil microarthropods are observed under organic management practices, thereby indicating an enhancement in the biological quality of agricultural soils (Coulis 2021; Doles et al. 2001; Vreeken-Buijs et al. 1998). Notwithstanding these ecological advantages, organic agriculture represents approximately 2% of the world's arable land and 0.8% of Japan's total arable land (MAFF 2025). To bridge this disparity, Japan's MIDORI-2021 Strategy aims to significantly reduce the use of chemical fertilizers and pesticides and to expand organic farmland to 25% of the total cultivated area by 2050. This emphasizes the necessity for regionally tailored organic farming systems that efficiently recycle local resources.

This chapter investigates a resource-recycling organic farming system implemented in Hakkayama, Toyama City, central Japan. The system integrates spent mushroom substrate (SMS) derived from commercial mushroom cultivation, wood chips from lumber mills and landscaping firms, and ash from a biomass power plant, which are subsequently applied to agricultural soils. SMS enhances microbial activity and soil nutrient availability, serving as a substrate for soil microarthropods (Li et al. 2017; Pintarič et al. 2024; Tang et al. 2024). Meanwhile, wood chips and biomass ash improve soil physical structure and regulate pH (Albuquerque et al. 2021; Bhandari et al. 2024; Fontana et al. 2023; Li et al. 2018). Furthermore, this system facilitates the managed coexistence of weeds and crops, potentially buffering soil temperature and moisture, thereby supporting belowground biodiversity amid changing climatic conditions. While prior research has assessed individual organic amendments, the combined effects of spent mushroom substrate, wood chips, biomass ash, and managed weed coexistence on soil properties and soil fauna remain poorly understood. Accordingly, this chapter assesses the effects of the integrated system, hereafter designated the SMS–CW treatment, on soil carbon and nitrogen concentrations and soil microarthropod densities. Additionally, it examines whether this system improves habitat conditions and biological activity relative to conventional and other organic farming practices.

7.2 Materials and Methods

7.2.1 Study Area

This study was conducted in Toyama Prefecture, located in central Japan. The primary study site, where vegetables were cultivated under the spent mushroom substrate–crop–weed coexistence (SMS–CW) treatment (Fig. 7.1a), is situated at the northeastern edge of the Kureha Hills in Hakkayama, Toyama City. For comparison, a total of 15 soybean croplands in Toyama, Nanto, and Oyabe, and 11 vegetable croplands in Namerikawa, Imizu, Toyama, Nanto, and Tateyama Town, were selected. These comparison sites included fields managed under both conventional and organic farming practices. Detailed information on site selection and management characteristics is provided in the supplementary material (Haque et al. 2025). The agricultural plains of the study region are predominantly characterized by Gleysols, Fluvisols, and Andosols (Yanai et al. 2012). The region exhibits a humid temperate climate, with a mean annual temperature of approximately 13–14 °C. The mean monthly temperature is approximately 26.8 °C in August and ranges from −1 to 2 °C in January. Annual precipitation averages approximately 2300 mm (Aiura et al. 2018).

7.2.2 Soil Sampling and Laboratory Analysis

This study involves a partial reanalysis of soil data from a previously published investigation (Haque et al. 2025). Soil samples were obtained from soybean fields (*Glycine max* (L.) Merr.; n = 15) and vegetable fields (including *Cucurbita moschata* (Duch.) Poir., *Solanum melongena* L., among others; n = 12) managed under conventional, organic, and SMS–CW treatment regimes. At each site, three soil cores (100 cm^3

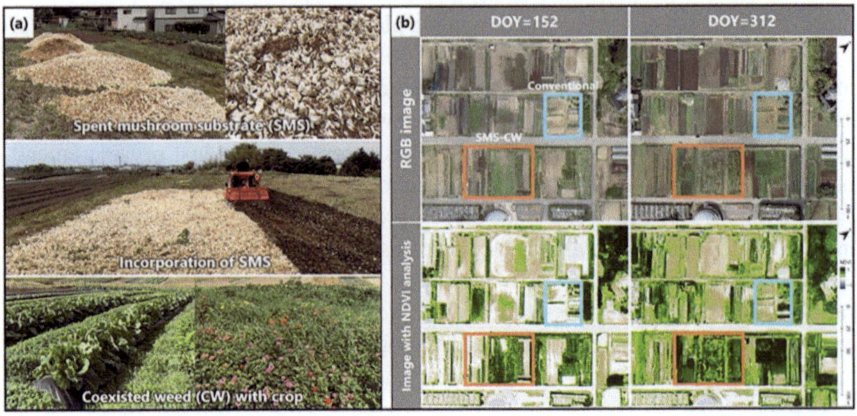

Fig. 7.1 **a** Spent mushroom substrate (SMS), treated as industrial waste, was collected and incorporated into a vegetable field where weeds were allowed to coexist with the crop **b** NDVI analysis illustrating seasonal effects on fields under different management practices (conventional and SMS–CW treated) at two day-of-year (DOY) periods (152 and 312). The RGB image comparison shows the conventional field outlined in sky blue (area: 0.1906 ha) and the SMS–CW treated field outlined in orange (area: 0.4488 ha)

each) were collected from 0–5 cm depth using a stainless-steel corer (5.0 cm in diameter × 5.1 cm in height). Sampling locations were separated by a minimum distance of 5 meters to ensure spatial independence. Soil microarthropods were extracted utilizing a Berlese–Tullgren funnel apparatus and preserved in 70% ethanol. Arthropods measuring less than 2 mm were identified and enumerated under a stereomicroscope (Carton DSZT-70IFL, MS-7833, Kanagawa, Japan), following established protocols (Briones 2014; Dombos et al. 2017). After extraction, soil samples were air-dried and sieved to remove particles larger than 2 mm. Subsequently, samples were oven-dried at 105 °C for 2 hours (ISUZU EPN-115, Niigata, Japan) to eliminate residual moisture prior to chemical analysis. Total carbon (TC) and total nitrogen (TN) concentrations were measured by dry combustion with a CN coder (MT-700, YANACO, Kyoto, Japan). Inorganic nitrogen (IN) forms, including ammonium (NH_4^+–N) and nitrate (NO_3^-–N), were quantified spectrophotometrically via the indophenol blue method for NH_4^+–N and the naphthylethylenediamine method following zinc reduction for NO_3^-–N (Hata et al. 1992). Inorganic nitrogen content was calculated as the sum of NH_4^+–N and NO_3^-–N (Dai et al. 2016), while organic nitrogen (ON) was estimated as the difference between TN and IN (Jickells et al. 2013; Yang et al. 2019).

7.2.3 Image Acquisition and Data Processing

Multispectral imagery was collected in 2024 from vegetable croplands in Hakkayama managed under conventional and SMS–CW practices to assess seasonal vegetation dynamics. Images were acquired using an unmanned aerial vehicle (UAV; DJI Phantom 4 Multispectral) operated with DJI GS Pro software at five time points (June, July, August, October, and November), with a spatial resolution of 0.06 m. Two representative vegetable fields planted with pumpkin (*Cucurbita moschata* (Duch.) Poir.) were analyzed to evaluate vegetation dynamics. One field was managed under the SMS–CW treatment and covered 0.4488 ha, whereas the other was managed conventionally and covered 0.1906 ha (see Fig. 7.1b).

The UAV was equipped with a multispectral sensor capturing five spectral bands: blue, green, red, red-edge, and near-infrared (NIR). Acquired images were processed and mosaicked using Agisoft Metashape software (version 2.0.4). The Normalized Difference Vegetation Index (NDVI) was calculated within ArcGIS Pro (version 3.5) utilizing the following equation:

$NDVI = (NIR - Red) / (NIR + Red)$.

The mean NDVI value and its standard error were computed for each image obtained on different days of the year (DOY 152, 184, 213, 280, and 312).

All statistical analyses and data visualizations were performed using R version 4.4.1 (R Core Team 2024). The R packages glmmTMB, MuMIn, effects, ggplot2, DHARMa, car, dplyr, MASS, and emmeans were utilized for data analysis and model evaluation.

7.3 Results

Total carbon (TC) content varied across crop types and management practices. The lowest mean value was recorded in soybean fields subjected to conventional management (2.75%), whereas the highest was observed in fields managed under the SMS–CW treatment (4.78%) (see Table 7.1). This table summarizes the total carbon (%) and organic nitrogen (%) contents measured across various agricultural fields, presenting mean values accompanied by standard deviations. Tukey's honestly significant difference (HSD) test was employed to evaluate differences among treatments; different letters indicate statistically significant differences ($P < 0.05$). The abbreviations Soy_Con, Soy_Org, Veg_Con, and Veg_Org denote soybean conventional, soybean organic, vegetable conventional, and vegetable organic management systems, respectively. The SMS–CW treatment refers to organic vegetable fields amended with spent mushroom substrate, wood chips, and combustion ash, in which weed coexistence was permitted.

Similarly, organic nitrogen (ON) content was highest under the SMS–CW treatment (0.343%) and lowest in soybean fields under conventional management

Table 7.1 A comparison of soil properties among different crops with managements

Crop with managements	Total carbon content (%)	Organic nitrogen content (%)	Microarthropod abundance (/100 cm^3)
Soy_Con (n = 18)	2.75 ± 0.66^{a}	0.232 ± 0.05^{a}	4.44 ± 4.70^{a}
Soy_Org (n = 24)	2.97 ± 1.56^{a}	0.247 ± 0.07^{a}	6.90 ± 5.16^{ab}
Veg_Con (n = 15)	3.75 ± 1.88^{a}	0.250 ± 0.06^{a}	12.50 ± 8.87^{ab}
Veg_Org (n = 18)	3.08 ± 1.18^{a}	0.272 ± 0.08^{a}	6.56 ± 3.54^{ab}
SMS-CW (n = 3)	4.78 ± 0.74^{a}	0.343 ± 0.06^{a}	31.70 ± 8.62^{b}

(0.232%) (Table 7.1). However, despite these observed trends, differences in TC and ON contents among management types were not statistically significant ($P > 0.05$).

In contrast, the abundance of soil microarthropods varied significantly across management practices. The SMS–CW treatment yielded the highest microarthropod density, averaging 31.7 individuals per 100 cm^3 of soil, which was markedly higher than those recorded under conventional and other organic management systems ($P < 0.05$; Table 7.1). In the size-scaled scatterplot (bubble plot), relatively elevated microarthropod abundances (24, 30, and 41 individuals per 100 cm^3) were observed under the SMS–CW treatment (solid circles in Fig. 7.2). In these fields, the total carbon content ranged from 4.13 to 5.59%, whereas the organic nitrogen content ranged from 0.29 to 0.40%. Although these values were above average, they did not reach the highest levels observed across all treatments examined (Fig. 7.2). A comparison of NDVI values between conventional vegetable fields and SMS–CW–treated fields across the day of year (DOY) demonstrated a broader range of NDVI in conventional fields (0.07–0.60). Conversely, NDVI values in the SMS–CW fields remained more consistent, ranging from 0.38 to 0.56 throughout the growing season (Fig. 7.3).

7.4 Discussion

The present study broadly concurs with prior research indicating that organic management systems generally sustain higher soil carbon and nitrogen levels compared to conventional systems (Lorenz et al. 2025; Wen et al. 2025). However, this pattern was less evident for carbon content in vegetable cultivation fields. Soil carbon, originating from plant residues, roots, and soil microorganisms, serves as an indicator of soil organic matter and is integral to soil fertility and structure (Bianchi et al. 2008). Furthermore, soil organic matter serves as the primary reservoir of nitrogen, with organic nitrogen accounting for over 90% of total nitrogen in numerous agricultural soils (Shu et al. 2025; Te et al. 2022). Although differences in management practices

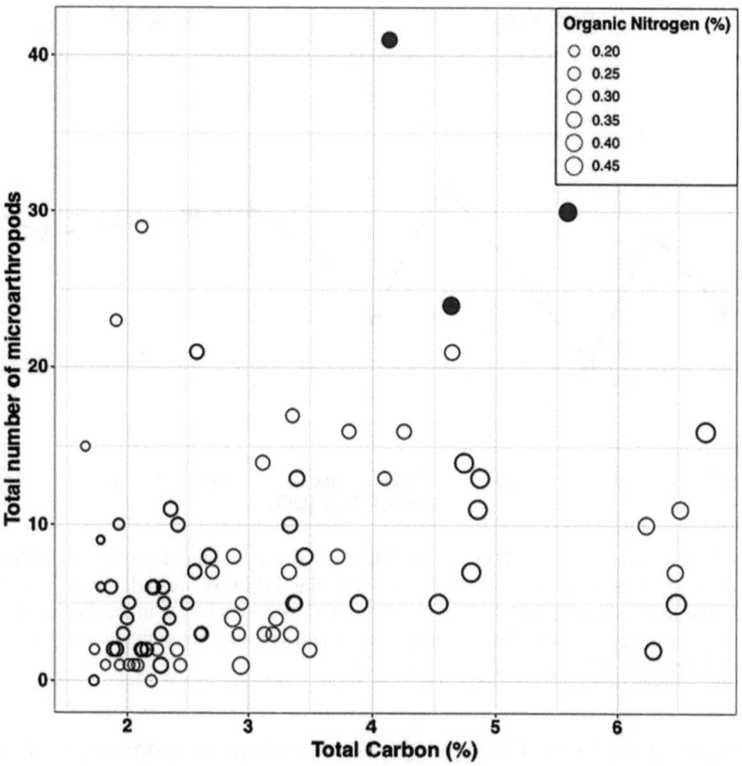

Fig. 7.2 Size-scaled scatterplot showing the relationship between total carbon content (%) and microarthropod abundance (modified from supporting information by Haque et al. (2025)). Bubble size represents organic nitrogen content (%), with larger circles indicating higher nitrogen levels. Black hollow circles denote organic management (soybean and vegetable), gray hollow circles denote conventional management (soybean and vegetable), and solid circles denote the SMS-CW treatment. The SMS-CW treatment had the highest microarthropod abundance among all treatments

did not reach statistical significance, the observed trends suggest that organic and resource-recycling management strategies may improve soil nutrient status.

Soil microarthropods are widely utilized as indicators of soil quality, as their abundance is closely correlated with the availability of organic matter, nitrogen cycling, and carbon-to-nitrogen ratios (Haque et al. 2025). In the current investigation, microarthropod abundance was significantly higher under the SMS–CW treatment than under conventional systems and alternative organic practices, indicating that this integrated approach provides particularly favorable conditions for soil fauna.

The utilization of spent mushroom substrates (SMS) appears to have played a significant role in enhancing the abundance of microarthropods through soil enrichment with organic material and nitrogen, as well as by stimulating microbial biomass and activity (Martín et al. 2023; Medina et al. 2012; Ngan and Riddech

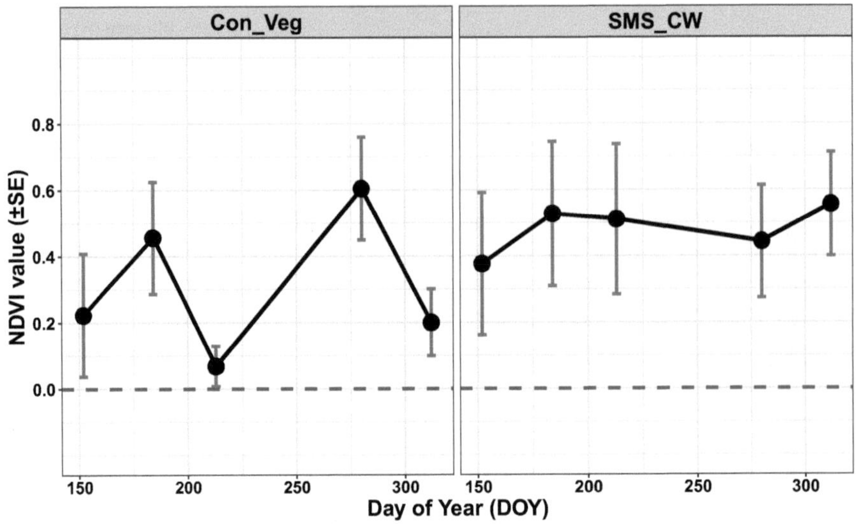

Fig. 7.3 Seasonal variation in NDVI values (Mean ± SE) across the day of year (DOY) in the conventional vegetable field (n = 529,644 pixels) and the SMS-CW–treated field (n = 1,246,902 pixels). In the conventional field, NDVI ranged from 0.07 to 0.60, whereas in the SMS-CW–treated field, it ranged from 0.38 to 0.56. The lowercase "n" denotes the number of pixels analyzed for each day of the year (DOY 152, 184, 213, 280, and 312)

2021; Wang et al. 2021). Considering microorganisms as a primary food source for microarthropods, increased microbial activity can lead to higher microarthropod densities (Kundel et al. 2020; Li et al. 2017). The incorporation of wood chips presumably further improved soil physical properties by elevating porosity, aeration, and water retention capacity, which are crucial for soil fauna inhabiting pore spaces (Fontana et al. 2023).

The regulation of soil chemistry also played a significant role in the observed patterns. Biomass ash, when applied in the SMS–CW treatment, likely functioned as a soil conditioner by moderating soil acidity and supplying essential mineral nutrients, including silicon, potassium, calcium, phosphorus, and magnesium. This intervention consequently enhanced nutrient availability and supported soil biological activity (Bhandari et al. 2024; Zhai et al. 2021). Maintaining an appropriate soil pH is particularly important, as soil acidification can adversely affect both plant growth and soil fauna (Du et al. 2024; Iloba and Odon 2006).

Managed weed coexistence constitutes a significant element of the SMS–CW system. While unmanaged weed–crop competition may reduce yields (Kaur et al. 2018), implementing appropriate weed management strategies can preserve biodiversity while minimizing yield losses (Karlsson et al. 2025; Santos et al. 2021). In this investigation, organically managed vegetable fields with weed coexistence maintained persistent vegetative cover throughout the growing season, whereas fields under conventional management did not (Figure 7.2). Persistent vegetation cover can mitigate fluctuations in the soil microclimate by enhancing soil moisture retention and lowering soil temperature, thereby reducing conditions that inhibit soil biological activity (Ni et al. 2019; Zhou et al. 2023). Overall, the synergy of recycled organic inputs (SMS, wood chips, and biomass ash) and managed weed coexistence within the SMS–CW treatment fostered optimal physical, chemical, and biological soil conditions that promote high microarthropod abundance. These results underscore the potential of integrated, resource-recycling organic farming systems to improve soil quality and belowground biodiversity, as well as to enhance soil resilience against climate-related stresses, including summer soil overheating and erosion caused by increasingly frequent heavy rainfall events.

7.5 Conclusion

This study demonstrated that the SMS–CW treatment—an organic farming methodology involving managed weed coexistence—resulted in the highest abundance of soil microarthropods. The use of spent mushroom substrate as a substrate, wood chips as habitat, and biomass ash as a soil conditioner likely fostered optimal conditions for microarthropod proliferation. Furthermore, continuous vegetation within the SMS–CW system had a possibility to maintain high soil moisture levels and more stable temperatures, thereby further supporting microarthropod populations. In the context of future climate change scenarios, which may lead to increased summer soil overheating and heightened erosion due to heavy precipitation, such vegetated and structurally enhanced soils could serve as protective measures against thermal stress and erosion. Future investigations into the relationships between soil microarthropods and parameters such as pH, moisture, and temperature are anticipated to deepen these insights.

References

Aiura H, Nakajima H, Ishida M (2018) Estimate average annual maximum snow depth with the temperature and the amount of precipitation of the mesh normal climatic data in Toyama Prefecture, Japan. J Jpn For Soc 100:174–177

Albuquerque AR, Angélica RS, Merino A, Paz SP (2021) Chemical and mineralogical characterization and potential use of ash from Amazonian biomasses as an agricultural fertilizer and for soil amendment. J Clean Prod 295:126472. https://doi.org/10.1016/j.jclepro.2021.126472

Assefa S, Tadesse S (2019) The principal role of organic fertilizer on soil properties and agricultural productivity: a review. Agric Res Technol: Open Access J 22(2):1–5. https://doi.org/10.19080/ARTOAJ.2019.22.556192

Barretto AG, Berndes G, Sparovek G, Wirsenius S (2013) Agricultural intensification in Brazil and its effects on land-use patterns: an analysis of the 1975–2006 period. Glob Chang Biol 19(6):1804–1815. https://doi.org/10.1111/gcb.12174

Bengtsson J, Ahnström J, Weibull AC (2005) The effects of organic agriculture on biodiversity and abundance: a meta-analysis. J Appl Ecol 42(2):261–269. https://doi.org/10.1111/j.1365-2664.2005.01005.x

Bergstrand KJ (2022) Organic fertilizers in greenhouse production systems: a review. Sci Hortic 295:110855. https://doi.org/10.1016/j.scienta.2021.110855

Bhandari D, Arlović S, Galić M, Bilandžija N, Leto J, Bilandžija D (2024) Influence of biomass ash fertilization on soil pH, $CaCO_3$, K_2O, and P_2O_5 under Miscanthus × giganteus. J Cent Eur Agric 25(3):833–841. https://doi.org/10.5513/JCEA01/25.3.4339

Briones MJI (2014) Soil fauna and soil functions: a jigsaw puzzle. Front Environ Sci 2:7. https://doi.org/10.3389/fenvs.2014.00007

Bünemann EK, Bongiorno G, Bai Z, Creamer RE, De Deyn G, De Goede R, Fleskens L, Geissen V, Kuyper TW, Mäder P, Pulleman M, Sukkel W, Van Groenigen JW, Brussaard L (2018) Soil quality: a critical review. Soil Biol Biochem 120:105–125. https://doi.org/10.1016/j.soilbio.2018.01.030

Bond W, Grundy AC (2001) Non-chemical weed management in organic farming systems. Weed Res 41(5):383–405. https://doi.org/10.1046/j.1365-3180.2001.00246.x

Bianchi SR, Miyazawa M, Oliveira ELD, Pavan MA (2008) Relationship between the mass of organic matter and carbon in soil. Braz Arch Biol Technol 51:263–269. https://doi.org/10.1590/S1516-89132008000200005

Coulis M (2021) Abundance, biomass and community composition of soil saprophagous macrofauna in conventional and organic sugarcane fields. Appl Soil Ecol 164:103923. https://doi.org/10.1016/j.apsoil.2021.103923

Cussans GW (1992) The implications for weeds and weed control of practices to reduce nitrate leaching. In: Aspects of applied biology (vol. 30), Nitrate and farming systems. Hortic Res Int, pp 351–358

Dai X, Yuan Y, Wang H (2016) Changes of anaerobic to aerobic conditions but not of crop type induced bulk soil microbial community variation in the initial conversion of paddy soils to drained soils. Catena 147:578–585. https://doi.org/10.1016/j.catena.2016.08.012

Dhaliwal SS, Sharma V, Mandal A, Naresh RK, Verma G (2021) Improving soil micronutrient availability under organic farming. In: Yadav M, Meena S, Purakayastha TJ (eds) Advances in organic farming. Woodhead Publishing, pp 93–114. https://doi.org/10.1016/B978-0-12-822358-1.00006-7

Doles JL, Zimmerman RJ, Moore JC (2001) Soil microarthropod community structure and dynamics in organic and conventionally managed apple orchards. Appl Soil Ecol 18:83–96. https://doi.org/10.1016/S0929-1393(01)00143-3

Dombos M, Kosztolányi A, Szlávecz K, Gedeon C, Flórián N, Groó Z, Dudás P, Banszegi O (2017) EDAPHOLOG monitoring system: Automatic, real-time detection of soil microarthropods. Methods Ecol Evo 8: 313–321. https://doi.org/10.1111/2041-210X.12662

Du L, Zhang Z, Chen Y, Wang Y, Zhou C, Yang H, Zhang W (2024) Heterogeneous impact of soil acidification on crop yield reduction: a global meta-analysis. Field Crop Res 319:109643. https://doi.org/10.1016/j.fcr.2024.109643

Erisman JW, Galloway JN, Seitzinger S, Bleeker A, Dise NB, Petrescu AMR, Leach AM, de Vries W (2013) Consequences of human modification of the global nitrogen cycle. Philos Trans R Soc B: Biol Sci 368(1621):20130116. https://doi.org/10.1098/rstb.2013.0116

Fontana M, Johannes A, Zaccone C, Weisskopf P, Guillaume T, Bragazza L, Elfouki S, Charles R, Sinaj S (2023) Improving crop nutrition, soil carbon storage and soil physical fertility using ramial wood chips. Environ Technol Innovation 31:103143. https://doi.org/10.1016/j.eti.2023.103143

Gergócs V, Flórián N, Tóth Z, Szili-Kovács T, Mucsi M, Dombos M (2022) Crop species and year affect soil-dwelling Collembola and Acari more strongly than fertilisation regime in an arable field. Appl Soil Ecol 173:104390. https://doi.org/10.1016/j.apsoil.2022.104390

Gu B, Zhang X, Lam SK, Yu Y, van Grinsven HJM, Zhang S, Wang X, Bodirsky BL, Wang S, Duan J, Ren C, Bouwman L, de Vries W, Xu J, Sutton MA, Chen D (2023) Cost-effective mitigation of nitrogen pollution from global croplands. Nature 613:77–84. https://doi.org/10.1038/s41586-022-05481-8

Haque MA, Minemura Y, Sazawa K, Kuramitz H, Wada N (2025) Interaction of organic nitrogen and C/N ratio enhances soil microarthropod abundance. Agric Environ Lett 10(2):e70034. https://doi.org/10.1002/ael2.70034

Hata N, Teraguchi K, Yamaguchi M, Kasahara I, Taguchi S, Goto K (1992) Spectrophotometric determination of ammonia-nitrogen after preconcentration as indothymol on a glass-fiber filter in the presence of a cationic surfactant. Microchim Acta 106:101–108. https://doi.org/10.1007/BF01242704

Hua W, Luo P, An N, Cai F, Zhang S, Chen K, Han X et al (2020) Manure application increased crop yields by promoting nitrogen use efficiency in the soils of 40-year soybean–maize rotation. Sci Rep 10:14882. https://doi.org/10.1038/s41598-020-71932-9

Iloba BN, Odon CF (2006) Studies on the biodiversity of soil microarthropods and their responses to precipitation regimes. African Scientist 7

Jickells T, Baker AR, Cape JN, Cornell SE, Nemitz E (2013) The cycling of organic nitrogen through the atmosphere. Philos Trans R Soc B: Biol Sci 368:20130115. https://doi.org/10.1098/rstb.2013.0115

Karlsson M, Carrié R, Wetterlind J, Bergkvist G, Ekroos J, Smith HG (2025) Weed–crop competition under improved nutrient management reveals trade-off between yields and weed diversity in organic farming. Biol Agric & Hortic 1–20. https://doi.org/10.1080/01448765.2025.2505896

Karlen DL, Mausbach MJ, Doran JW, Cline RG, Harris RF, Schuman GE (1997) Soil quality: a concept, definition, and framework for evaluation. Soil Sci Soc Am J 61(1):4–10. https://doi.org/10.2136/sssaj1997.03615995006100010001x

Kaur S, Kaur R, Chauhan BS (2018) Understanding crop–weed–fertilizer–water interactions and their implications for weed management in agricultural systems. Crop Prot 103:65–72. https://doi.org/10.1016/j.cropro.2017.09.011

Kundel D, Bodenhausen N, Jørgensen HB, Truu J, Birkhofer K, Hedlund K, Mäder P, Fliessbach A (2020) Effects of simulated drought on biological soil quality, microbial diversity and yields under long-term conventional and organic agriculture. FEMS Microbiol Ecol 96:fiaa205. https://doi.org/10.1093/femsec/fiaa205

Li R, Tao R, Ling N, Chu G (2017) Chemical, organic and bio-fertilizer management practices effect on soil physicochemical property and antagonistic bacteria abundance of a cotton field: implications for soil biological quality. Soil Tillage Res 167:30–38. https://doi.org/10.1016/j.still.2016.11.001

Li Z, Schneider RL, Morreale SJ, Xie Y, Li C, Li J (2018) Woody organic amendments for retaining soil water, improving soil properties and enhancing plant growth in desertified soils of Ningxia, China. Geoderma 310:143–152. https://doi.org/10.1016/j.geoderma.2017.09.009

Lorenz K, Omondi E, Lal R, Das S, Smith A (2025) Soil organic carbon and total nitrogen after 34 years under conventional and organic management practices at the Rodale institute farming systems trial. Soil Sci Soc Am J 89(1):e14165. https://doi.org/10.1002/saj2.70000

Lori M, Symnaczik S, Mäder P, De Deyn G, Gattinger A (2017) Organic farming enhances soil microbial abundance and activity a meta-analysis and meta-regression. PLoS One 12(7):e0180442. https://doi.org/10.1371/journal.pone.0180442

Martín C, Zervakis GI, Xiong S, Koutrotsios G, Strætkvern KO (2023) Spent substrate from mushroom cultivation: exploitation potential toward value-added products. Bioengineered 14(1):2252138. https://doi.org/10.1080/21655979.2023.2252138

Medina E, Paredes C, Bustamante MA, Moral R, Moreno-Caselles J (2012) Relationships between soil physico-chemical and biological properties in a soil amended with spent mushroom substrate. Geoderma 173:152–161. https://doi.org/10.1016/j.geoderma.2011.12.011

Menta C, Remelli S (2020) Soil health and arthropods: from complex system to worthwhile investigation. Insects 11:54. https://doi.org/10.3390/insects11010054

Ministry of Agriculture, Forestry and Fisheries (MAFF: Sustainable Agriculture Division, Crop Production Bureau) (2025) The situation surrounding organic agriculture in Japan. https://www.maff.go.jp/e/policies/env/sustainagri/TheSituationSurroundingOrganicAgricultureinJapan_2025.09.pdf

Ngan NM, Riddech N (2021) Use of spent mushroom substrate as an inoculant carrier and an organic fertilizer and their impacts on roselle growth (Hibiscus sabdariffa L.) and soil quality. Waste Biomass Valorization 12(7):3801–3811. https://doi.org/10.1007/s12649-020-01278-w

Ni J, Cheng Y, Wang Q, Ng CWW, Garg A (2019) Effects of vegetation on soil temperature and water content: field monitoring and numerical modelling. J Hydrol 571:494–502. https://doi.org/10.1016/j.jhydrol.2019.02.009

Pintarič M, Štuhec A, Tratnik E, Langerholc T (2024) Spent mushroom substrate improves microbial quantities and enzymatic activity in soils of different farming systems. Microorganisms 12(8):1521. https://doi.org/10.3390/microorganisms12081521

R Core Team (2024) R: A Language and Environment for Statistical Computing (Version 4.4.1) [Computer software]. CRAN. https://www.r-project.org/

Sahu H, Kumar U, Mariappan S, Mishra AP, Kumar S (2024) Impact of organic and inorganic farming on soil quality and crop productivity for agricultural fields: a comparative assessment. Environ Chall 15:100903. https://doi.org/10.1016/j.envc.2024.100903

Santos RNV, Santos LLO, Marques LJP, Rodrigues AAC, Correa MJP, Mesquita MLR, Silva MRM (2021) Influence of periods of coexistence of weeds on organic okra production. Aust J Crop Sci 15(1):41–47

Shu Y, Gong X, Wu J (2025) Characteristics of soil organic nitrogen fractions under vegetation restoration in karst areas. Sci Rep 15(1):28180. https://doi.org/10.1038/s41598-025-08232-7

Stewart WM, Roberts TL (2012) Food security and the role of fertilizer in supporting it. Procedia Eng 46:76–82. https://doi.org/10.1016/j.proeng.2012.09.448

Tang Q, Liu W, Huang H, Peng Z, Deng L (2024) Responses of crop yield, soil fertility, and heavy metals to spent mushroom residues application. Plants 13(5):663. https://doi.org/10.3390/plants13050663

Te X, Hassan MJ, Cui K, Xiao J, Aslam MN, Saeed A, Yang W, Ali S (2022) Effect of different planting pattern arrangements on soil organic matter and soil nitrogen content under a maize/soybean strip relay intercropping system. Front Plant Sci 13:995750. https://doi.org/10.3389/fpls.2022.995750

Vreeken-Buijs MJ, Hassink J, Brussaard L (1998) Relationships of soil microarthropod biomass with organic matter and pore size distribution in soils under different land use. Soil Biol Biochem 30(1):97–106. https://doi.org/10.1016/S0038-0717(97)00064-3

Wang HW, Xu M, Cai XY, Tian F (2021) Evaluation of soil microbial communities and enzyme activities in cucumber continuous cropping soil treated with spent mushroom (Flammulina velutipes) substrate. J Soils Sediments 21(8):2938–2951. https://doi.org/10.1007/s11368-021-02989-w

Wang L, Lyons J, Kanehl P, Bannerman R, Emmons E (2000) Watershed urbanization and changes in fish communities in southeastern Wisconsin streams. JAWRA. J Am Water Resour Assoc 36(5):1173–1189. https://doi.org/10.1111/j.1752-1688.2000.tb05719.x

Wen Y, Yao W, Yu T, Cheng L, Zhang Q, Yang J, Lin F, Zhu H, Gunina A, Yang Y, Mganga KZ (2025) Long-term organic farming improves red soil quality and microbial diversity in subtropics. Agric Ecosyst Environ 381:109410. https://doi.org/10.1016/j.agee.2024.109410

Xiang T, Malik TH, Nielsen K (2020) The impact of population pressure on global fertiliser use intensity, 1970–2011: An analysis of policy-induced mediation. Technol Forecast Soc Chang 152:119895. https://doi.org/10.1016/j.techfore.2019.119895

Yanai J, Okada T, Yamada H (2012) Elemental composition of agricultural soils in Japan in relation to soil type land use and region. Soil Sci Plant Nutr 58:1–10. https://doi.org/10.1080/00380768.2012.658349

Yang XM, Drury CF, Xu W, Reeb MA, Oloya T (2019) Rapid determinations of dissolved inorganic and organic nitrogen in soil leachate using mid-infrared spectroscopy. Can J Soil Sci 99:579–583. https://doi.org/10.1139/cjss-2019-0116

Zhai J, Burke IT, Stewart DI (2021) Beneficial management of biomass combustion ashes. Renew Sust Energ Rev 151:111555. https://doi.org/10.1016/j.rser.2021.111555

Zhou D, Si J, He X, Jia B, Zhao C, Wang C, Qin J, Zhu X, Liu Z (2023) The relationship between vegetation and soil moisture reveals the vegetation carrying capacity threshold: a case study of a Haloxylon ammodendron plantation in the Alxa Desert. China Front Environ Sci 11:1054195. https://doi.org/10.3389/fenvs.2023.1054195

Chapter 8
Carbon and Energy Footprints in Conservation Agriculture: Short- and Long-Term Effects

Prithwiraj Dey, Soham Rana, Priya Bhattacharya, and Manindra Barman

Abstract Agriculture, covering more than 40% of the land surface, underpins human survival yet contributes significantly to climate change, accounting for 10–12% of global greenhouse gas (GHG) emissions. Conservation agriculture (CA) has emerged as a sustainable solution, promoting productivity while reducing environmental impacts. CA practices like minimum tillage, crop residue retention, and diversification minimize carbon emissions and energy consumption compared to conventional farming methods. Controlled tillage decreases soil disturbance, preserves soil organic carbon (SOC), and lowers energy needs for field operations. Practices such as no tillage and minimum tillage are especially effective in reducing the carbon footprint by keeping soil structure intact throughout the crop cycle. Residue retention enhances carbon sequestration, improves soil health, and reduces the reliance on synthetic fertilizers, cutting down energy use in fertilizer production and application. Over time, CA practices contribute significantly to carbon sequestration, improvement in SOC, water retention, and nutrient cycling. This results in more resilient agroecosystems to climate variability. Additionally, CA's reduction of fossil fuel-based inputs curbs direct GHG emissions along the agricultural landscapes. However, a number of variables, including crop rotation, climate, soil type, and management techniques, affect how successfully CA practices are being implemented. Long-term CA adoption offers a path toward climate change mitigation, resource conservation, and enhanced resilience, underscoring its importance in sustainable land management and transition to carbon friendly agriculture.

Keywords Carbon sequestration · Resource conservation technologies · Sustainability · Carbon footprint · Energy footprint

P. Dey (✉) · S. Rana · P. Bhattacharya · M. Barman
Agricultural and Food Engineering Department, Indian Institute of Technology Kharagpur, Kharagpur, West Bengal, India
e-mail: prithwi@agfe.iitkgp.ac.in

D. Kumar Swain et al. (eds.), *Climate-Smart Technologies for Sustainable Agricultural Production*, https://doi.org/10.1007/978-3-032-25666-9_8

8.1 Introduction

Unsustainable management and exploitation of natural resources in contemporary agricultural practices have undermined the entitlement of future generations to a cleaner environment (Jat et al. 2019). The expansion of energy-intensive mechanized farming remains appealing, particularly as developing nations pursue modernization (Dey et al. 2024). Yet, such modernization is often accompanied by increased energy demand and carbon inputs in crop production systems, while the efficiency of energy and carbon use continues to decline (Mandal et al. 2002). Food security for a rapidly expanding population depends not just on increasing output but also on implementing it in a way that is environmentally sustainable and resource-efficient, and thus this falling efficiency is a matter of serious concern. Approximately 691 million people experienced food insecurity worldwide in December 2019, a number that escalated to 820 million by December 2020, highlighting the urgency of achieving Sustainable Development Goal 2 (Zero Hunger) by 2030 (UNICEF, FAO 2021). Global food insecurity is, in fact, a complicated issue which not only depends on yield gaps but is also influenced by the environmental costs of agricultural intensification as well as production and distributional inequalities and sustainability issues.

To confront these multidimensional challenges, agriculture has become increasingly dependent on inputs and processes with substantially higher energy and carbon footprints, thereby raising concerns about the long-term sustainability of the system (Choudhary et al. 2017; Dey et al. 2021; Dwivedi et al. 2024). India, for example, has achieved remarkable progress in food grain production, nearly tripling output between the 1960s and the 2020s (Das et al. 2023). However, these gains have come at significant environmental costs, including soil degradation, excessive energy use, and substantial greenhouse gas (GHG) emissions, all of which have contributed in accelerating climate change, soil health problems, yield stagnation and so on (Ghorbani et al. 2011). Regarding climate change, the average global temperature has increased by 1.09 °C during the previous few decades relative to the pre-industrial reference level (IPCC 2023). This increase in temperature by 1 °C can potentially reduce the productivity of cereals and vegetable crops by 3–7% globally (Goswami et al. 2006; Pandit et al. 2021). According to the Paris Agreement (2015), an action plan was aimed to limit the global warming to <2 °C which can potentially be achieved by implementing emission reduction technologies, carbon sequestration and use of renewable energy sources (Rhodes 2016).

Conventional agriculture has mostly been defined by tillage, which makes the land susceptible to erosion (Montgomery 2007). Long-term intensive farming accelerates GHG emissions, stagnates crop yields, and depletes SOC. Open-field burning is still the most common method of managing crop residues, but it is also the least sustainable, making a significant contribution to emissions from biomass burning worldwide. Such burning releases significant amounts of CO, CO_2, NO_n, SO_2 and volatile organic compounds (VOCs) in addition to air polluting particulate matters (Grover and Chaudhry 2019). It is also reported that biomass burning and the associated emissions are more prevalent in rice growing areas of South-Eastern Asia, mostly for the crops harvested in the kharif season (June–November) across the Indo-Gangetic plains (IGP) (Ravindra et al. 2019; Raj et al. 2025). The adverse impacts of climate change are further exacerbated by soil degradation diminishing the inherent capacity of soil to deliver essential ecosystem services which include ecological regulatory provisioning, and cultural benefits that are vital for human well-being (Sharma and Malaviya 2023). Soil erosion, being one form of degradation, increases the mineralization of SOC pool, which in turn is affected by climate change through a positive feedback loop and increases radiative forcing (Lal 2012; Bhatt et al. 2024). Climate change, along with land degradation, is reported to have adverse effects on multiple cropping systems, which make up over 80% of all the cereals produced on more than 100 mha in South Asian countries (Hobbs et al. 2008).

Conservation agriculture (CA) has emerged as a potential concept in contemporary agricultural systems with the primary goal of addressing the dual difficulties of maintaining food security and reducing environmental degradation. Even though conventional farming methods have been effective in increasing agricultural output in recent decades, these caused environmental pollution.. Such ways of intensive farming systems for crop establishment and residue management are responsible for being the major contributors to agricultural GHG emissions (Saha et al. 2023; Samanta et al. 2025). Given this, the empirical ideas of energy footprint (EF) and carbon footprint (CF), which measure resource consumption and life cycle emissions in agricultural production systems, are now commonly acknowledged as important markers of environmental sustainability. The energy and carbon flow in agricultural systems is depicted in Fig. 8.1, where the balance of inputs and outputs establishes whether the systems function as net carbon sources or sinks. In the end, reducing carbon emissions and maximising the ratio of carbon input to carbon output are critical to sustainability of agricultural systems (Bista et al. 2024). Resource conservation technologies (RCTs), which can reduce climate change, improve ecosystem sustainability, and protect biodiversity, must be adopted in order to accomplish this (Kundu et al. 2024; Rani et al. 2025).

CA's efficacy in minimizing environmental footprints is highly context-dependent, varying across soil types, climatic conditions, cropping systems, and time scales, which raises questions about short-term trade-offs and long-term benefits despite its potential (Villagrán et al. 2023). Short-term adoption may be linked to yield reductions, slower nutrient mineralisation, and higher residue management costs, all of which have an impact on footprint assessments, even though long-term improvements in soil health, water-use efficiency (WUE), and biological activity can significantly reduce energy and carbon footprints (Pramanick et al. 2023; Sahoo et al. 2023; Srivastava 2025). Therefore, it is critical to assess how CA affects energy and carbon footprints over a range of time periods and to think about how this data can guide policy interventions and decision-making at the farm level. This chapter offers a thorough analysis of CA's contribution to sustainable agricultural development by looking at both its short-term and long-term effects on CF and energy footprints EF. Ultimately, advancing CA requires not only scientific evidence but also practical actions that secure resilient and sustainable farming systems for future generations.

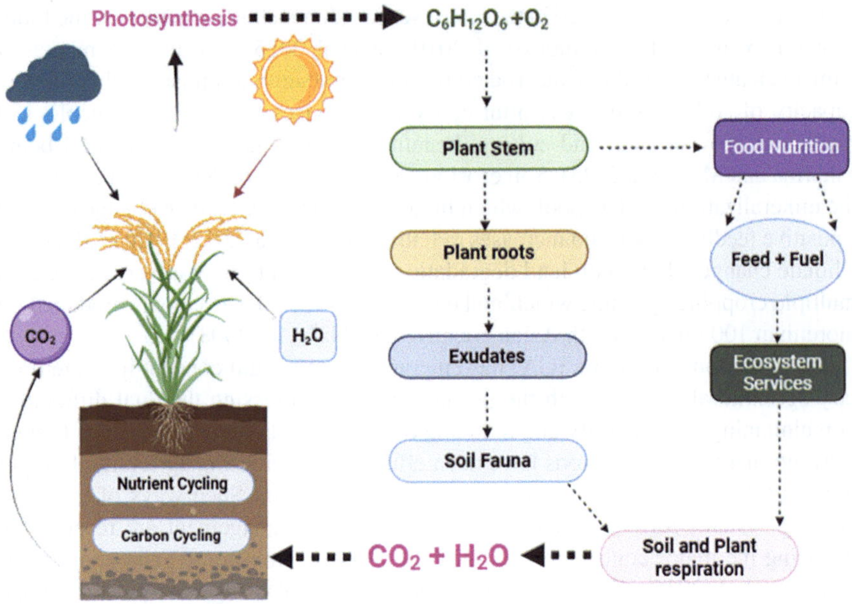

Fig. 8.1 Carbon and energy flow in the agricultural systems

8.2 Understanding Carbon and Energy Footprint in Agriculture

Pandey et al. (2011) defined CF as, "the quantity of GHGs expressed in terms of CO_2-e, emitted into the atmosphere by an individual, organization, process, product, or event from within a specified boundary". According to Odum (2007), energy is the quantity of work done or available energy utilized in the production of goods or services from ecological point of view. From a footprint perspective, the EF represents the land area required to offset the CO_2 emissions generated (Banerjee et al. 2021a). Three interconnected strategies serve as the foundation for our understanding of agriculture's CF and EF (Kamyab et al. 2024). First, mitigating climate change entails measuring emission sources like CO_2 from various farm operations, N_2O emissions from fertilizer application, and CH_4 emissions from anaerobic decomposition in rice fields and livestock farms. The second area is energy and resource-use efficiency, which focuses on identifying practices that use a lot of energy and resources and improving input efficiency through technological advancements like the use of renewable energy, better irrigation, and less reliance on artificial agrochemicals. Third, assessing the effects of various establishment techniques on soil health, SOC dynamics and carbon fractions is necessary for soil fertility, sustainability and carbon sequestration. In the present climate change scenario, CF is the most crucial parameter to measure especially from system sustsainability point of view. It takes into account the quantity of GHG emissions caused by human activity, either directly or indirectly, or the quantity estimated for a product's life cycle. The following formula (Eq. 8.1) is used to calculate CF (Li et al. 2016).

$$CF = \sum (I^{*}EF) \tag{8.1}$$

where, I = Quanity of resources used as input; EF = "Emission factor" for a specific resource.

According to Bartocci et al. (2016), CF can be defined as the "net GHG emissions over the complete lifecycle of any product." More broadly, it measures the total GHGs emitted within a defined system boundary by an individual, entity, process, product, or event, expressed as CO_2-e (Singh et al. 2022). An alternative formulation (Eq. 8.2) expresses CF as the sum of emissions of individual constituents weighted by their global warming potential (GWP) (Holka et al. 2022):

$$CF = \sum_{i}^{n} m(i) \times GWP(a, i) \tag{8.2}$$

where, m(i) = quantity of ith constituents emitted (in kg per functional unit); GWP (a, i) = Global Warming Potential for constituents 'i' over time horizon a (expressed as CO_2 per kg 'i').

Figure 8.2 illustrates a conceptual framework of how carbon footprints are assessed and managed within agricultural systems. It reflects the balance of carbon stocks and flows emphasizing the importance of quantifying sources and sinks of CO_2 within different subsystems. When it comes to carbon credit programmes (CCP) and their impact on land management, most attention has been given to forest and tree-based biomass for carbon storage. However, there is significant untapped potential in agricultural soils for storing carbon. FAO states that agricultural practices focused on building soil carbon could account for nearly 90% of agriculture's ability to help fight climate change (Mitra et al. 2022; Bhatt et al. 2024).

The EF represents the amount of energy consumed by an individual, product, organization, or nation (Banerjee et al. 2021b). EF is further divided into sub-footprints viz. renewable (Chen and Lin 2008), nuclear (Stoeglehner et al. 2005), fossil (Stoeglehner and Narodoslawsky 2009) and solar (Browne et al. 2009), collectively referred to as the energy mix. Globally, agriculture accounts for approximately 5% of total energy use (Ghisellini et al. 2016), with the majority derived from non-renewable sources, leading to substantial GHG emissions. In the context of the ongoing energy crisis, EF estimation has become increasingly critical. Advances in agrotechnology and mechanization have raised the energy demands of agroecosystems, resulting in significant GHG emissions and higher energy subsidies in the agricultural sector. Consequently, assessing EF across different cropping systems has become a vital priority for sustainable agricultural management.

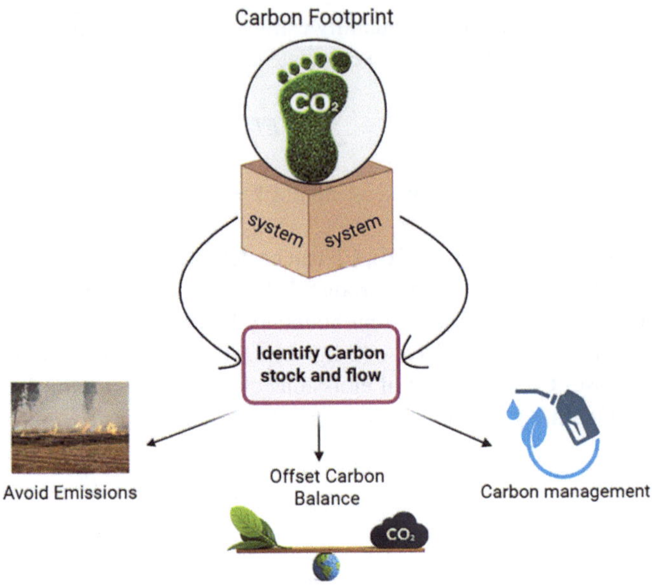

Fig. 8.2 Conceptual framework of carbon footprints in agriculture systems

8.3 Factors Affecting Agricultural Carbon and Energy Footprints

8.3.1 Crop Establishment, Tillage and Crop Rotation

Generally, no tillage (NT) practices have smaller CF than conventional tillage (CT) owing to less soil disturbance, more carbon sequestration, and less GHG emissions. Intensive soil disturbance caused by CT can result in lower SOC and considerable CO_2 emissions (Nisar et al. 2021; Saraswat et al. 2024; Rana et al. 2025). An experiment was conducted by Dachraoui and Sombrero (2020), to calculate the CF of CT and NT systems in continuous irrigated maize using varying rates of N fertilization. The findings revealed that the maize CF was mainly driven by direct and indirect N_2O emissions from synthetic fertilizer use, ranging between 3.3–4.2 tCO_2-e ha^{-1} under CT and 3.4–4.4 tCO_2-e ha^{-1} under NT management. As compared to CT, conservation tillage typically has smaller EF too since it uses less fuel and machinery. CT uses more energy since it disturbs the soil more thoroughly and necessitates more passes with tractors, other machinery as well as manpower (Rusu 2014; Mahapatra and Dey 2022). Several researchers have indicated that farming practices that do not properly diversify crop choices, generally have increased energy consumption and negatively impact the environment due to higher emissions (Al-Kaisi and Yin 2005; Knudsen et al. 2014). Crop diversification, particularly with legumes, can significantly reduce the need for N fertilizers, reducing emissions of

N_2O and lowering carbon emissions considerably (Patra et al. 2024). Meena et al. (2022) conducted a field experiment in the semi-arid plains of India using nine cropping systems comprising 14 different crops, aiming to evaluate productivity, profitability, and environmental sustainability through CF and EF analysis. The results showed that, compared to the cereal-based system, a legume-integrated rotation viz. groundnut–wheat–cluster bean–onion over two years, increased system productivity by 44.5% and profitability by 60.8%, while simultaneously reducing energy use by 5.2% and CF by 21.8%. In addition to lowering production costs, conservation tillage techniques have long-term implications on soil qualities that includes increase and stabilization of soil organic matter (SOM). Increased soil biological activity and organic matter as a result of CA improves the productivity of agroecosystems by improving the structural stability and overall improvement of soil health parameters (Sapkota 2012).

8.3.2 Crop Residue Management and Carbon Sequestration

The SOM is one of the most important active carbon sources existing in soil and terrestrial ecosystems. It plays an important role in maintaining the balance of the global carbon cycle through the processes of carbon sequestration and release. With an estimated 3.8 to 4.5 billion tonnes produced worldwide each year, crop residues are a significant source of organic inputs into agricultural soils. They contain around 1.5 to 2.0 billion tonnes of carbon (Lal 2007). However, during microbial decomposition, a large portion of this carbon is released back into the atmosphere as CO_2, meaning that only a small portion of it is stabilised in soils. Environmental factors, management techniques, and residue quality all affect how stable and release should be balanced (Ussiri and Lal 2009). The dynamics of soil carbon cycling are drastically changed by residue addition. Legume leftovers and other fresh organic matter with a low C:N ratio decompose more quickly, resulting in a rapid evolution of CO_2 and the release of nutrients. Cereal straws and other residues with larger C:N ratios, on the other hand, break down more slowly and have a higher potential to contribute to stable SOM pools (Six et al. 2002). Microbial assimilation of labile carbon fractions dominates the early stages of decomposition, while the formation of humified organic matter and mineral-associated organic carbon leads to long-term stabilisation (Cotrufo et al. 2012).

Short-term increases in microbial respiration, soil respiration, and nutrient cycling brought about by residue addition frequently result in sharp bursts of CO_2 emissions (Song et al. 2025). These immediate impacts include improvement in soil aggregation and water retention while also reflecting the faster mineralisation of labile carbon (LC) pools. However, by accumulating more recalcitrant carbon (RC) fractions, ongoing residue incorporation improves structural stability, SOC sequestration, and increases agro-ecosystem resilience over time (Lehmann and Kleber 2015). Interactions with soil minerals, aggregation processes, and microbial byproducts which protect different SOC fractions from quick decomposition, are necessary for the stability of residue-derived carbon. According to estimates, optimaly incorporating residues could sequester 0.4–1.2 Pg C per year, offsetting a significant portion of GHG emissions from agriculture (Paustian et al. 2016). However, the efficacy hugely differs depending on climate, soil type, and location. For instance, tropical soils may exhibit faster turnover rates because of higher microbial activity and temperature sensitivity, whereas temperate soils frequently experience significant long-term stabilisation when residue incorporation occurs (Zhou et al. 2014). As a result, crop residue management ought to be seen as both an important strategy in global carbon sequestration initiatives and a method for improving soil health. Long-term advantages include stabilising carbon pools, lowering net emissions, and maintaining soil health, whereas short-term effects are mostly linked to decomposition and CO_2 fluxes.

8.3.3 Regional and Contextual Factors

Regional and contextual conditions significantly influence the magnitude and direction of benefits or trade-offs of CA adoption. Because agroecosystems are inherently heterogeneous, responses to CA practices vary across agroecological zones, soil types, cropping systems, and socio-economic context (Corbeels et al. 2020). For example, in tropical and subtropical regions, high temperatures tend to accelerate SOC mineralization, which may constrain the carbon sequestration benefits of CA in tropical and subtropical regions as compared to temperate systems (Powlson et al. 2016). Even in semi-arid and arid environments, trade-offs tend to emerge. Agricultural residue retention improves overall soil moisture, which reduces the energy that must be expended for irrigation and tillage. Among agro-climatic moderators, temperature and rainfall patterns are crucial in determining CA impacts. In tropical environments, high soil temperatures promote organic matter degradation, decreasing retention and persistence of SOC gains than for colder regions (Powlson et al. 2016). In contrast, semi-arid and arid areas get more pronounced benefits from retaining residues that improve soil water storage and irrigation energy requirements. Soils at field level being mostly heterogeneous, play primary role in influencing carbon pool stability, energy requirement and resource use efficiency of agricultural practices. Sandy soils have little water-holding capacity and poor soil aggregation, which limits SOC sequestration potential under NT systems, while clay-rich soils help to stabilize SOC but often have negative N_2O losses when occurring in anaerobic processes (Pittelkow et al. 2014).

Legumes and cover crops in diversified rotations improve energy use efficiency and carbon sequestration by increasing SOC inputs, decreasing reliance on synthetic fertilisers, and increasing nitrogen use efficiency in contrast to unsustainable monoculture systems (Wang et al. 2025). The viability and sustainability of CA systems are also greatly influenced by socio-economic parameters viz. labour availability, competitive demands for residue/organic ammendments, accessibility to farm machines, market incentives etc. Residues are frequently diverted to domestic energy or livestock feed in smallholder contexts in South Asia and Sub-Saharan Africa, which limits their ability to protect or improve soil (Giller et al. 2015). Adoption may be hampered and short-term costs involvement especially due to restricted access to CA-specific high end equipment. CA significantly lowers labour costs and the use of fossil fuels in mechanised large-scale farms, but policy support, subsidies, and information sharing are necessary for profitability. At microlevel, farmers' knowledge base, access to extension services, and institutional capacity all have an impact on adoption outcomes. CA's environmental benefits may be undermined by improper residue management or fertilizer use during the transition period, which can raise weed-pest pressure and N_2O emissions (Knowler and Bradshaw 2006). On the other hand, adoption that is knowledge-intensive and backed by networks for extension and participatory research improves resource efficiency and optimises long-term sustainability benefits.

8.4 Short-Term Effects of CA on Footprints

Proper implementation and adoption of CA practices have quantifiable short-term impacts on CF and EFs; however, they often have associated transitional costs. The LCA concept allows estimation of CO_2 emissions by considering emissions from all on-farm and off-farm activities and reflects the comprehensive perspective on short-term effects caused by the change from conventional to RT/ZT systems (Liu et al. 2016).

Tillage breaks soil aggregate integrity, increases aeration, and exposes SOC to microbial oxidation, thus increasing the rate of CO_2 release from the soil (Lascalajr et al. 2008). CA, on the other hand, reduces aggregate disturbance, limiting short-term carbon losses. Al-Kaisi and Yin (2005) performed a three-year rotating grain, corn, and soybean cropping at loamy soils and found that the conventional cultivator and plough tilled systems released 415 and 511 kg CO_2 ha^{-1} respectively, in 20 days, whereas the ZT system with residue cover released only 300 kg CO_2 ha^{-1}. Likewise, Lascalajr et al. (2008) measured ~250 kg CO_2 ha^{-1} in the first 25 days in NT systems. These studies demonstrate the capacity for CA to decrease early season carbon fluxes compared to CT.

Short-term trade-offs are also evident. Don and Jantz (2013) provided a meta-analysis that showed that NT systems enhances N_2O emissions by ~40% when compared to tillage practices, indicating that adopting NT alone, without consideration for residue and nutrient management, may not reduce footprints. Xiao et al. (2021) also indicated that retention of residue had a greater effect of building up SOC than any reductions in tillage alone. Similarly, Govaerts et al. (2009) underlined the need to complement cover residue with crop diversification and retention in order to stabilize SOC and minimize transitional emissions from agricultural systems.

Similarly system energy dynamics also change considerably in the short-term. Choudhary et al. (2017) stated that total energy utilized by CT wheat production systems (119,413 MJ ha^{-1}) were nearly six times that of energy used in systems of CA (19,225 MJ ha^{-1}). The difference is largely attributed to energy requirements for fuel, machinery, and labour due to repeating tillage and weed management. Energy budgeting provides input-output and efficiency relationships based on cropping system, residue management, and yield reports. Estimates in rice-wheat systems indicate that CT systems have increased inputs for field preparation and irrigation pumping alone which can double energy footprints (Yadav et al. 2018; Kumara et al. 2020).

CA based RT or ZT systems are frequently associated with increased weed pressure and pest incidence, which then involves higher pesticide usage and, in some circumstances, a larger EF and ecotoxicological footprints (Yadav et al. 2018). However, in majority of the studies worldwide, various authors have emphasized energy efficiency benefits of CA, even in the short term. Singh et al. (2022) found the highest energy use efficiency (3.39) with CA, while Jat et al. (2018) found that minimal till options, like NT that is coupled with rational residue management in a cereal-based system, improves energy use efficiency, but also stabilizes transitional yield variability.

Some of the short term positive and negative effects of CA as reported by multiple researchers worldwide has been summarized in Table 8.1. CA systems may deliver a period of adjustment in yield, variable weed management obstacles, and varied N_2O responses; nevertheless, they appear to provide immediate gains in CO_2 emissions and energy use relative to CT. The transition period, still a challenge, and opportunity, especially since strategies that retain residues, diversify crop rotations, and enhance nutrient management, should be investigated closely at integrating and stabilizing performance because if these practices in fact help to reduce agricultural carbon and energy footprints.

8.5 Long-Term Impacts of Conservation Practices

By reducing fossil fuel use and emissions from agricultural systems, CA can potentially minimize GHG outputs in terms of total global warming potentials (GWP) while simultaneously improving soil carbon storage and energy efficiency. ZT and reduced tillage, when applied consistently and optimally, are particularly effective in lowering energy input from farm operations, leading to reduced fuel consumption and labour requirements and thereby decreasing the overall GHG footprint. Importantly, CA alters the fluxes of CO_2, CH_4, and N_2O from soils in addition to lowering emissions associated with reduced power use. Under anaerobic conditions, soils release CH_4, while N_2O arises from nitrification and denitrification processes. Sapkota et al. (2014) demonstrated that direct-seeded rice under CA emitted substantially lower CH_4 (< 50 mg m^{-2} d^{-1}) compared with puddled transplanted rice under conventional tillage (50–250 mg m^{-2} d^{-1}) specifically when it is practiced under long-terms. Jat et al. (2020) reported a substantial 39% reduction in CH_4 emissions under CA-based rice systems relative to conventional puddled transplanted systems.

Table 8.1 Short-term effects of onservation agriculture over diverse cropping systems

Aspect	Positive effect	Negative effect	Adoption year of CA (years)	Reference
Energy and fuel usage	Tillage operation reduction lowers fossil fuel demand and machinery usage	Residue handling and herbicide application lead to higher initial energy demand	3	Jat et al. (2018); Jat et al. (2019)
Soil carbon dynamics	Surface carbon input is contributed by residue retention	SOC sequestration requires time to stabilize	3	Bhattacharya, et al., (2020); Dash et al. (2025) 2020
Nutrient management	Partial recycling of nutrients	N_2O emissions may increase due to slower nutrient mineralization	4	Fuentes-Ponce et al. (2022)
Crop yields	Soil moiture retention reduces early season stress thus maintaining crop yields	Yield fluctuations are observed during initial years	2	Zizinga et al. (2022)
Water use efficiency	WUE increases due to mulch retention	Residue retention initially interfers with seeding operations	2	Yang et al. (2021)
Footprint impact	Immediate reduction in energy usage and fossil fuel dependencies	Carbon balance improvements in the short term are limited and vary with contexts	3	Singh et al. (2022)

Regular adoption of CA based farming also enhances long-term soil health through improved organic matter retention and sustained decomposition of SOC. Effective residue management under NT or reduced tillage regimes has been shown to improve SOC in long term and nutrient availability, thereby improving multiple soil physico-chemical and biological properties (Mishra et al. 2022). However, SOC accumulation under NT is often concentrated in the upper 0–10 cm layer, resulting in higher stratification within the soil profile (Mondal et al. 2023). Long-term NT further enhances soil structure, infiltration, and aggregate stability. Blanco-Canqui and Ruis (2018) observed that NT increased water infiltration rates by 17–86%, improved available water by up to 44%, and raised wet aggregate stability by 1–97% relative to mold-board ploughing. SOC levels under NT were also higher, with contents 4.2% and 2.3% greater compared with rotational tillage (RT) and conventional tillage (CT), respectively. Parallel improvements were reported in soil nitrogen, with NT achieving 1.54 t ha^{-1} compared with 1.30 t ha^{-1} under RT and 1.42 t ha^{-1} under CT (Sithole and Magwaza 2019). Energy savings and emission reductions are another significant result of practicing CA in long-term. NT practices have been estimated to conserve approximately 2500 MJ ha^{-1} of energy compared with conventional methods. Equivalent carbon emissions (ECE) also vary significantly across tillage systems. Naorem et al. (2023) reported highest ECE in CT (145 kg ha^{-1}) followed by reduced tillage (72 kg ha^{-1}), and lowest under NT (31 kg ha^{-1}) from long term experiments. Integrated crop management systems aligned with CA principles demonstrate further long-term advantages. Over nine consecutive years, CA-based ICMs were shown to increase productivity by 16.5–22.9% relative to CT-based systems, while also enhancing renewable energy contributions, as 76–80% of inputs were derived from crop residues (Barman et al. 2024). Recent evidence also highlights trade-offs that require careful management. For instance, Kundu et al. (2024) reported that NT increased bulk density (BD) compared with CT and initially reduced system productivity by 10.3%. However, mulching strategies mitigated these effects, raising yields by 16.4–25.2% compared with no mulch. Importantly, both NT and RT reduced input energy requirements by 22.9% and 13.5%, respectively, compared with CT, while NT reduced the CF of rice and sweet corn production by 41.5% and 22.2%, respectively. The cumulative evidence clearly indicates that CA, when integrated with appropriate residue and nutrient management, constitutes a robust strategy for reducing carbon and energy footprints in the long term. Some of the long term effects of CA that has been reported by a range of researchers globaly are represented in Table 8.2.

8.6 Optimization of Energy Use in Conservation Agriculture

The Patterns of agricultural energy use have a significant impact on soil productivity and sustainability over the long term. Excessive use of heavy machinery, intensive tillage, and chemical fertilisers often leads to problems like higer energy use, soil compaction, erosion, and salinisation, as well as the loss of SOC and nutrients (Zhang et al. 2024). On the other hand, CA is founded on the principles of permanent soil cover, diversified cropping, and minimal soil disturbance which essentially optimise energy flows within crop production systems while also enhancing soil health, biodiversity, and resource use efficiency. The implementation of CA techniques can enhance the characteristics of soil, boost its dearth resistance and enhance the utilization of water and nutrients. Among the many benefits of conservation systems are energy savings (from the NT system), decreased soil erosion, and increased soil absorptivity by the release of agricultural waste. Tillage uses around 30% of the energy in the field. As we use conservation tillage or minimum tillage in CA, conservation tillage can reduce fuel use by up to 77% as compared to conventional methods (Moradi et al. 2018). Reduced intensity and frequency of tillage are most obvious energy-saving advantages of practicing CA. Thus CA can provide a more sustainable way to achieve energy-efficient and climate-resilient agriculture by lowering down the need for input energy and improving ecosystem services.

Table 8.2 Long-term effects of CA on carbon and energy footprints

Study	Location	Crops	Effect on footprint	Remarks
Bansal et al. (2022)	Eastern gulf Coastal Plains	Continuous corn Soyabean-cotton	28–31% reduction in carbon footprint	Long term CA alog with suitable crop rotation results in carbon sequestration and lower emissions
Sun et al. (2022)	China	Winter wheat-summer maize	Energy efficiency increased by 47.8%	Ridge cultivation along with long term CA was used
Yadav et al. (2018)	India	Rice, mustard	Energy footprint reduced by 39% GHG emissions reduced by ~13%	Adopted NT with residue retention in a long term field experiment
Chaudhary et al. (2024)	India	Wheat	Carbon footprint reduced by 10–20%	Long term study reported ZT with residue as sustainable
Gan et al. (2012)	Canada	Spring wheat	38–55% gain in SOC and decrease in carbon foot print was observed	Effect of freaquency of fallow and continuous wheat was studied in a long term experiment
Hu et al. (2015)	China	Wheat-maize	6.7% decrease in carbon footprint	Intercropping with reduced tillage was studied

CA systems perform better than conventional systems on a number of efficiency metrics, according to energy audits. An energy audit comparing conventional wheat production and conservation farming systems revealed substantial differences in energy efficiency indicators. Under conservation agriculture, the mean energy ratio was 4.31, with a net energy gain of 137,656 MJ ha^{-1}, a specific energy requirement of 5.56 MJ kg^{-1}, and an energy productivity of 0.18 kg MJ^{-1}. By contrast, conventional systems exhibited lower efficiency, with corresponding values of 3.03, 90,101 MJ ha^{-1}, 7.69 MJ kg^{-1}, and 0.13 kg MJ^{-1}, respectively (Moradi et al. 2018). These enhancements are made possible by a combination of crop diversification, residue recycling, and improved nutrient cycling, in addition to the direct energy savings from less tillage (Meena et al. 2019). Increased soil health, lower production costs, increased yields, and increased resilience are just a few of the long-term advantages of adopting CA, even though it frequently necessitates larger initial investments in specialised seeding and residue management equipment (Pannell et al. 2014). From a wider angle, energy use optimisation in CA offers both short and long-term benefits. Short-term effects include reductions in fuel and irrigation demands resulting in lower energy inputs. Long-term increases in SOC accumulation, improved overall resource use efficiency, and system resilience may result in better system sustainability and profitability while lowering the carbon intensity of agricultural production (Lal et al. 2021; Soni et al. 2024).

8.7 Modern Assessment of Carbon and Energy Footprints

Matrices viz. CF and EF are essential for assessing the potential for climate mitigation of CA based practices which are being used more and more to quantify the environmental performance of agricultural systems. In contrast to traditional accounting methods, improved recent frameworks generally use life cycle assessment (LCA) based assessment techniques that provide a better view of environmental impacts by inclusion of upstream (input manufacturing), on-farm (operations, emissions), and downstream (post-harvest handling) processes (Fan et al. 2022). The temporal dynamics of GHG emissions in CA systems can be captured by CF analysis and CF based system modelling. For example, breakdown of LC fractions, short-term techniques like residue retention and reduced tillage frequently promote CO_2 efflux (Rigon and Calonego 2020). Over longer timeframes, the accumulation of stable SOC, improved nutrient-use efficiencies, and lower fuel consumption all work together to reduce emissions footprints and contribute to net sequestration. EF evaluations can also show short-term trade-offs, such as the potential increases in energy demands from mechanisation and irrigation, but emphasise long-term benefits, such as higher energy productivity, decreased dependency on fossil fuels, improved energy-use efficiency and changes in energy mix of the systems. Real time mapping of crop biomass, soil cover, and water stress is now possible with the use of advancements in sensing technologies (proximal and remote), such as high-resolution satellite data and UAV-based imagery, which can be directly connected to emission and energy models (Dong et al. 2024). Large scale datasets are used by AI-ML algorithms to model different management scenarios, accurately forecasting emission trajectories and energy efficiency results. Near real-time carbon and energy accounting is made possible by the integration of IoT-enabled sensor networks, which further improve precision by continuously monitoring variables like soil respiration, nutrient fluxes, and machinery fuel use (Sharma et al. 2021).

8.8 Conclusion

The CA is better understood today as a central practice for sustainable farming systems with a huge potential for emission reductions, energy savings, and soil health improvements. The challenges of measuring CF and EF in CA lie within the specialized and variable nature of the agricultural landscapes. The variability that results from spatially and temporally heterogeneous farming systems, along with the multitude of management techniques, creates complex uncertainties. There is a paucity problem whereby reliable, sustained carbon, energy inputs and outputs records are difficult to obtain. Models, such as that from the IPCC, are overly simplistic and highly generalized to the extent of relying on global averages which are disconnected from local realities, particularly in complex cropping systems under small holder conditions. There are also unintended negative consequences from the assessments that CA seeks to lower (trade-offs), such as the reduction in tillage which in turn, increases usage of chemical fertilizer, herbicides to a large extent. Despite these challenges, the long-term trajectory of CA is highly promising. By minimizing soil disturbance, maintaining permanent soil cover, and diversifying crop rotations, CA enhances SOC storage, improves resource-use efficiency, and reduces reliance on fossil energy. Energy audits consistently show that CA outperforms conventional systems in terms of net energy gain, energy productivity, and fuel savings, even though adoption may initially involve higher equipment costs and transitional yield fluctuations. Over time, however, these barriers diminish as soils stabilize and farmers adapt management strategies to reduced-disturbance systems. Future progress in the field of CA will largely depend on integrating advanced tools and site-specific methodologies into footprint assessments. Emerging technologies such as drones, LiDAR, hyperspectral sensors, and GPS-linked fuel meters can provide high-resolution data on soil, biomass, and energy use, while stratified soil sampling and long-term monitoring sites can improve the accuracy of carbon stock measurements. The development of localized emission factors, region-specific models, and farmer-accessible footprint calculators will further strengthen both scientific assessments and on-farm decision-making. CA should be viewed not only as a farming practice but also as a strategic response to global challenges of climate change, resource scarcity, and food security. With appropriate policy support, investment in monitoring technologies, and farmer-centered advisory services, CA can evolve into a robust, scalable, and climate-resilient production system that secures both environmental sustainability and agricultural productivity for future generations.

References

Al-Kaisi MM, Yin X (2005) Tillage and crop residue effects on soil carbon and carbon dioxide emission in corn–soybean rotations. J Environ Qual 34(2):437–445. https://doi.org/10.2134/jeq2005.0437

Banerjee A, Jhariya MK, Meena RS, Yadav DK (2021a) Ecological footprints in agroecosystem: an overview. In: Agroecological footprints management for sustainable food system. Springer Singapore, pp 1–23. https://doi.org/10.1007/978-981-15-9496-0_1

Banerjee A, Jhariya MK, Raj A, Yadav DK, Khan N, Meena RS (2021b) Energy and climate footprint towards the environmental sustainability. In: Agroecological footprints management for sustainable food system. Springer Singapore, pp 415–443. https://doi.org/10.1007/978-981-15-9496-0_14

Bansal S, Yin X, Schneider L, Sykes V, Jagadamma S, Lee J (2022) Carbon footprint and net carbon gain of major long-term cropping systems under no-tillage. J Environ Manag 307:114505. https://doi.org/10.1016/j.jenvman.2022.114505

Barman A, Pooniya V, Zhiipao RR, Biswakarma N, Kumar D, Das TK, Shivay YS, Rathore SS, Das K, Babu S, Saikia N, Meena MC, Bhatia A (2024) Integrated crop management for long-term sustainability of maize-wheat rotation focusing on productivity, energy and carbon footprints. Energy 311:133304. https://doi.org/10.1016/j.energy.2024.133304

Bartocci P, Bidini G, Saputo P, Fantozzi F (2016) Biochar pellet carbon footprint. Chem Eng Trans 50:217–222. https://doi.org/10.3303/CET1650037

Bhatt MK, Singh DK, Raverkar KP, Chandra R, Pareek N, Dey P, Pramanick B, Joshi HC, Kumar M, Gaber A, Alsuhaibani AM, Hossain A (2024) Corrigendum: effects of varied nutrient regimes on soil health and long-term productivity in a rice–wheat system: insights from a 29-year study in the mollisols of the Himalayan Tarai region. Front Sustain Food Syst 8. https://doi.org/10.3389/fsufs.2024.1373475

Bhattacharya P, Maity PP, Mowrer J, Maity A, Ray M, Das S, Chakrabarti B, Ghosh T, Krishnan P (2020) Assessment of soil health parameters and application of the sustainability index to fields under conservation agriculture for 3, 6, and 9 years in India. Heliyon 6(12)

Bista D, Sapkota S, Acharya P, Acharya R, Ghimire R (2024) Reducing energy and carbon footprint in semi-arid irrigated cropping systems through crop diversification. Heliyon 10(6):e27904. https://doi.org/10.1016/j.heliyon.2024.e27904

Blanco-Canqui H, Ruis SJ (2018) No-tillage and soil physical environment. Geoderma 326:164–200

Browne D, O'Regan B, Moles R (2009) Use of ecological footprinting to explore alternative domestic energy and electricity policy scenarios in an Irish city-region. Energy Policy 37(6):2205–2213. https://doi.org/10.1016/j.enpol.2009.01.039

Chen CZ, Lin ZS (2008) Multiple timescale analysis and factor analysis of energy ecological footprint growth in China 1953–2006. Energy Policy 36(5):1666–1678

Choudhary M, Rana KS, Bana RS, Ghasal PC, Choudhary GL, Jakhar P, Verma RK (2017) Energy budgeting and carbon footprint of pearl millet—mustard cropping system under conventional and conservation agriculture in rainfed semi-arid agro-ecosystem. Energy 141:1052–1058. https://doi.org/10.1016/j.energy.2017.09.136

Chaudhary VP, Sawant CP, Chaudhary R, Gautam R, Wakchaure GC (2024) Conservation tillage enhances energy efficiency and mitigates carbon footprint and greenhouse gas emissions in long-term wheat production trials in the Western indo-Gangetic plain of India. Int J Plant Prod 18(4):531–548. https://doi.org/10.1007/s42106-024-00308-0

Corbeels M, Naudin K, Whitbread AM, Kühne R, Letourmy P (2020) Limits of conservation agriculture to overcome low crop yields in sub-Saharan Africa. Nat Food 1(7):447–454. https://doi.org/10.1038/s43016-020-0114-x

Cotrufo MF, Wallenstein MD, Boot CM, Denef K, Paul E (2012) The microbial efficiency-matrix stabilization (MEMS) framework integrates plant litter decomposition with soil organic matter stabilization: do labile plant inputs form stable soil organic matter? Glob Chang Biol 19(4):988–995. https://doi.org/10.1111/gcb.12113

Dachraoui M, Sombrero A (2020) Effect of tillage systems and different rates of nitrogen fertilisation on the carbon footprint of irrigated maize in a semiarid area of castile and Leon. Spain Soil Tillage Res 196:104472. https://doi.org/10.1016/j.still.2019.104472

Das A, Saha S, Layek J, Babu S, Saxena R, Ramkrushna GI (2023) Agricultural Technologies. In: Trajectory of 75 years of Indian agriculture after Independence. Springer Nature Singapore, pp 57–78. https://doi.org/10.1007/978-981-19-7997-2_4

Dash AK, Meena MC, Das S, Dey A, Raza MB, Tripathy S, Kumar A, Panda D, Divyadarshan A (2025) Temporal effects of conservation agriculture-based Rice–wheat cropping system on soil aggregation and organic carbon dynamics in Northwestern indo-Gangetic Plains. J Soil Sci Plant Nutr:1–17

Dey P, Mahapatra BS, Mitra B, Sarvadamana AK, Chowdhury AR, Rana A, Mandal D (2024) Potential nexus approach for sustainable soil productivity resource management. In: Environmental nexus for resource management. CRC Press. https://doi.org/10.1201/9781003358169-12

Dey P, Mahapatra BS, Pramanick B, Kumar A, Negi MS, Paul J, Shukla DK, Singh SP (2021) Quality optimization of flax fibre through durational management of water retting technology under sub-tropical climate. Ind Crop Prod 162. https://doi.org/10.1016/j.indcrop.2021.113277

Don A, Jantz M (2013) Impact of reduced tillage on the greenhouse gas balance-a meta-analysis. In: EGU general assembly conference abstracts, pp EGU2013–EG10658

Dong H, Dong J, Sun S, Bai T, Zhao D, Yin Y, Shen X, Wang Y, Zhang Z, Wang Y (2024) Crop water stress detection based on UAV remote sensing systems. Agric Water Manag 303:109059. https://doi.org/10.1016/j.agwat.2024.109059

Dwivedi S, Srivastava A, Gangwar SP, Dey P, Dey P, Bhatt MK, Sarkar S, Bhattacharya P, Mandal D, Alotaibi M, Seleiman MF (2024) Soil-test-crop-response based nutrient scheduling can improve soybean-wheat productivity and system sustainability. BMC Plant Biol 24(1). https://doi.org/10.1186/s12870-024-05890-z

Fan J, Liu C, Xie J, Han L, Zhang C, Guo D, Niu J, Jin H, McConkey BG (2022) Life cycle assessment on agricultural production: a mini review on methodology, application, and challenges. Int J Environ Res Public Health 19(16):9817. https://doi.org/10.3390/ijerph19169817

Fuentes-Ponce MH, Gutiérrez-Díaz J, Flores-Macías A, González-Ortega E, Mendoza AP, Sánchez LMR, Novotny I, Espíndola IPM (2022) Direct and indirect greenhouse gas emissions under conventional, organic, and conservation agriculture. Agric Ecosyst Environ 340:108148

Gan Y, Liang C, Campbell CA, Zentner RP, Lemke RL, Wang H, Yang C (2012) Carbon footprint of spring wheat in response to fallow frequency and soil carbon changes over 25 years on the semiarid Canadian prairie. Eur J Agron 43:175–184. https://doi.org/10.1016/j.eja.2012.07.004

Ghisellini P, Setti M, Ulgiati S (2016) Energy and land use in worldwide agriculture: an application of life cycle energy and cluster analysis. Environ Dev Sustain 18(3):799–837. https://doi.org/10.1007/s10668-015-9678-2

Ghorbani R, Mondani F, Amirmoradi S, Feizi H, Khorramdel S, Teimouri M, Sanjani S, Anvarkhah S, Aghel H (2011) A case study of energy use and economical analysis of irrigated and dryland wheat production systems. Appl Energy 88(1):283–288. https://doi.org/10.1016/j.apenergy.2010.04.028

Giller KE, Andersson JA, Corbeels M, Kirkegaard J, Mortensen D, Erenstein O, Vanlauwe B (2015) Beyond conservation agriculture. Front Plant Sci 6. https://doi.org/10.3389/fpls.2015.00870

Goswami BN, Venugopal V, Sengupta D, Madhusoodanan MS, Xavier PK (2006) Increasing trend of extreme rain events over India in a warming environment. Science 314(5804):1442–1445. https://doi.org/10.1126/science.1132027

Govaerts B, Verhulst N, Castellanos-Navarrete A, Sayre KD, Dixon J, Dendooven L (2009) Conservation agriculture and soil carbon sequestration: between myth and farmer reality. CritAl Rev Plant Sci 28(3):97–122

Grover D, Chaudhry S (2019) Ambient air quality changes after stubble burning in rice–wheat system in an agricultural state of India. Environ Sci Pollut Res 26(20):20550–20559

Hobbs PR, Sayre K, Gupta R (2008) The role of conservation agriculture in sustainable agriculture. Philos Trans R Soc B Biol Sci 363(1491):543–555. https://doi.org/10.1098/rstb.2007.2169

Holka M, Kowalska J, Jakubowska M (2022) Reducing carbon footprint of agriculture—can organic farming help to mitigate climate change? Agriculture 12(9):1383. https://doi.org/10.3390/agriculture12091383

Hu F, Chai Q, Yu A, Yin W, Cui H, Gan Y (2015) Less carbon emissions of wheat–maize intercropping under reduced tillage in arid areas. Agron Sustain Dev 35(2):701–711. https://doi.org/10.1007/s13593-014-0257-y

IPCC (2023) Summary for policymakers. In: Climate change 2021—the physical science basis, working group I contribution to the sixth assessment report of the intergovernmental panel on climate change. Cambridge University Press, pp 3–32. https://doi.org/10.1017/9781009157896.001

Jat RD, Jat HS, Nanwal RK, Yadav AK, Bana A, Choudhary KM, Kakraliya SK, Sutaliya JM, Sapkota TB, Jat ML (2018) Conservation agriculture and precision nutrient management practices in maize-wheat system: effects on crop and water productivity and economic profitability. Field Crop Res 222:111–120. https://doi.org/10.1016/j.fcr.2018.03.025

Jat SL, Parihar CM, Singh AK, Kumar B, Choudhary M, Nayak HS, Parihar MD, Parihar N, Meena BR (2019) Energy auditing and carbon footprint under long-term conservation agriculture-based intensive maize systems with diverse inorganic nitrogen management options. Sci Total Environ 664:659–668. https://doi.org/10.1016/j.scitotenv.2019.01.425

Jat HS, Kumar V, Datta A, Choudhary M, Kakraliya SK, Poonia T, McDonald AJ, Jat ML, Sharma PC (2020) Designing profitable, resource use efficient and environmentally sound cereal-based systems for the Western indo-Gangetic plains. Sci Rep 10(1):19267

Kamyab H, SaberiKamarposhti M, Hashim H, Yusuf M (2024) Carbon dynamics in agricultural greenhouse gas emissions and removals: a comprehensive review. Carbon Lett 34(1):265–289. https://doi.org/10.1007/s42823-023-00647-4

Knowler D, Bradshaw B (2006) Farmers' adoption of conservation agriculture: a review and synthesis of recent research. Food Policy 32(1):25–48. https://doi.org/10.1016/j.foodpol.2006.01.003

Knudsen MT, Meyer-Aurich A, Olesen JE, Chirinda N, Hermansen JE (2014) Carbon footprints of crops from organic and conventional arable crop rotations—using a life cycle assessment approach. J Clean Prod 64:609–618. https://doi.org/10.1016/j.jclepro.2013.07.009

Kumara TK, Kandpal A, Pal S (2020) A meta-analysis of economic and environmental benefits of conservation agriculture in South Asia. J Environ Manag 269:110773. https://doi.org/10.1016/j.jenvman.2020.110773

Kundu A, Kundu CK, Dey P, Rana S, Majumder J, Bera A, Paramanik B, Patra PS, Uddin MG, Abioui M, Saraswat A (2024) Management of soil cover and tillage regimes in upland rice-sweet corn systems for better system performance, energy use and carbon footprints. Heliyon 10(4). https://doi.org/10.1016/j.heliyon.2024.e26524

Lal R (2007) Soil science and the carbon civilization. Soil Sci Soc Am J 71(5):1425–1437

Lal R (2012) Climate change and soil degradation mitigation by sustainable Management of Soils and Other Natural Resources. Agric Res 1(3):199–212. https://doi.org/10.1007/s40003-012-0031-9

Lal R, Monger C, Nave L, Smith P (2021) The role of soil in regulation of climate. Philos Trans R Soc B 376(1834):20210084. https://doi.org/10.1098/rstb.2021.0084

Lascalajr N, Lopes AF, Spokas K, Bolonhezi DE, Archer DW, Reicosky DC (2008) Short-term temporal changes of soil carbon losses after tillage described by a first-order decay model. Soil Tillage Res 99(1):108–118. https://doi.org/10.1016/j.still.2008.01.006

Lehmann J, Kleber M (2015) The contentious nature of soil organic matter. Nature 528(7580):60–68. https://doi.org/10.1038/nature16069

Li J, Liu Z, He C, Tu W, Sun Z (2016) Are the drylands in northern China sustainable? A perspective from ecological footprint dynamics from 1990 to 2010. Sci Total Environ 553:223–231

Liu C, Cutforth H, Chai Q, Gan Y (2016) Farming tactics to reduce the carbon footprint of crop cultivation in semiarid areas. A Rev Agron Sustain Dev 36(4):69. https://doi.org/10.1007/s13593-016-0404-8

Mahapatra BS, Dey P (2022) Integrated management practices for incremental wheat productivity. In: New horizons in wheat and barley research: crop protection and resource management. Springer Nature. https://doi.org/10.1007/978-981-16-4134-3_13

Mandal KG, Saha KP, Ghosh PK, Hati KM, Bandyopadhyay KK (2002) Bioenergy and economic analysis of soybean-based crop production systems in Central India. Biomass Bioenergy 23(5):337–345. https://doi.org/10.1016/S0961-9534(02)00058-2

Meena OP, Sammauria R, Gupta AK, Gupta KC, Behera B, Saxena R, Yadav MR, Singh P, Meena RK, Raza MB, Anil AS, Lal MK (2022) Energy-carbon footprint Vis-à-Vis system productivity and profitability of diversified crop rotations in Semi-Arid Plains of north-West India. J Soil Sci Plant Nutr 22(2):2026–2041. https://doi.org/10.1007/s42729-022-00791-2

Meena RS, Kumar S, Yadav GS (2019) Soil carbon sequestration in crop production. In: Nutrient dynamics for sustainable crop production. Singapore, Springer Singapore, pp 1–39

Mishra AK, Roohi R, Sheoran HS, Mishra S, Pandey A, Sah D, Bhat MA, Sharma S (2022) Effect of conservation agriculture on energy consumption and carbon emission. In: Agriculture, livestock production and aquaculture: advances for smallholder farming systems, vol 2. Springer International Publishing, Cham, pp 75–96

Mitra B, Chowdhury AR, Dey P, Hazra KK, Sinha AK, Hossain A, Meena RS (2022) Use of agrochemicals in agriculture: alarming issues and solutions. In: Input use efficiency for food and environmental security. Wiley. https://doi.org/10.1007/978-981-16-5199-1_4

Mondal S, Chakraborty D, Paul RK, Mondal A, Ladha J (2023) No-till is more of sustaining the soil than a climate change mitigation option. Agric Ecosyst Environ 352:108498. https://doi.org/10.1016/j.agee.2023.108498

Montgomery DR (2007) Soil erosion and agricultural sustainability. Proc Natl Acad Sci 104(33):13268–13272. https://doi.org/10.1073/pnas.0611508104

Moradi M, Nematollahi MA, Mousavi Khaneghah A, Pishgar-Komleh SH, Rajabi MR (2018) Comparison of energy consumption of wheat production in conservation and conventional agriculture using DEA. Environ Sci Pollut Res 25(35):35200–35209. https://doi.org/10.1007/s11356-018-3424-x

Naorem A, Jayaraman S, Sinha NK, Mohanty M, Chaudhary RS, Hati KM, Mandal A, Thakur JK, Patra AK, Srinivasarao C, Chaudhari SK, Dalal RC, Lal R (2023) Eight-year impacts of conservation agriculture on soil quality, carbon storage, and carbon emission footprint. Soil Tillage Res 232:105748. https://doi.org/10.1016/j.still.2023.105748

Nisar S, Benbi DK, Toor AS (2021) Energy budgeting and carbon footprints of three tillage systems in maize-wheat sequence of north-western indo-Gangetic Plains. Energy 229:120661. https://doi.org/10.1016/j.energy.2021.120661

Odum HT (2007) Environment, power, and society for the twenty-first century: the hierarchy of energy. Columbia University Press

Pandey D, Agrawal M, Pandey JS (2011) Carbon footprint: current methods of estimation. Environ Monit Assess 178(1–4):135–160. https://doi.org/10.1007/s10661-010-1678-y

Pandit P, Dey P, Krishnamurthy KN (2021) Comparative assessment of multiple linear regression and fuzzy linear regression models. SN Comput 2(2). https://doi.org/10.1007/s42979-021-00473-3

Pannell DJ, Llewellyn RS, Corbeels M (2014) The farm-level economics of conservation agriculture for resource-poor farmers. Agric Ecosyst Environ 187:52–64. https://doi.org/10.1016/j.agee.2013.10.014

Patra PS, Saha R, Ahmed AS, Kanjilal B, Debnath MK, Paramanik B, Hoque A, Kundu A, Adhikary P, Biswas A, Dey P, Biswas A (2024) Enhancing aromatic rice production through agronomic and nutritional management for improved yield and quality. Sci Rep 14(1). https://doi.org/10.1038/s41598-024-65476-5

Paustian K, Lehmann J, Ogle S, Reay D, Robertson GP, Smith P (2016) Climate-smart soils. Nature 532(7597):49–57. https://doi.org/10.1038/nature17174

Pittelkow CM, Liang X, Linquist BA et al (2014) Productivity limits and potentials of the principles of conservation agriculture. Nature 517(7534):365–368. https://doi.org/10.1038/nature13809

Powlson DS, Stirling CM, Thierfelder C, White RP, Jat M (2016) Does conservation agriculture deliver climate change mitigation through soil carbon sequestration in tropical agro-ecosystems? Agric Ecosyst Environ 220:164–174. https://doi.org/10.1016/j.agee.2016.01.005

Pramanick B, Mahapatra BS, Datta D, Dey P, Singh SP, Kumar A, Paramanik B, Awasthi N (2023) An innovative approach to improve oil production and quality of mustard (Brassica juncea L.) with multi-nutrient-rich polyhalite. Heliyon 9(3). https://doi.org/10.1016/j.heliyon.2023.e13997

Raj A, Singh SP, Singh VP, Pratap T, Bhattacharya P, Dey P (2025) Combining site-specific nutrient allocation tools can decrease input demands in no-till wheat farming. Commun Soil Sci Plant Anal 56(1):90–107. https://doi.org/10.1080/00103624.2024.2406475

Rana S, Swain DK, Bhattacharya P, Roy A, Samanta S, Seleiman MF, Ali N, Dey P (2025) Soil biochemical quality indices can capture transitional changes of tillage and residue regime in lateritic soils. BMC Plant Biol 25(1). https://doi.org/10.1186/s12870-025-07212-3

Rani J, Gulia V, Sangwan A, Dhull SS, Mandzhieva S (2025) Synergies of traditional ecological knowledge in biodiversity conservation: a paradigm for sustainable food security, pp 27–49. https://doi.org/10.1007/978-981-96-2413-3_2

Ravindra K, Singh T, Mor S (2019) Emissions of air pollutants from primary crop residue burning in India and their mitigation strategies for cleaner emissions. J Clean Prod 208:261–273. https://doi.org/10.1016/j.jclepro.2018.10.031

Rhodes CJ (2016) The 2015 Paris climate change conference: Cop21. Sci Prog 99(1):97–104. https://doi.org/10.3184/003685016X14528569315192

Rigon JPG, Calonego JC (2020) Soil carbon fluxes and balances of crop rotations under long-term no-till. Carbon Balance Manag 15(1). https://doi.org/10.1186/s13021-020-00154-3

Rusu T (2014) Energy efficiency and soil conservation in conventional, minimum tillage and no-tillage. Int Soil Water Conserv Res 2(4):42–49. https://doi.org/10.1016/S2095-6339(15)30057-5

Saha A, Samanta S, Dey P, Halder R, Sinha AC (2023) Choice of varieties and organic-inorganic nutrient integrations in rainfed buckwheat can affect the performance of succeeding green gram grown on residual fertility. Arch Agron Soil Sci 69(14):3030–3043. https://doi.org/10.1080/03650340.2023.2196413

Sahoo SK, Mishra KN, Panda N, Sethi D, Dey P (2023) Carbon and nitrogen mineralization kinetics in soil under differential nutrient management regimes in Inceptisols of coastal India. Commun Soil Sci Plant Anal 54(20):2782–2799. https://doi.org/10.1080/00103624.2023.2241500

Samanta S, Saha A, Saha A, Mandal D, Pandit P, Dey P (2025) Optimization of planting geometry can manifest performance, energy balance and marketability of table potato. Potato Res. https://doi.org/10.1007/s11540-024-09841-x

Sapkota TB (2012) Conservation tillage impact on soil aggregation, organic matter turnover and biodiversity, pp 141–160. https://doi.org/10.1007/978-94-007-4113-3_6

Sapkota TB, Majumdar K, Jat ML, Kumar A, Bishnoi DK, McDonald AJ, Pampolino M (2014) Precision nutrient management in conservation agriculture-based wheat production of Northwest India: profitability, nutrient use efficiency and environmental footprint. Field Crop Res 155:233–244. https://doi.org/10.1016/j.fcr.2013.09.001

Saraswat A, Ram S, Sharma S, Chawla R, Khardia N, Chauhan D, Vishwakarma DK, Raza MB, Rajput VD, Dey P, Meena RS, Behera B (2024) EDCs exposure-induced alteration in the germination, growth, and physiological trait of the plant. In: Emerging contaminants: sustainable agriculture and the environment. Elsevier. https://doi.org/10.1016/B978-0-443-18985-2.00001-8

Scopel E, Findeling A, Guerra EC, Corbeels M (2005) Impact of direct sowing mulch-based cropping systems on soil carbon, soil erosion and maize yield. Agron Sustain Dev 25(4):425–432. https://doi.org/10.1051/agro:2005041

Sharma A, Kumar S, Iqbal Mir J, Kumar R, Chand Sharma O, Lal S, Lal Kumawat K, Ahmed N, Beer Singh D, Ahmad Ganie M, Razvi S (2021) Restricting depletion of soil organic carbon by amending nutrient-N input to soils. Land Degrad Dev 32(11):3204–3219. https://doi.org/10.1002/ldr.3974

Sharma R, Malaviya P (2023) Ecosystem services and climate action from a circular bioeconomy perspective. Renew Sust Energ Rev 175:113164. https://doi.org/10.1016/j.rser.2023.113164

Singh R, Singh A, Sheoran P, Fagodiya RK, Rai AK, Chandra P, Rani S, Yadav RK, Sharma PC (2022) Energy efficiency and carbon footprints of rice-wheat system under long-term tillage and residue management practices in western indo-Gangetic Plains in India. Energy 244:122655

Sithole NJ, Magwaza LS (2019) Long-term changes of soil chemical characteristics and maize yield in no-till conservation agriculture in a semi-arid environment of South Africa. Soil Tillage Res 194:104317. https://doi.org/10.1016/j.still.2019.104317

Six J, Conant RT, Paul EA et al (2002) Stabilization mechanisms of soil organic matter: implications for C-saturation of soils. Plant Soil 241:155–176. https://doi.org/10.1023/A:1016125726789

Song W, Xiang X, Wang J, Radosevich M, Lin Y (2025) Editorial: anthropogenic effects on the microbial communities of terrestrial ecosystems. Front Microbiol 16. https://doi.org/10.3389/fmicb.2025.1636546

Soni P, Zaya F, Kumar R, Kumari A, Gupta R, Singh N, Ghoshal N (2024) Variation in soil quality in tropical dryland agroecosystem: impact of different crop sequences. Land Degrad Dev 35(14):4213–4223

Srivastava RK (2025) Conservation tillage practices on GHG emissions, soil health and overall agricultural sustainability. Soil Use Manag 41(2). https://doi.org/10.1111/sum.70096

Stoeglehner G, Levy JK, Neugebauer GC (2005) Improving the ecological footprint of nuclear energy: a risk-based lifecycle assessment approach for critical infrastructure systems. Int J CritAl InfrastructS 1(4):394–403. https://doi.org/10.1016/j.biortech.2009.01.059

Stoeglehner G, Narodoslawsky M (2009) How sustainable are biofuels? Answers and further questions arising from an ecological footprint perspective. Bioresour Technol 100(16):3825–3830

Sun J, Wang Z, Du Y, Zhang E, Gan H, Sun D, Niu W (2022) Optimized tillage improves yield and energy efficiency while reducing carbon footprint in winter wheat-summer maize rotation systems. Sci Total Environ 820:153278. https://doi.org/10.1016/j.scitotenv.2022.153278

UNICEF (2021) The state of food security and nutrition in the world 2021. FAO. https://doi.org/10.4060/cb4474en

Ussiri DA, Lal R (2009) Long-term tillage effects on soil carbon storage and carbon dioxide emissions in continuous corn cropping system from an alfisol in Ohio. Soil Tillage Res 104(1):39–47

Villagrán E, Romero-Perdomo F, Numa-Vergel S, Galindo-Pacheco JR, Salinas-Velandia DA (2023) Life cycle assessment in protected agriculture: where are we now, and where should we go next? Horticulturae 10(1):15. https://doi.org/10.3390/horticulturae10010015

Wang L, Garland GM, Ge T et al (2025) Integrated strategies for enhancing agrifood productivity, lowering greenhouse gas emissions, and improving soil health. Int J Hydrogen Energ:101006. https://doi.org/10.1016/j.xinn.2025.101006

Yadav GS, Das A, Lal R, Babu S, Meena RS, Saha P, Singh R, Datta M (2018) Energy budget and carbon footprint in a no-till and mulch-based rice–mustard cropping system. J Clean Prod 191:144–157. https://doi.org/10.1016/j.jclepro.2018.04.173

Yang Y, Wu J, Du YL, Gao C, Pan X, Tang DW, van der Ploeg M (2021) Short-and long-term straw mulching and subsoiling affect soil water, photosynthesis, and water use of wheat and maize. Front Agron 3:708075. https://doi.org/10.3389/fagro.2021.708075

Xiao L, Kuhn NJ, Zhao R, Cao L (2021) Net effects of conservation agriculture principles on sustainable land use: a synthesis. Glob Chang Biol 27(24):6321–6330

Zhang B, Jia Y, Fan H, Guo C, Fu J, Li S, Li M, Liu B, Ma R (2024) Soil compaction due to agricultural machinery impact: a systematic review. Land Degrad Dev 35(10):3256–3273. https://doi.org/10.1002/ldr.5144

Zhou W, Hui D, Shen W (2014) Effects of soil moisture on the temperature sensitivity of soil heterotrophic respiration: a laboratory incubation study. PLoS One 9(3):e92531. https://doi.org/10.1371/journal.pone.0092531

Zizinga A, Mwanjalolo JGM, Tietjen B, Bedadi B, Gabiri G, Luswata KC (2022) Impacts of climate smart agriculture practices on soil water conservation and maize productivity in rainfed cropping systems of Uganda. Front Sustain Food Syst 6:889830. https://doi.org/10.3389/fsufs.2022.889830

[illegible]

Chapter 9
Organic Farming—A Climate Resilient Technology

K. Ashok Kumar

Abstract Agriculture is both a contributor to and a victim of climate change, with rising temperatures, erratic rainfall, and resource degradation posing significant threats to food security and rural livelihoods. Organic farming, grounded in ecological principles, biodiversity, and closed nutrient cycles, has emerged as a promising climate-resilient production system. This chapter synthesizes the current understanding of organic agriculture as both a mitigation and adaptation pathway in the face of climate variability. It highlights how organic practices—such as the use of farmyard manure, crop rotations, legume integration, composting, biological pest management, and minimal external inputs—reduce greenhouse gas emissions, enhance carbon sequestration, and promote energy efficiency. Evidence from global and Indian contexts is reviewed to assess benefits for soil fertility, water quality, biodiversity conservation, and farm incomes. The chapter further explores nutrient, pest, and weed management strategies within organic systems, and their implications for crop yield stability, nutritional quality, and value chains. Special attention is given to India, where organic farming aligns with smallholder needs and policy priorities but faces barriers of lower initial yields, knowledge intensity, certification hurdles, and market constraints. Interfaces with conservation agriculture are discussed, showing complementarities in resource conservation and resilience building. The chapter concludes by outlining a research and policy agenda to mainstream climate-smart organic transitions, emphasizing supportive institutional mechanisms, farmer training, participatory research, and accessible certification and market systems. Collectively, organic farming represents both an adaptive strategy and a low-carbon development pathway for sustainable agri-food systems.

Keywords Organic agriculture · Climate resilience · Conservation agriculture · Nutrient management · Carbon sequestration · Agroecology

K. Ashok Kumar (✉)
Acharya N.G.Ranga Agricultural University (ANGRAU), Guntur, Andhra Pradesh, India
e-mail: k.ashokkumar@angrau.ac.in

D. Kumar Swain et al. (eds.), *Climate-Smart Technologies for Sustainable Agricultural Production*, https://doi.org/10.1007/978-3-032-25666-9_9

9.1 Introduction

Climate change is reshaping agricultural risk profiles through rising temperatures, erratic precipitation, increased frequency and intensity of droughts and floods, and heightened pest and disease pressures. Approximately, 22% of global GHG emissions are from Agriculture, forestry and other land use (AFOLU) (IPCC 2023). Livestock enteric fermentation, rice methane emissions, and nitrous oxide (N_2O) from synthetic nitrogen fertilizer use are prominent sources. In many regions, especially in the Global South, smallholders face compounding vulnerabilities due to degraded soils, imbalanced input use, and market volatility.

Organic farming has emerged as a comprehensive production and management paradigm that seeks to sustain soil, ecosystem, and human health by relying on ecological processes, biodiversity, and cycles adapted to local conditions in lieu of synthetic inputs. The International Federation of Organic Agriculture Movements (IFOAM) articulates the principles of health, ecology, fairness, and care as the foundation of organic agriculture (IFOAM 2018). These principles translate into practices such as diversified crop rotations, green manuring, farmyard manure (FYM) and compost application, biological nitrogen fixation (BNF), botanical and biological pest management, and reduced tillage.

Positioned within climate policy, organic agriculture offers a dual dividend: (i) mitigation through reduced reliance on energy-intensive synthetic fertilizers and pesticides, lower N_2O emissions via improved nitrogen management, and carbon sequestration in soils and biomass; and (ii) adaptation through improved soil structure, water retention, agro-biodiversity, and farm system diversification that buffer climatic shocks. The Indian experience—with its ecological heterogeneity, post-Green Revolution nutrient imbalances, and growing consumer demand for residue-free food—provides a salient case for understanding organic transitions at scale.

This chapter surveys scientific and policy evidence on the climate resilience of organic farming. It provides an analytical framework and practical guidance for researchers, practitioners, and policymakers, with particular emphasis on smallholder-oriented strategies and the integration of organic and conservation-agriculture principles.

9.2 Concepts and Principles of Organic Farming

9.2.1 Definition and Scope

Organic farming is a holistic production system that promotes and enhances agro-ecosystem health, including biodiversity, biological cycles, and soil biological activity. It emphasizes the use of management practices over external inputs, acknowledging that regional conditions require locally adapted solutions. Synthetic nitrogen fertilizers, most synthetic pesticides, and genetically modified organisms (GMOs) are excluded, while off-farm organic inputs are used judiciously to close nutrient loops (IFOAM 2018; FAO 2020).

9.2.2 *Core Principles*

- **Health**: Sustain and enhance the health of soil, plants, animals, humans, and the planet as one and indivisible.
- **Ecology**: Base production on living ecological systems and cycles, work with them, emulate them, and help sustain them.
- **Fairness**: Build relationships that ensure fairness among farmers (equitable access to resources, livelihoods, and decision making) and for consumers (access to safe, nutritious food), while safeguarding the common environment and life opportunities.
- **Care**: Manage in a precautionary and responsible manner to protect the health and well-being of current and future generations and the environment.

9.2.3 *Systemic Features*

Organic systems rely on interacting components rather than single-factor solutions. Diversified rotations, intercrops, cover crops, legumes, and agroforestry enhance functional biodiversity; organic amendments and composts build soil organic matter (SOM) and support microbial activity; and mechanical, cultural, and biological controls suppress pests and weeds. The result is a system that is less dependent on fossil fuel-derived inputs and more resilient to climatic and market shocks (Reganold and Wachter 2016).

9.3 Organic Farming and Climate Change

9.3.1 *Greenhouse Gas Emissions*

Agriculture emits significant N_2O owing to nitrification and denitrification in soils, especially following application of synthetic N fertilizers and manure; CH_4 from ruminants and flooded rice; and CO_2 from soil organic carbon (SOC) loss and fossil fuel use. The global emissions attributed to agriculture and livestock have been estimated at ~ 7.8 Gt CO_2-eq annually, with trends sensitive to fertilizer use and livestock population growth (IFOAM 2018; FAO 2024). Organic systems can reduce these emissions by:

(i) **Eliminating Synthetic N Fertilizers**: Avoiding industrial N reduces upstream emissions from the Haber–Bosch process and curtails field N_2O emissions due to lower overall N surpluses (Skinner et al. 2013).
(ii) **Closing Nutrient Cycles**: Legume BNF, green manures, and composting synchronize N supply with crop demand, decreasing losses via leaching, runoff, and volatilization (Drinkwater and Snapp 2007).
(iii) **Improving Soil Carbon Stocks**: Recycling residues and adding organic matter enhance SOC, offsetting CO_2 emissions (Gattinger et al. 2012).
(iv) **Energy Savings**: Reduced reliance on synthetic inputs and more on-farm nutrient sources can cut energy use by 30–70% per unit area compared to input-intensive systems (Kumar et al. 2018a; Gomiero et al. 2019; Lucie et al. 2024).

9.3.2 Carbon Sequestration and Soil Health

Organic practices—regular organic amendments, cover crops, reduced tillage, and diversified rotations—tend to increase SOC, aggregate stability, and soil biodiversity (microbial biomass, mycorrhizae, mesofauna). Enhanced SOC improves water-holding capacity, infiltration, and nutrient buffering, thereby stabilizing yields under variable rainfall. While sequestration rates vary with climate, soil type, and management, meta-analyses report higher SOC stocks in organically managed topsoils relative to conventional counterparts (Leifeld and Fuhrer 2010; Gattinger et al. 2012).

9.3.3 Adaptation: Resilience to Climate Extremes

Soils in organic systems typically have higher porosity and better structure due to higher SOM and biological activity, translating into reduced runoff and erosion, greater infiltration during heavy rains, and sustained moisture during dry spells. Functional agrobiodiversity (e.g., multi-species cover crops, varietal mixtures, agroforestry) diversifies risk and provides ecological services such as natural pest regulation and pollination (Winans et al. 2015; Altieri and Nicholls 2017; Gururani et al. 2021). At farm level, diversified enterprises (crops-livestock-trees) and embedded knowledge (local/indigenous practices) enhance adaptive capacity.

9.3.4 Limits and Trade-Offs

Mitigation/adaptation gains are not automatic. Poorly timed manure applications may elevate N_2O; intensive mechanical weeding can increase CO_2 emissions; and yield gaps—when present—could shift pressure to land expansion. Therefore, context-specific best practices, nutrient budgeting, and integration with reduced-tillage and precision organic management-defined as the targeted, site-specific application of organic inputs and cultural practices based on soil, crop and spatial variability- are crucial (Ponisio et al. 2015; Asanda et al. 2017).

9.4 The Indian Context

9.4.1 *Post-Green Revolution Imperatives*

India's Green Revolution boosted staple production but also precipitated second-generation problems: soil structure deterioration, organic matter decline, multi-nutrient deficiencies (N, P, K, S, Zn, Fe, Mn), groundwater depletion and contamination, herbicide resistance, and stagnant or declining total factor productivity in some regions (Ladha et al. 2003; GOI 2014). Despite high fertilizer use in many districts (> 100 kg NPK ha^{-1}), soils often remain low in available N and P due to imbalances, poor timing, and limited organic matter return.

9.4.2 *Policy Landscape and Initiatives*

Government programs such as Paramparagat Krishi Vikas Yojana (PKVY), Mission Organic Value Chain Development for North Eastern Region (MOVCDNER), and state-level missions (e.g., Sikkim Organic Mission) promote organic and natural farming through cluster formation, certification (PGS-India), and value-chain support. Public procurement, crop insurance designs, and price information systems are being explored to reduce market risk for organic producers (DAC and FW 2020).

9.4.3 *Opportunities for Smallholders*

Organic farming is knowledge-intensive but input-light in external chemicals, aligning with smallholder contexts where family labor and locally available biomass can be mobilized. Benefits include reduced input costs, access to premium markets, improved soil health, and diversified incomes (e.g., composting enterprises, seed production, livestock integration). Participatory guarantee systems (PGS) can lower certification barriers for local markets while third-party certification targets exports. Farmer Producer Organizations (FPOs) support aggregation, branding, and logistics.

9.4.4 Constraints

Yield gaps relative to conventional systems may persist during transition years (1–3 seasons), with reported reductions of approximately 10–25% depending on crop, soils and nutrient cycles re-equilibrate (Seufert et al. 2012; Ponisio et al. 2015). Organic input availability (quality compost, green manure seed), knowledge on nutrient budgeting, pest management, and market linkages remain limiting in several states. Infrastructure for separate handling, storage, and transportation is often inadequate. Robust extension, credit, and risk-sharing mechanisms are needed to ensure equitable participation of small and marginal farmers; for example, public extension systems (e.g., Agricultural Technology Management Agency (ATMA) and Krishi Vigyan Kendras), institutional credit through Kisan Credit Cards, Farmer Producer Organizations (FPOs), and risk-sharing instruments such as Pradhan Mantri Fasal Bima Yojana provide enabling policy and institutional support in the Indian context.

9.5 Nutrient Management in Organic Systems

9.5.1 Principles and Sources

Nutrient management aims to optimize on-farm resource use, minimize losses, and maintain soil health. Major nutrient sources include FYM, composts (aerobic vermicompost), green manures (e.g., Sesbania, Crotalaria), crop residues, animal manures, oil-cakes (neem, castor), biofertilizers (Rhizobium, Azotobacter, Azospirillum, phosphate-solubilizing bacteria), and safe off-farm organic wastes, which together support soil fertility, nutrient availability, and crop performance in organic systems (Kumar et al. 2024; Kumar and Swain 2025). Moreover, combining sources rather than relying on a single input improves nutrient synchronization and system resilience.

9.5.2 Nitrogen: Synchrony and Biological Fixation

In the absence of synthetic N, synchronizing soil N supply with crop demand is pivotal. Strategies include:

- **Legume Rotations/Intercrops**: Pulses or green manure legumes add N via BNF (30–200 kg N ha^{-1} depending on species and management) about 1.0 kg N ha^{-1} day^{-1} (Singh and Chahal 2020), break pest/disease cycles (Maitra and Ray 2019), and improve soil structure (Lal 2017).
- **Compost and FYM**: Provide slow-release N through soil organic matter mineralization, enhancing microbial activity and cation exchange capacity; however,

under field conditions, nitrogen release may be insufficient or poorly synchronized with peak crop demand Kumar et al. 2018b, particularly during early growth stages or under cool or dry conditions, potentially leading to temporary N limitations.

- **Mulches and Cover Crops**: Reduce volatilization and leaching, moderate soil temperature, and feed soil biota.
- **Nutrient Budgeting**: Farm-scale N budgets track inflows (BNF, feeds, bedding, purchased organics) and outflows (harvest, losses), guiding application timing and rates.

9.5.3 Phosphorus and Potassium: Mobilization and Recycling

Organic systems rely on P mobilization from sparingly soluble pools through organic acids, microbial solubilization (PSB), mycorrhizal associations, and manure/compost applications. Recycled biomass (ash, rock phosphate where permitted) and residue retention conserve P and K. Composting with mineral additives (e.g., rock phosphate co-composting) enhances P availability by increasing water soluble P from 0.37% to 0.89% (Sonali and Vallabh 2023); and plant-available P through organic acid production and microbial activity during composting, while simultaneously reducing P fixation in soil (Bustamante et al. 2016).

9.5.4 Micronutrients and Soil Biology

Multi-nutrient deficiencies common in intensively farmed soils can be alleviated by increasing SOC—which chelates and buffers micronutrients—and by using diverse organic inputs (FYM, composts, oil-cakes) that supply Zn, Fe, Mn, B, etc. For example, under subtropical conditions, application of vermicompost @16.8 t ha^{-1} significantly improved SOC content, which resulted in significant increase in soil micronutrients (Fe, Zn and Mn) content (Kumar and Swain 2021). Mycorrhizae and microbial consortia expand the effective root zone and nutrient access. Routine soil testing, including SOC and micronutrients, is recommended to tailor inputs.

9.5.5 Practical Example: Rice–Wheat Rotation

For a rice–wheat system, a 30-t ha^{-1} FYM equivalent annual nutrient need can be met by combining a summer green manure (Chander et al., 2023) (Sesbania; ~ 60–100 kg N ha^{-1}), FYM (5–10 t ha^{-1}), vermicompost (2–3 t ha^{-1}), and oil-cakes (neem 0.5–1 t ha^{-1}), supported by biofertilizers and residue retention (Singh et al., 2025).

Such combinations supply macro- and micronutrients, improve tilth, and reduce pest incidence while building SOC.

9.6 Pest, Disease, and Weed Management

9.6.1 Ecological Pest Management (EPM)

Organic systems prioritize prevention through diversity and habitat management, supported by monitoring and thresholds. Key components include:

- **Cultural Controls**: Crop rotations, resistant varieties, sanitation, timely planting, balanced nutrition.
- **Habitat Management**: Flowering strips, hedgerows, beetle banks to conserve natural enemies (Coccinellids, Chrysopids, parasitoids like Trichogramma, Campoletis).
- **Biological Control Agents**: Bacillus-based bio-pesticides, entomopathogenic fungi, NPV formulations; microbial antagonists for diseases (e.g., Trichoderma spp.).
- **Botanical Extracts**: Neem (azadirachtin), pyrethrum, rotenone (restricted/region-specific), sabadilla, ryania, essential oils; used judiciously under organic standards.
- **Bio-rational Insecticides**: Insect growth regulators (IGRs) such as juvenoids and ecdysteroids were permitted under organic regulations.

Some fermented and microbial-derived products (e.g., spinosad from *Saccharopolyspora spinosa*) may be allowed under strict conditions as last-resort measures for specific pests (e.g., leaf miners, thrips) consistent with organic certification frameworks.

9.6.2 Disease Management

Disease suppression arises from enhanced soil biology, organic amendments that induce systemic resistance, crop diversity, and hygiene. Seed treatments with Trichoderma and plant extracts, compost teas (where evidence supports), and resistant varieties are integral. Avoiding excess available N curtails disease susceptibility.

9.6.3 *Weed Management*

Weeds remain a major challenge in organic systems due to the prohibition of most synthetic herbicides. Integrated strategies include:

- **Preventive**: Clean seed, sanitation, minimizing seedbank replenishment.
- **Cultural**: Competitive cultivars, narrow row spacing, stale seedbed technique, cover crops, living mulches.
- **Mechanical/Physical**: Timely hand/mechanical weeding, inter-row cultivators, flame weeding (where feasible), mulching with residues or biodegradable films.
- **Biological**: Emerging approaches with fungal pathogens and grazing were integrated with livestock.

Residue retention and cover cropping also suppress weeds while protecting soil moisture and temperature—synergizing with conservation agriculture.

9.7 Crop Quality and Nutritional Outcomes

9.7.1 *Quality Dimensions*

Crop quality encompasses macronutrients (protein, oil), micronutrients (Fe, Zn), phytochemicals (polyphenols, antioxidants), nitrates, sensory attributes, and post-harvest behavior. Nutrient sources and soil management influence these dimensions.

9.7.2 *Evidence across Commodity Groups*

- **Cereals**: Organic management can improve organoleptic properties and, in some cases, protein quality. In rice, organic manures have been associated with higher milling and hulling percentages and altered amylose content (Kumar et al. 2018a). In bread wheat, bone meal applied before stem extension has been shown to increase grain protein and improve loaf volume compared to solely chemical fertilization (Worthington 2001; Mäder et al. 2002).
- **Pulses**: Vermicompost at ~ 5 t ha^{-1} in chickpea has improved dry matter, yield, and grain protein while enhancing soil N and P and microbial counts relative to mineral fertilization in select studies (Kaur et al. 2023).
- **Oilseeds**: Biofertilization with composted residues (e.g., water hyacinth, rice straw) has increased seed P, Zn, Mn, Fe, and Cu in peanut and improved protein and oil content; FYM enhances sunflower photosynthetic area, oil content, and yield (Ramesh et al. 2010).
- **Tubers**: Meta-analyses indicate significantly lower nitrate concentrations in organic carrot, lettuce, and potato; organic potato tubers often exhibit higher dry

matter suitable for processing, without significant differences in K, Ca, Mg versus conventional (Smith-Spangler et al. 2012).

- **Vegetables**: Organically managed tomatoes can show higher salicylic acid and soluble solids but sometimes lower ascorbic acid and total phenolics, reflecting management and varietal interactions; organic nutrient strategies (e.g., FYM substitution) can reduce post-harvest onion losses by lowering rotting and sprouting. Foliar applications such as diluted vermiwash have been reported to enhance chilli ascorbic acid content in specific trials.
- **Fruit Crops**: Grapes under organic manures may show higher ascorbic acid and total sugars with lower acidity; organic strawberries have displayed higher ascorbate: dehydroascorbate ratios and, in some assays, greater antiproliferative activity against cell lines, suggestive of higher secondary metabolite content (Olsson et al. 2006; Sergi and Bermejo 2024).
- **Sericulture**: Organic nutrient and biopesticide packages in mulberry enhance leaf quality, which translates to improved silkworm growth and cocoon traits; reduced chemical pesticide exposure also protects beneficials and handlers.

9.7.3 Health Considerations

Lower nitrate accumulation in leafy and tuber crops under organic management is relevant to public health due to risks of methemoglobinemia and potential carcinogenic nitrosamines from excessive nitrate intake. Reduced pesticide residues in organic produce further strengthen consumer confidence—though rigorous, residue-aware post-harvest handling and traceability remain essential irrespective of production system (EU 2018).

9.8 Interfaces with Conservation Agriculture

9.8.1 Shared and Distinct Principles

Conservation Agriculture (CA) rests on three pillars: (i) minimal mechanical soil disturbance; (ii) permanent soil cover with residues/cover crops; and (iii) diversified crop rotations. Organic agriculture shares the latter two strongly and the first to a degree—recognizing that reduced tillage is desirable but often constrained by reliance on mechanical weed control. Integrating organic and CA principles can maximize synergistic benefits for soil structure, SOC accrual, and water dynamics (Kassam et al. 2018).

9.8.2 Practical RCTs (Resource-Conserving Technologies)

No- or low-till organic systems that use roller-crimped cover crops to form surface mulch enable direct seeding of cash crops, effectively suppressing weeds, conserving soil moisture, and reducing the number of tillages passes. Maintaining crop residues to ensure at least 30% soil cover after planting further enhances soil infiltration, reduces evaporative losses, moderates soil temperature, and limits weed germination. In addition, adopting diverse crop rotations and intercropping systems involving two to three species disrupts pest and disease cycles, distributes nutrient demand more evenly, and contributes to improved yield stability.

9.8.3 Soil Biology as "Biological Tillage"

Enhanced soil biota—fungal hyphae, earthworms, and micro-aggregates—create stable pore networks that mimic tillage functions, enabling gas exchange, water infiltration, and root penetration. Mechanical tillage can disrupt these biological structures; therefore, organic systems increasingly aim to minimize disturbance while maintaining weed control through mulches, timing, and mechanical precision.

9.9 Economic and Social Dimensions

9.9.1 Productivity and Profitability

Yield outcomes under organic management are heterogeneous. Meta-analyses report average organic yield gaps of 8–25% relative to conventional systems, but gaps narrow with diversified rotations and optimized management; in water-limited conditions or years with pest/disease pressure, organic yields can be comparable or superior due to resilience effects (Ponisio et al. 2015). Profitability often improves through lower input costs and price premiums, particularly where supply chains and certification support are strong.

9.9.2 Markets, Certification, and Trust

Certification assures process compliance but can be costly and complex. Participatory Guarantee Systems (PGS) reduce transaction costs for local markets via peer review and transparency. For exports or large retail chains, third-party certification remains standard. Where feasible, traceability, residue testing, and group certification through FPOs can maintain credibility and reduce per-farmer costs.

9.9.3 Gender and Inclusion

Women frequently manage on-farm composting, seed saving, and kitchen gardens; targeted training and access to resources can amplify benefits and agency. Inclusive models should consider landless laborers and tenant farmers through service enterprises (e.g., composting units, biocontrol production, weeding services) and fair contracts.

9.9.4 Risk Management

Transition risks can be mitigated via staggered conversion, transitional labeling, forward contracts, climate risk insurance tailored to diversified organic systems, and working capital for group-owned input units (vermicompost, biopesticides). Digital advisory tools and community knowledge platforms strengthen adaptive management.

9.10 Implementation Pathways and Case Illustrations

9.10.1 Cluster-Based Approaches

Cluster formation at village or watershed scale enables shared learning, input production, residue logistics, common certification, and collective marketing. Demonstration plots and farmer-researcher partnerships can validate locally adapted packages for dominant cropping systems (rice–wheat, cotton-based rotations, millets, pulses, horticulture).

9.10.2 On-Farm Input Enterprises

Community-level composting, vermicompost, enriched compost (e.g., rock phosphate co-composting), and biofertilizer multiplication can generate employment while ensuring reliable supply and quality. Quality protocols (C:N ratios, maturation indices, pathogen suppression) are essential.

9.10.3 Knowledge Systems and Extension

Experiential learning (farmer field schools), where available decision support tools for nutrient budgeting, and advisory on cover crop selection, timing, and residue management accelerate adoption. Linking traditional knowledge with modern soil and plant diagnostics enhances credibility and outcomes.

9.11 Challenges and Research Priorities

9.11.1 Agronomic and Ecological Research

Yield stability remains a critical research priority, particularly under increasing climate variability. Future studies should focus on multi-location and multi-year trials that quantify yield stability under extreme climatic conditions such as heat stress, drought, and excess rainfall. Special emphasis is needed on identifying soil organic carbon thresholds that buffer crops against these stresses, thereby enhancing resilience and sustaining productivity across diverse agro-ecological zones.

Advances in nitrogen dynamics are essential for improving nutrient use efficiency in organic and low-input systems. Refinement of biological nitrogen fixation (BNF) estimates, coupled with better understanding of synchrony between nitrogen release from organic sources and crop uptake is required. Additionally, research should address mitigation of nitrous oxide (N_2O) emissions through optimized timing, placement, and forms of organic amendments, aligning productivity goals with climate-smart agriculture.

In the domain of weed ecology, future efforts should evaluate innovative non-chemical weed management strategies, including mechanical precision tools, thermal and steam-based weeding technologies, and optimized mulching practices. These approaches need to be tested across contrasting agro-ecological conditions to develop context-specific, scalable weed management solutions that reduce labour intensity while maintaining crop competitiveness.

Soil biology research should move beyond bulk indicators and focus on functional biological metrics such as enzyme activities, microbial functional groups, soil microbial network structures. Linking these indicators to specific management practices and measurable yield outcomes will improve mechanistic understanding of soil-plant interactions and support the development of biologically driven soil health benchmarks.

9.11.2 Socio-Economic and Policy Research

From a socio-economic and policy perspective, transition economics warrants detailed investigation, particularly the time-path of yields, production costs, and market premiums during shifts to sustainable and organic systems. Evidence-based design of transition support mechanisms, including risk-sharing instruments and targeted incentives, is needed to reduce adoption barriers for farmers.

Innovation in certification systems is another priority area. Scalable Participatory Guarantee Systems (PGS), supported by digital traceability tools and residue-aware supply chains, can enhance transparency, reduce certification costs, and improve consumer trust while maintaining integrity in organic and natural farming markets.

Finally, landscape-level approaches should be promoted to enhance ecological services and resource efficiency. Research on organic corridors, pollinator habitat restoration, and integrated water management at the catchment scale can help align farm-level practices with broader environmental outcomes, supporting biodiversity conservation, climate resilience, and sustainable intensification.

9.11.3 Data and Metrics

Standardized indicators for climate mitigation (SOC change, N_2O intensity), adaptation (yield stability indices, water productivity), biodiversity (functional groups), and livelihoods (income diversification, gender equity) will enable robust monitoring and comparative analysis.

9.12 Conclusions

Organic farming, grounded in ecological principles and local knowledge, provides a credible pathway to climate mitigation and adaptation while delivering co-benefits for soil health, biodiversity, nutrition, and rural livelihoods. Its non-reliance on synthetic nitrogen fertilizers reduces upstream and field emissions; closed nutrient cycles enhance nutrient-use efficiency; organic amendments and diversified rotations build SOC and improve hydrological functioning; and ecological pest and weed management strengthen system resilience.

In India and similar contexts, organic farming can help address second-generation challenges stemming from input-intensive agriculture, including nutrient imbalances, soil degradation, and groundwater stress. However, realizing its full potential requires addressing constraints: bridging yield gaps through research and innovation, ensuring quality and availability of organic inputs, strengthening extension and farmer-led knowledge systems, and building robust, transparent markets and certification systems. Integration with conservation agriculture and resource-conserving technologies can further amplify climate benefits by reducing soil disturbance while maintaining effective weed control.

A coherent policy and investment agenda—supporting transitional finance, cluster-based models, on-farm input enterprises, gender-responsive capacity building, and data-driven monitoring—can mainstream climate-resilient organic systems at scale. With thoughtful design and inclusive governance, organic agriculture can potentially be a cornerstone of sustainable, climate-smart agri-food transitions.

References

Altieri MA, Nicholls CI (2017) The adaptation and mitigation potential of traditional agriculture in a changing climate. Clim Change 140(1):33–45

Asanda M, Lembe SM, Samson ZT, Nokwazi M (2017) Postharvest quality and composition of organically and conventionally produced fruits: A review. Sci Hortic 216:148–159

Bustamante MA, Ceglie FG, Aly A, Mihreteab HT, Ciaccia C, Tittarelli F (2016) Phosphorus availability from rock phosphate: Combined effect of green waste composting and sulfur addition. J Environ Manage 182:557–563

Chander S, Tripathi S, Venkatesh K, Neeraj K, Raj M, Chhokar R, Kamboj N, Kumar N (2023) Sustainability assessment of rice-wheat system through organic fertilizers and green manuring in sub-tropical humid climate. Int J Plant Prod 17. https://doi.org/10.1007/s42106-023-00232-9

DAC&FW (2020) Paramparagat Krishi Vikas Yojana (PKVY) Guidelines. Department of Agriculture, Cooperation & Farmers Welfare, Government of India

Drinkwater LE, Snapp S (2007) Nutrients in agroecosystems: rethinking the management paradigm. Agric Ecosyst Environ 119(1–2):1–3

EU (2018) Regulation (EU) 2018/848 on organic production and labelling of organic products. European Union

FAO (2020) The state of the world's biodiversity for food and agriculture. Food and Agriculture Organization of the United Nations

FAO (2024) Greenhouse gas emissions from agrifood systems—global, regional and country trends, 2000–2022. FAOSTAT Analytical Brief Series, No. 94. Rome

Gattinger A, Muller A, Haeni M, Skinner C, Fliessbach A, Buchmann N, Mäder P, Stolze M, Smith P, Scialabba H, Niggli U (2012) Enhanced topsoil carbon stocks under organic farming. Proc Natl Acad Sci 109(44):18226–18231

GOI (2014) State of Indian agriculture 2013–14. Ministry of Agriculture, Government of India

Gomiero T, Pimental D, Paoletti MG (2019) Energy and environmental issues in organic and conventional agriculture. Energy Policy 36(2):446–557

Gururani K, Sood S, Kumar A, Joshi DC, Pandey D (2021) Mainstreaming Barahnaja cultivation for food and nutritional security in the Indian Himalayas. Biodivers Conserv 30:3561–3580

IFOAM (2018) Principles of organic agriculture. IFOAM—Organics International

IPCC (2023) Summary for policymakers. In: Lee H, Romero J (eds) Climate change 2023: synthesis report. Contribution of working groups I, II and III to the sixth assessment report of the intergovernmental panel on climate change. IPCC, Geneva, Switzerland, pp 1–34. https://doi.org/10.59327/IPCC/AR6-9789291691647.001

Kassam A, Friedrich T, Derpsch R (2018) Global spread of conservation agriculture. Int J Environ Stud. https://doi.org/10.1080/00207233.2018.1494927

Kaur G, Vats A, Singh R, Behl R, Singh A (2023) Effect of vermicompost, biofertilizers and in organic fertilizers on nutrient uptake and physico-chemical properties of soil: a case study of

Chickpea (Cicer arietinum L.). Int J Environ Clim Chang 13:4519–4525. https://doi.org/10.9734/ijecc/2023/v13i103130

Kumar P, Singh S, Yadav VK, Yadav J, Singh S, Singh RP, Pratap M, Abhimanyu M, Pandy A (2024) Effect of integrated nutrient management through organic sources on growth parameters and yield of chickpea (Cicer arietinum L.). Int J Res Agron 7(12S):270–274

Kumar KA, Swain DK (2021) Nutrient management strategy in rice-chickpea for improving chemical and biological properties of lateritic soil. Agron J 113:135–146

Kumar KA, Swain DK (2025) Impacts of split application of organic fertilizer on nitrogen uptake and use efficiency in direct seeded rice. Bangladesh J Bot 54(2):311–318

Kumar KA, Swain DK, Bhadoria PBS (2018a) Split application of organic nutrient improved productivity, nutritional quality and economics of rice-chickpea cropping system in lateritic soil. Field Crop Res 223:125–136

Kumar KA, Swain DK, Pallavi, Ghosh BC (2018b) Effect of organic and inorganic nutrient management on soil nutrient dynamics and productivity of rice-chickpea system in lateritic soil. Org Agric 8:15–28

Ladha JK, Dawe D, Pathak H, Padre AT, Yadav RL, Bijay S, Yadvinder S, Singh Y, Singh P, Kundu AL, Sakal R, Ram N, Regmi AP, Gami SK, Bhandari AL, Amin R, Yadav CR, Bhattarai EM, Das S, Aggarwal HP, Gupta RK, Hobbs PR (2003) How extensive are yield declines in long-term rice-wheat experiments in Asia? Field Crop Res 81(2–3):159–180

Lal R (2017) Improving soil health and human protein nutrition by pulses-based cropping systems. Adv Agron 145:167–120

Leifeld J, Fuhrer J (2010) Organic farming and soil carbon sequestration: what do we really know about the benefits? Ambio 39(8):585–599

Lucie C, Schmid H, Anke S, Hülsbergen K (2024) Energy-use efficiency of organic and conventional plant production systems in Germany. Scientific Reports. 14. https://doi.org/10.1038/s41598-024-51768-3

Mäder P, Fliessbach A, Dubois D, Gunst L, Fried P, Niggli U (2002) Soil fertility and biodiversity in organic farming. Science 296(5573):1694–1697

Maitra S, Ray DP (2019) Enrichment of biodiversity, influence in microbial population dynamics of soil and nutrient utilization in cereal-legume inter-cropping systems: a review. Int J Bioresour Sci 6:11–19

Olsson M, Staffan CA, Stina MO, Rakel HB, Karl EG (2006) Antioxidant levels and inhibition of cancer cell proliferation in vitro by extracts from organically and conventionally cultivated strawberries. J Agric Food Chem 54:248–255

Ponisio LC, M'Gonigle LK, Mace KC, Palomino J, Valpine P, Kremen C (2015) Diversification practices reduce organic to conventional yield gap. Proc R Soc B: Biol Sci 282(1799):20,141,396

Ramesh P, Panwar NR, Singh AB, Ramana S, Yadav SK, Rao AS (2010) Status of organic farming in India. Curr Sci 98(9):1190–1194

Reganold JP, Wachter JM (2016) Organic agriculture in the twenty-first century. Nat Plants 2(15):221

Sergi MB, Bermejo NF (2024) Fruit quality in organic and conventional farming: advantages and limitations. Trends Plant Sci 29(8):878–894

Seufert V, Ramankutty N, Foley JA (2012) Comparing the yields of organic and conventional agriculture. Nature 485:229–232

Singh A, Chahal HS (2020) Organic grain legumes in India: potential production strategies, perspective, and relevance. In: Legume crop-prospects, production and uses. IntechOpen

Singh V, Gupta RK, Seema S, Singh G, Batra C (2025) Assessing rice residue and nitrogen management strategies on productivity and nitrogen use efficiency of wheat in rice-wheat cropping system in Indo-Gangetic Plains. Field Crop Res 325:109826

Skinner C, Gattinger A, Muller A, Paul M, Andreas F, Stolze M, Ruser R, Niggli U (2013) Greenhouse gas fluxes from agricultural soils under organic and non-organic management—a global meta-analysis. Sci Total Environ 468–469c:553–563

Smith-Spangler C, Brandeau ML, Hunter GE, Bavinger JC, Pearson M, Eschbach PJ, Sundaram V, Liu H, Schirmer P, Stave C, Olkin I, Bravata DM (2012) Are organic foods safer or healthier? A systematic review. Ann Intern Med 157(5):348–366

Sonali C, Vallabh Z (2023) The impact of conditioning periods and rock phosphate on phosphorus enrichment, total nutrient content and microbial count in compost. Biol Forum Int J 15(11):564–573

Winans K, Sahrawat K, Wani S, Rao M, Reddy K, Hanlon E (2015) Effects of diversification of rainfed cropping systems to enhance sustained yields for semi-arid tropical agroecosystems: effects of 11–years of farmers' practice. J Indian Soc Soil Sci 63:166–172

Worthington V (2001) Nutritional quality of organic versus conventional fruits, vegetables, and grains. J Altern Complement Med 7(2):161–173

Chapter 10
Conservation Agriculture Through Watershed Planning and Management

Susanta Kumar Jena and Kalpesh Borse

Abstract Conservation agriculture (CA) has emerged as a globally recognized paradigm for achieving sustainable intensification of farming system. The key tenets of CA which are minimum soil disturbance, permanent organic soil cover, and diversified crop rotations have proven very effective in restoring degraded lands, enhancing soil fertility, and improving water-use efficiency. However, implementation at the farm scale often fails to capture upstream–downstream interactions in resource management. This chapter emphasizes the importance of integrating CA within watershed planning and management frameworks to maximize benefits across ecological and socio-economic dimensions. The chapter details the role of Land Resource Inventory (LRI) for characterizing soils, terrain, and fertility parameters, and the integration of hydrological assessment for evaluating water availability, runoff, and groundwater recharge. Case examples from rice-based systems in India demonstrate how CA reduces water demand, improves soil carbon, and enhances crop resilience to climatic extremes. The REWARD (Rejuvenating Watersheds for Agricultural Resilience through innovative Development) initiative further illustrates participatory watershed management approaches were scientific tools guide interventions. By embedding CA within watershed planning, agricultural landscapes can be rejuvenated through reduced erosion, improved ecosystem services, and higher resource productivity. The integration of modern tools like remote sensing, GIS, and hydrological modeling with participatory governance holds the key to scaling CA for climate-resilient agriculture.

Keywords Conservation agriculture · Climate resilience · Watershed management · Hydrological assessment · LRI

S. K. Jena (✉) · K. Borse
ICAR-Indian Institute of Water Management, Bhubaneswar, India
e-mail: skjena.icar@gmail.com

D. Kumar Swain et al. (eds.), *Climate-Smart Technologies for Sustainable Agricultural Production*, https://doi.org/10.1007/978-3-032-25666-9_10

Fig. 10.1 Schematic diagram of conservation agriculture (CA)

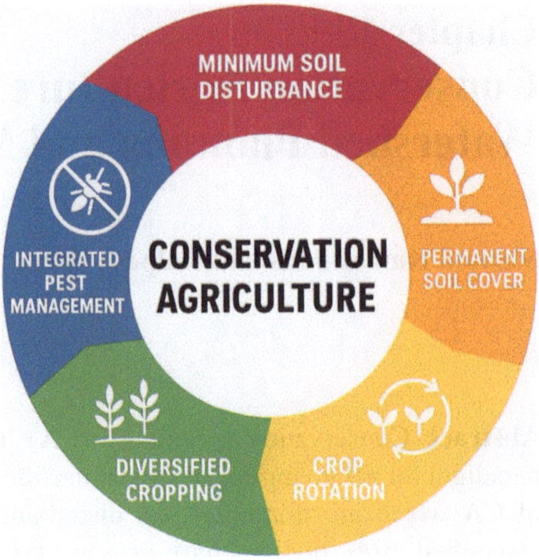

10.1 Introduction

Agriculture in South Asia is under significant stress due to land degradation, declining water resources, and increasing climatic variability. India loses nearly 5.3 billion tonnes of soil annually due to erosion, which severely undermines productivity and ecosystem stability (Mandal et al. 2021). Intensive tillage, monocropping, and unsustainable input use have accelerated soil fertility decline and reduced water-use efficiency. Conservation agriculture (CA) offers a paradigm shift by focusing on soil health restoration, water conservation, and energy efficiency (Hobbs et al. 2008). While individual farm-level benefits are well-documented, scaling CA practices to watershed units provides synergistic advantages. Watershed planning ensures that soil–water conservation measures account for topography, hydrology, and social contexts, thereby enhancing resource-use efficiency and equity (Wani et al. 2003). This chapter aims to (i) discuss the scientific principles of CA, (ii) explain the role of Land Resource Inventory (LRI) and hydrological assessment, (iii) present the integration of CA in rice-based systems, and (iv) highlight the importance of watershed planning for resilient agriculture (Fig. 10.1).

10.2 Principles of Conservation Agriculture

The three universally accepted principles of CA are:

1. **Minimum Soil Disturbance**
   - Reduces breakdown of soil aggregates and loss of organic carbon
   - Promotes biological activity and improves infiltration
   - Zero or reduced tillage decreases energy and labour costs (Kassam et al. 2009)

2. **Permanent Soil Cover**
   - Crop residues or cover crops act as a protective layer
   - Reduces erosion, moderates soil temperature, and conserves moisture
   - Enhances soil organic matter and carbon sequestration (Lal 2015)
3. **Crop Rotation and Diversification**
   - Reduces pest and disease cycles
   - Improves soil nutrient cycling and resilience to climate shocks
   - Integrating legumes enhances nitrogen fixation and system sustainability (Thierfelder and Wall 2010)

When embedded within a watershed, these practices collectively improve upstream resource conservation and downstream water quality, enhancing ecosystem services.

10.3 Land Resource Inventory (LRI)

LRI is a geospatially referenced, field-based assessment of land characteristics. It provides the scientific foundation for site-specific CA planning. Parameters typically assessed include:

- Morphological: geology, soil depth, texture, gravelliness, and colour
- Topographic: slope, erosion status, and flood proneness
- Hydrological: drainage, available water capacity, and infiltration rates
- Fertility: organic carbon, soil reaction (pH and EC), macro- and micronutrient status.

In India, LRI has been applied under the National Bureau of Soil Survey and Land Use Planning (NBSS&LUP) initiatives to support watershed development projects (Velayutham et al. 1999). When combined with GIS tools, LRI enables high-resolution mapping for precise planning of CA interventions.

10.4 Hydrology Assessment and Integration with LRI

Hydrological processes govern the distribution and productivity of water resources in watersheds. Hydrology assessment includes rainfall analysis, runoff estimation, infiltration capacity, and groundwater dynamics. Key tools include the Soil and Water Assessment Tool (SWAT), Water Evaluation and Planning System (WEAP), and remote sensing-based rainfall–runoff models (Arnold et al. 2012).

Integration of LRI and hydrology enables:

- Identification of critical erosion-prone areas
- Planning of check dams, farm ponds, and recharge structures
- Estimation of crop water requirements aligned with soil water-holding capacity

- Improved irrigation scheduling and drainage planning

Thus, LRI–hydrology integration ensures that CA practices are adapted not only to soil fertility but also to water availability and watershed-scale dynamics.

10.5 Conservation Agriculture in Cereal Based Cropping Systems

Rice-based systems predominate in eastern India, where waterlogging and groundwater depletion often coexist. Traditional puddled transplanted rice consumes ~ 2500–3000 L of water per kg of grain (Bouman and Tuong 2001).

CA offers alternatives:

- Zero tillage wheat after rice: reduces turnaround time and conserves soil moisture.
- Direct-seeded rice (DSR): reduces water demand by 25–30% compared to transplanted rice.
- Residue management: prevents burning, enhances soil carbon, and moderates microclimate.
- Diversification with legumes and oilseeds: improves profitability and soil fertility.

Field trials in Odisha under ICAR-IIWM demonstrated that DSR combined with mulching and crop rotation increased water productivity by 20% and reduced erosion (Jena et al. 2022).

The study finds that CA practices increase soil organic matter content and enhance soil physical, chemical, and biological properties. Furthermore, enhancing water use efficiency involves optimizing nutrient management. CA also improves nutrient cycling and use efficiency, potentially reducing the need for external fertilizer inputs. The adoption of conservation agriculture-based practices, such as zero-tillage or reduced tillage with residue retention, can lead to comparable or even higher crop yields, while reducing input costs (e.g., labour, fuel for tillage), thus increasing farm income. By reducing greenhouse gas emissions (compared to puddling for rice cultivation) and improving resilience to climate variability, CA contributes to more climate-smart agriculture.

ICAR-Flexi Check Dam (Rubber Dam) for Conservation Agriculture- A Case Study:

ICAR-rubber Check Dams (Rubber dams), which are inflatable and flexible hydraulic structures, provide an innovative alternative to conventional rigid check dams by enabling dynamic water storage and controlled water release. Their ability to inflate during dry periods to store water, and deflate fully during floods to allow safe passage of excess flow makes them ideal for monsoon-dependent river systems. Rubber dams operate on a simple yet highly effective engineering principle. The dam body is made of tough, rubberized fabric reinforced with high-tensile synthetic fibers. This membrane is securely anchored to a concrete foundation, ensuring stability even under high hydraulic loads. When water is injected into the dam body, it rises and acts as a barrier, impounding water upstream. During heavy rainfall or potential floods, the dam can be deflated completely, allowing unobstructed discharge downstream to prevent structural damage and flooding. This adaptability makes rubber dams superior to masonry structures, which cannot respond dynamically to changing hydrological conditions.

Odisha was the first state where indigenous rubber dams were installed and rigorously field-tested. Locations such as Baghamari, Badapokharia, Mendhasal, and Chandeswar of Odiaha were selected based on stream characteristics, community demand, and topography. These early installations demonstrated that rubber dams could perform effectively under diverse hydrological conditions, providing valuable data on sedimentation patterns, structural wear, community acceptance, and maintenance requirements (ICAR-IIWM 2016).

Agricultural Impacts

- Yield Increases in *kharif* rice

The supplemental irrigation provided by rubber dams proved highly beneficial for monsoon rice. Yield increases of 14% at Baghamari and 19–23% at Chandeswar were recorded due to reduced moisture stress during key phenological stages (ICAR-IIWM 2023). These improvements are consistent with broader research indicating that even small amounts of supplemental irrigation can significantly improve yields in rainfed rice systems (Mishra and Panigrahi 2015).

The average productivity of *kharif*, *rabi,* and summer crops pre-installation and post-installation period in rubber dam command area at Chandeswar, Khurdha district, Odisha is given in Table 10.1.

- Yield Increases in Rabi Crops and Vegetables

The availability of stored water during *rabi* and summer seasons encouraged crop diversification. Farmers expanded into high-value vegetables such as brinjal, ridge gourd, watermelon, and cowpea, with yield increases ranging from 23 to 49% (ICAR-IIWM 2023). This shift not only enhanced household nutrition but also increased income, as vegetable crops typically fetch higher market prices compared to cereals.

- Socio-Economic Impacts and Benefits

Rubber dams significantly increased net farm income by enabling multi-season cultivation and higher productivity. Farmers practicing rice-only systems saw their net returns nearly double, while those adopting diversified systems such as rice–green gram and rice–sunflower achieved even higher gains (ICAR-IIWM 2023). The improved water availability reduced seasonal migration by up to 28.5%, as families could engage in year-round agriculture instead of seeking wage labor elsewhere.

Pradhan et al. (2016) in their study, "Conservation Agriculture Practices in Rainfed Uplands of India: Improved Maize-Based System Productivity and Profitability" evaluated the impact of conservation agriculture practices such as reduced tillage, maize–cowpea intercropping, and cover cropping on productivity and profitability in

Table 10.1 Productivity of *kharif*, *rabi*, and summer crops pre-installation and post-installation period in rubber dam command area at Chandeswar, Khurdha district, Odisha

Crop	Yield (t ha^{-1})		Percent increase in yield
	Pre-installation stage	Post-installation stage	
Kharif season			
Rice	4.12	4.72	15
Rabi season			
Greengram	0.71	0.90	27
Brinjal	7.40	9.70	31
Ridgegourd	6.10	8.40	38
Summer season			
Brinjal	4.80	6.30	31
Watermelon	9.20	11.70	27
Cowpea	5.20	6.90	33
Ridgegourd	5.50	7.40	35
Cucumber	7.80	10.20	31

rainfed uplands of Odisha, India. Conducted over three years (2011–2014), the field experiment compared conventional and minimum tillage systems with and without intercropping and different cover crops (mustard, horsegram, and fallow). Results showed that maize yields remained stable across treatments, while intercropping with cowpea provided additional yield and income without affecting maize performance. Subsequent mustard and horsegram crops recorded 25 and 37% higher yields, respectively, following maize–cowpea intercropping, due to improved soil nitrogen and fertility. Overall system productivity, expressed as maize equivalent yield, and economic returns were highest under minimum tillage combined with maize–cowpea intercropping and mustard residue retention, achieving a net benefit of \$959 ha- 1 and a marginal rate of return of 228%. The study concluded that integrating these conservation practices enhances crop productivity, profitability, and soil health while reducing labor and input costs, making them highly suitable and adoptable for smallholder and tribal farmers in rainfed regions.

10.6 Watershed Planning for Agricultural Resilience

Watershed management integrates soil–water–crop–livelihood strategies. The REWARD (Rejuvenating Watersheds for Agricultural Resilience through Innovative Development) project provides a robust example. Using LRI and hydrological models, REWARD promotes participatory micro-watershed interventions in Odisha and Karnataka, combining CA practices with livelihood diversification (World Bank 2022). Under the REWARD initiative, the Soil and Water Conservation plan, and

Drainage line treatment plans have been proposed to overcome the climate change problems in the rainfed areas of Odisha and Karnataka. The following figures (Figs. 10.2, 10.3, and 10.4) show the Soil and Water Conservation plan, and drainage line treatment plans proposed in one of the micro-watersheds of Odisha under the REWARD project. Figure 10.2 represents the soil conservation plan proposed in the Bhandar MWS of Odagaon block of Nayagarh district of Odisha. Considering the slope, soil depth and land capability a staggered trenches are recommended for moderately sloping and moderately deep soil phases with a view that plantation may be adopted in these land parcels. Trench-cum-diversion drain has been recommended along the contour to resist the eroded soil and to harvest possible runoff for the coming from relatively higher slopes. Any breach in field bunds may lead to erosion of the top fertile soil over years ultimately converting the lands as barren and unsuitable for cultivation. To avoid the problem, it is recommended to strengthen the field bunds with surplus weir at relatively higher slopes under the SWC plan. Further, strengthening of bunds would allow the generated runoff to be detained in the fields for more time leading to maintaining of adequate soil moisture for a longer period as well as facilitating groundwater recharge. There are several waste and scrub land pockets present in the MWS which also leads to high erosion potential sites, to reduce this erosion and to cover the open or scrub land, Agroforestry cum plantation and Fodder plantation has been suggested for 71.39 and 9.00 ha respectively under SCP plan.

Figure 10.3 shows the Water Storage plan proposed for the Gunduribadi micro-watershed of Daspalla block of Nayagarh district of Odisha. To harvest the proposed

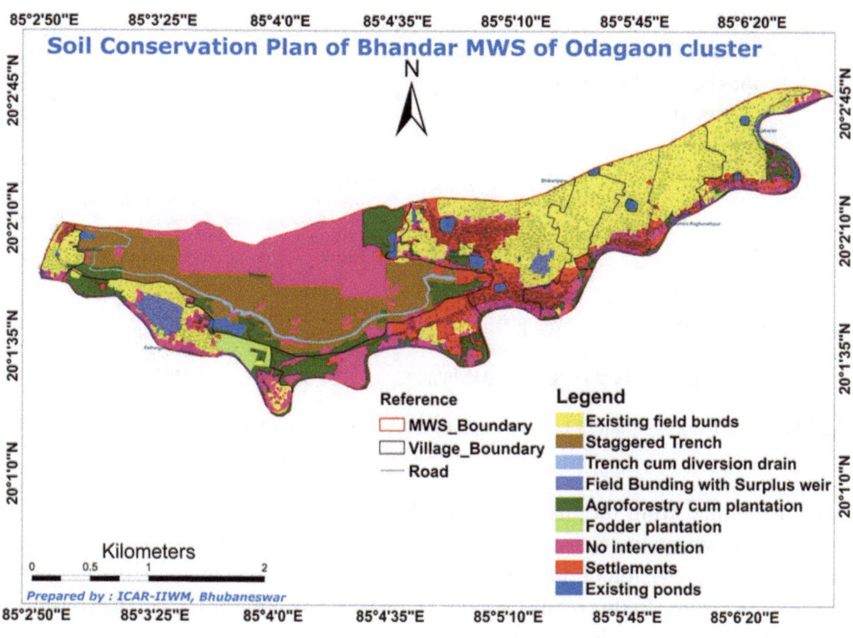

Fig. 10.2 Soil Conservation plan of Bhandar micro-watershed of Odisha

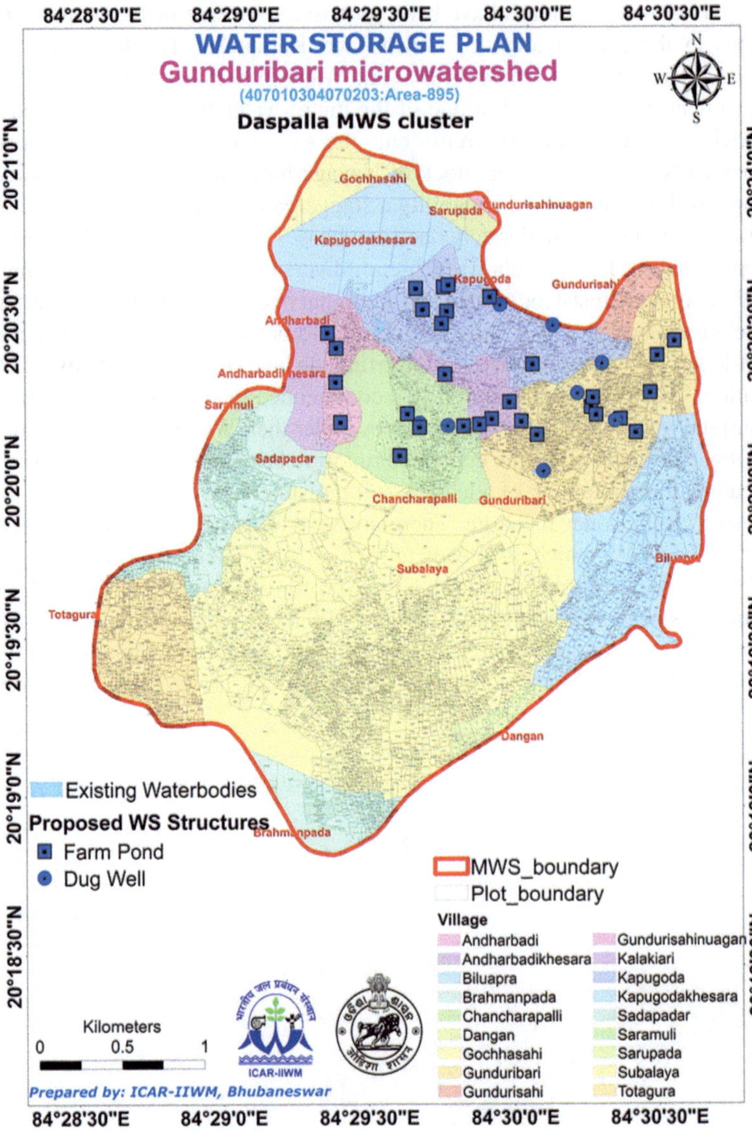

Fig. 10.3 Water storage plan of Gunduribadi micro-watershed of Odisha

runoff generated from the mws, 31 Farm Ponds (20mx20mx3m) in private plot is proposed. However, to ensure the irrigation water supply during rabi and subsequent season there is provision of dug wells which can meet the demand of irrigation water during dry seasons. In this regard, total 8 Dug Wells (Minimum 12 feet dia) has been proposed to dig in the whole Gunduribadi mws.

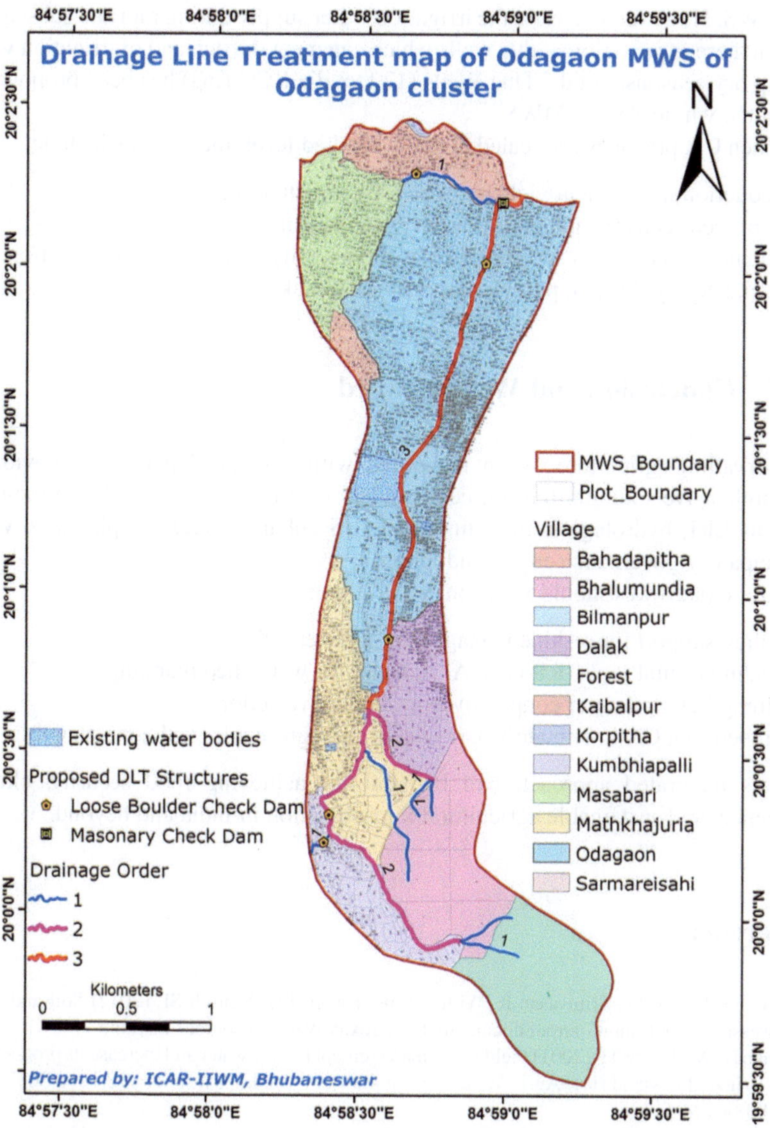

Fig. 10.4 Drainage line treatment plan of Odagaon micro-watershed of Odisha

Figure 10.4 shows the DLT plan proposed for the Odagaon MWS of Odagaon block of Nayagarh district of Odisha. Basing upon the criteria table developed for drainage line treatment, 2 No. of Loose Boulder Check Dams (LBCD) are recommended for the 1st order streams. Similarly, 01 No. of LBCD are suggested for the treatment of 2nd order streams in the MWS. For 3rd order and above streams, as per the criteria table, 02 LBCD and 01 Masonry Check Dam (MCD) is recommended for the MWS. However, to ensure the irrigation water supply during rabi and subsequent season there is provision of dug wells which can meet the demand of irrigation water during dry seasons. Total 3 Dug Wells (12 feet dia RCC ring) has been proposed to dig in the whole Rabara MWS.

When CA practices are scaled at the watershed level, the benefits include:

- Reduction in sediment yield and downstream siltation.
- Improved recharge and groundwater sustainability.
- Enhanced ecosystem services such as biodiversity and carbon sequestration.
- Social equity through participatory decision-making.

10.7 Conclusion and Way Forward

Conservation agriculture, when integrated with watershed planning, provides a comprehensive solution to resource degradation and climate stress. Scientific tools such as LRI, hydrological modeling, and GIS enhance precision planning, while participatory governance ensures adoption.

Future pathways should focus on:

- Policy support for residue management and zero tillage
- Scaling digital tools (RS-GIS, AI models) for watershed planning
- Strengthening farmer cooperatives for collective action
- Embedding CA in climate-smart village and agro-ecological zone planning

This integrated approach will be vital for achieving food security, climate resilience, and sustainable agricultural intensification in India and beyond.

References

Arnold JG, Kiniry JR, Srinivasan R, Williams JR, Haney EB, Neitsch SL (2012) Soil and water assessment tool: input/output documentation. Texas Water Resources Institute

Bouman BAM, Tuong TP (2001) Field water management to save water and increase its productivity in irrigated lowland rice. Agric Water Manag 49(1):11–30. https://doi.org/10.1016/S0378-3774(00)00128-1

Hobbs PR, Sayre K, Gupta R (2008) The role of conservation agriculture in sustainable agriculture. Philosoph Trans Royal Soc b: Biol Sci 363(1491):543–555. https://doi.org/10.1098/rstb.2007.2169

ICAR-IIWM (2016) Water harvesting and supplemental irrigation technologies for rainfed agriculture. ICAR–IIWM

ICAR-IIWM (2023) Assessment of agricultural impacts of rubber dams in Eastern and Western India. ICAR–IIWM

Jena SK, Behera B, Rath P (2022) Conservation agriculture-based water and nutrient management in rice-based systems of Eastern India. Indian J Soil Conserv 50(3):205–213

Kassam A, Friedrich T, Shaxson F, Pretty J (2009) The spread of conservation agriculture: justification, sustainability and uptake. Int J Agric Sustain 7(4):292–320. https://doi.org/10.3763/ijas.2009.0477

Lal R (2015) Restoring soil quality to mitigate soil degradation. Sustainability 7(5):5875–5895. https://doi.org/10.3390/su7055875

Mandal D et al (2021) Assessment of soil erosion in India using revised universal soil loss equation (RUSLE) and geospatial technology. CATENA 200:105174. https://doi.org/10.1016/j.catena.2021.105174

Mishra A, Panigrahi B (2015) Rainwater harvesting and small water storage interventions in eastern India: a review. Agricult Eng Rev 37(3):15–28

Pradhan A, Idol T, Roul PK (2016) Conservation agriculture practices in rainfed uplands of india improve maize-based system productivity and profitability. Front Plant Sci. 15(7):1008. https://doi.org/10.3389/fpls.2016.01008

Thierfelder C, Wall PC (2010) Rotation in conservation agriculture systems of Zambia: effects on soil quality and maize productivity. Field Crop Res 123(3):241–248. https://doi.org/10.1016/j.fcr.2009.12.013

Velayutham M, Pal DK, Mandal DK (1999) Land resources of India for agricultural development. NBSS Publ. 43. National Bureau of Soil Survey and Land Use Planning, Nagpur

Wani SP, Rockström J, Oweis T (2003) Rainfed agriculture: unlocking the potential. comprehensive assessment of water management in agriculture series. CABI Publishing

World Bank (2022) Rejuvenating Watersheds for Agricultural Resilience through Innovative Development (REWARD): Project Appraisal Document. Washington, DC

Chapter 11
Farmer-Centric Digital Transformation for Agri-Food System

Mukund Patil, Pranuthi Gogumalla, Cuba Perumal, Sreevani Minatala, Kishore Kumar, Sukanya Burry, Nagendra Katakamsetti, Pushpajeet Choudhari, Shalander Kumar, and Rebbie Harawa

Abstract Digital agriculture plays a key role in helping farmers deal with the challenges of food security, climate change, and sustainable livelihoods. For more than three decades, ICRISAT has been working at the intersection of science and technology to bring digital innovations to farmers, especially smallholders in the semi-arid tropics. The journey has evolved from early experiments in using digital tools for knowledge sharing to the development of farmer-centered, data-driven solutions that provide timely and location-specific advice. These efforts represent a shift from technology for information delivery to building integrated digital ecosystems that support farmers across the entire agricultural value chain. ICRISAT's farmer-centric digital transformation strategy is built on the idea that digital tools must empower farmers to make better decisions, use resources more efficiently, and become more resilient to climate risks. By using advances in artificial intelligence, climate analytics, remote sensing, and mobile technologies, farmers can access advisories on which crops to grow, when to sow, how much fertilizer and water to apply, how to protect crops from pests and diseases, and how to manage harvests and markets. These tools also support new opportunities such as carbon farming, where farmers can adopt regenerative practices that store carbon in soils and reduce greenhouse gas emissions, while benefiting from climate financing.

Keywords Digital agriculture · Data-driven advisories · Farmer-centric digital tools

M. Patil (✉) · P. Gogumalla · C. Perumal · S. Minatala · K. Kumar · S. Burry · N. Katakamsetti · P. Choudhari · S. Kumar · R. Harawa
International Crops Research Institute for the Semi-Arid Tropics, Hyderabad, India
e-mail: Mukund.patil@icrisat.org

D. Kumar Swain et al. (eds.), *Climate-Smart Technologies for Sustainable Agricultural Production*, https://doi.org/10.1007/978-3-032-25666-9_11

11.1 Introduction

Ensuring global food security for an estimated nine billion people by 2050, while improving rural livelihoods, remains one of the greatest challenges of our time. This challenge is compounded by degrading land resources, shrinking water reserves, and the growing uncertainties of climate change. It is particularly acute in developing countries, where agricultural growth is constrained by limited investment, weak infrastructure, and low adoption of science-led solutions. Smallholder farmers, who form the backbone of food systems in these regions, often depend on traditional practices that are increasingly inadequate to cope with changing environmental and socioeconomic realities.

At the core of this challenge lies the issue of knowledge gaps. Reliable and accessible knowledge is essential for effective agricultural planning, yet it is often fragmented, difficult to interpret, or siloed within institutions. Digital innovations can bridge this gap by democratizing access to timely, actionable insights. For instance, satellite imagery, once restricted to research institutions, is now widely available through user-friendly platforms like Google Earth (Butler 2006). However, vast datasets on soil fertility, crop health, and climate risks remain underutilized, largely because of limited capacity to integrate them into decision-making. Building interoperable data infrastructures for agriculture is therefore a critical step toward inclusive and efficient systems for planning, monitoring, and implementation.

Digital technologies offer immense potential to transform agriculture by enhancing access to information, strengthening decision-making, and improving service delivery. Tools such as geospatial analytics, sensors, artificial intelligence, and mobile-based applications can deliver localized, real-time advisories to farmers (Fabregas et al., 2019). Yet, despite significant advancements, a disconnect persists between the availability of sophisticated digital tools and their accessibility, usability, and affordability for smallholders (Mehrabi et al. 2021; Fairbairn et al. 2025). Bridging this gap requires more than deploying new technologies, it calls for a farmer-centric approach that ensures digital innovations are tailored to farmers' needs and realities.

Such transformation depends on collaborative models that engage governments, private enterprises, civil society, and farmers themselves. Co-creating the solutions, building local capacities, and scaling innovations through public–private partnerships, agriculture can transition toward inclusivity and resilience. With the rapid spread of mobile technologies, the rise of agri-tech startups, and the institutionalization of digital innovation hubs, there is an unprecedented opportunity to empower rural communities and strengthen food, land, and water systems.

ICRISAT's experience demonstrates the transformative potential of integrating digital tools into agricultural research and development. Over five decades of work in dryland regions have shown how genetic improvement, natural resource management, and inclusive value chain development can be amplified by digital innovations. Mobile-based agro-advisories, AI-driven plant health tools, and digital monitoring systems have delivered scalable solutions for millions of smallholders. Importantly, these initiatives were co-designed with farmers and stakeholders, ensuring local adaptation, gender inclusion, and equitable access.

This chapter explores how farmer-centric digital transformation can address the intertwined challenges of food security, climate resilience, and sustainable livelihoods. It draws lessons from past and ongoing initiatives, examines pathways for integrating big data and digital platforms into smallholder systems, and outlines strategies for accelerating an inclusive and sustainable digital future in agriculture.

11.2 Evolution of Farmer-Centric Digital Agriculture at ICRISAT

ICRISAT's journey in digital agriculture has evolved through distinct phases, shaped by advances in technology and the changing needs of smallholder farmers in the semi-arid tropics. Early efforts in the 2000s emphasized the use of information and communication technologies (ICTs) to share climate information and improve preparedness for drought and other risks (Navarro 2004). These pioneering initiatives established the foundation for ICT-enabled extension, demonstrating how digital tools could strengthen access to knowledge, support risk management, and enhance decision-making for dryland farming communities.

In the following decade, the scope of ICT applications widened through mobile-based advisories and interactive knowledge platforms, which provided farmers with broader access to agricultural information. Farmer-to-farmer digital knowledge exchange highlighted the value of participatory learning and locally relevant solutions (Patil et al. 2016). However, these efforts also revealed the limitations of conventional ICT models: information often remained generic and insufficient to address the growing complexity of farming systems affected by climate variability, soil degradation, and volatile markets.

The mid-2010s marked a pivotal shift toward farmer-centric digital transformation. ICRISAT redefined its strategy from delivering static information to creating intelligent, adaptive systems capable of generating personalized, predictive, and actionable insights. Advances in artificial intelligence, machine learning, remote sensing, and mobile-based analytics enabled real-time climate forecasting, soil and plant health diagnostics, and precision farming interventions. This transition recognizes farmers need for tailored solutions, not one-size-fits-all advisories, to build resilience and improve productivity.

Farmer-centric digital agriculture at ICRISAT rests on embedding innovations across the entire production cycle (Fig. 11.1). At the planning stage, decision-support systems help farmers identify crops that align with soil conditions, climate forecasts, water availability, and market signals. Climate analytics then inform the most suitable sowing windows, while access to mechanization through custom-hiring services reduces labor burdens and improves efficiency. During the crop cycle, site-specific nutrient management, smart irrigation systems, and AI-powered pest and

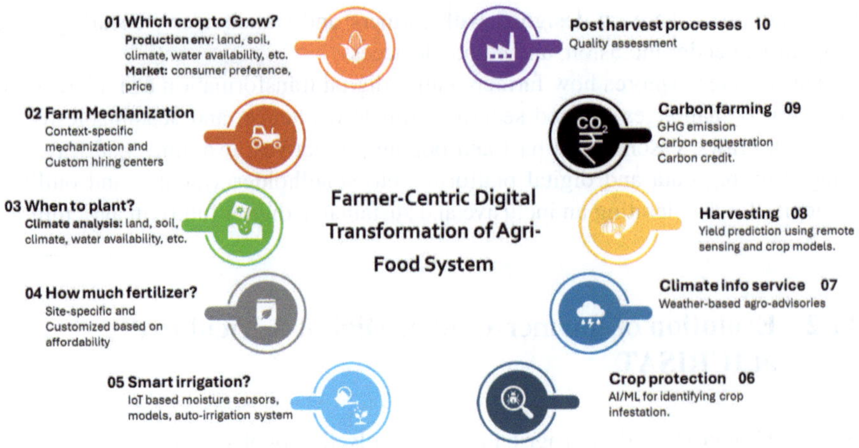

Fig. 11.1 Ways to farmer-centric digital transformation at critical stages in agri-food production cycle

disease detection tools provide actionable recommendations that safeguard yields and promote sustainability.

Post-harvest, digital solutions for quality assessment, market intelligence, and logistics enable farmers to enhance value capture and strengthen market integration. At the same time, emerging digital monitoring, reporting and verification (MRV) systems for carbon farming and climate finance open opportunities for new income streams while contributing to global sustainability goals. The integration of these tools aligns with ICRISAT's vision of positioning smallholders as active participants in data-driven farming systems rather than passive recipients of information. This approach ensures that smallholder farmers not only gain access to localized and actionable insights but also benefit from opportunities that strengthen livelihoods and contribute to sustainable agriculture in the face of climate and market challenges.

11.3 Climate Information Services

Climate variability remains one of the greatest risks for smallholder farmers in the drylands, where erratic rainfall, shifting seasons, and frequent climate shocks often lead to substantial yield losses. One of the most visible impacts of climate change is the increasing variability of rainfall patterns, with onset dates shifting while duration and intensity are becoming more unpredictable. Accurate and accessible rainfall information helps farmers decide not only when to sow and harvest, but also when to dry crops and prepare for outbreaks of pests and diseases that thrive under certain weather conditions. By working with such information, farmers can boost productivity and reduce post-harvest losses. In the wider context, climate information services also support governments and markets to prepare for extreme weather events such as storms, droughts, and floods. To address these challenges, ICRISAT has developed a suite of Digital Climate Information Services (CIS) that integrate advanced weather forecasting, agro-advisory generation, and localized delivery systems to support risk-informed farming decisions. These services provide farmers with timely and actionable advisories on critical questions such as which crop to grow, when to sow, and how to plan key operations like irrigation, fertilization, or harvesting.

In an early effort, ICRISAT along with Microsoft India had developed one of the most impactful innovations, Sowing App, which used cloud-based machine learning, satellite imagery, and historical rainfall data to identify the optimal sowing window through the moisture adequacy index (Rao and Dixit 2019). Delivered via SMS to feature phones, this app had demonstrated yield gains of 10–30% across crops like groundnut, maize, and cotton, with no need for sensor installation or capital investment. Complementing this, the Meghdoot mobile app, developed in partnership with the India Meteorological Department and Indian Institute of Tropical Meteorology, ensures last-mile delivery of district-level weather forecasts and crop advisories under the national program (Dhulipala et al. 2021). Available in 12 Indian languages, Meghdoot has reached hundreds of thousands of farmers, bridging information gaps in regions with limited digital infrastructure.

ICRISAT's flagship innovation, the Intelligent Agricultural Systems Advisory Tool (ISAT), represents the next generation of AI-enabled climate services. Designed with national and international partners, ISAT integrates weather forecasts, soil and crop data, and expert agronomic rules to deliver hyperlocal, crop-specific advisories throughout the season (Ramaraj et al. 2023; Kumar et al. 2024). Its hybrid AI-rule-based framework provides pre-season planning guidance and dynamic in-season recommendations, covering sowing, irrigation, nutrient management, and harvest timing (Fig. 11.2). Pilots in India have shown yield gains of up to 56%, and international deployments in Kenya and Senegal have demonstrated the adaptability of this approach for South-South collaboration.

Through these innovations, ICRISAT's climate information services are able to transform fragmented weather data into farmer-friendly, actionable insights. By aligning cutting-edge analytics with participatory design and multilingual delivery systems, these platforms empower farmers to make informed decisions, reduce climate risk, and enhance resilience across diverse dryland farming systems. Looking ahead, the integration of AI copilots, digital twins of farming systems, and conversational multilingual advisory agents will further strengthen CIS, ensuring that smallholder farmers remain at the center of a digitally empowered and climate-resilient future.

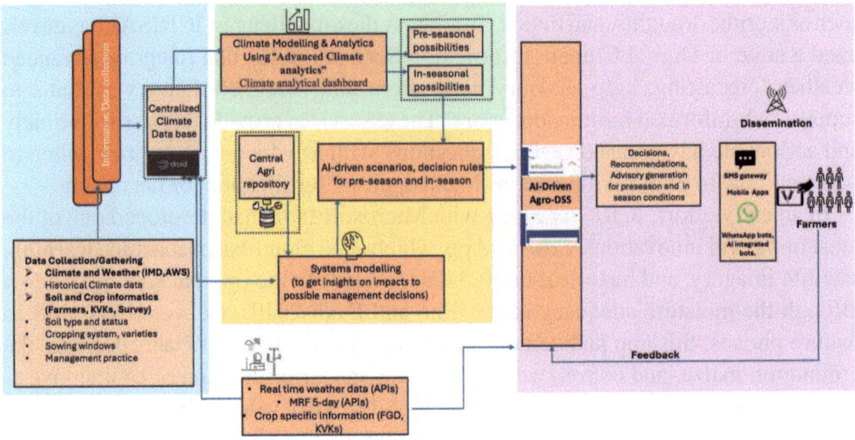

Fig. 11.2 Workflow of the ISAT platform for AI-driven agro-advisories

11.4 Digital Tools for Soil Health Monitoring and Precision Advisory

While climate information services equip farmers to manage weather-related risks, another critical gap lies in soil fertility management. Soil health is a cornerstone of agricultural productivity and sustainability, particularly in the semi-arid tropics, where nutrient imbalances, soil degradation, and inefficient fertilizer use persistently constrain yields (Snapp et al. 2025). Smallholder farmers often rely on blanket fertilizer recommendations, which rarely account for local soil variability or crop-specific requirements, leading to low nutrient use efficiency, rising input costs, and long-term soil depletion. To address this challenge, ICRISAT has developed a suite of digital platforms that combine soil intelligence with precision decision support to deliver site- and context-specific nutrient management solutions.

11.4.1 MSOIL: Digital Platform for Soil Health Monitoring and Management

The MSOIL platform represents a comprehensive approach to soil data management and farmer advisories (Patil et al. 2024). It integrates mobile applications, web-based portals, and real-time analytics to streamline the entire soil information lifecycle, from sample collection to advisory dissemination. The mobile app, designed for offline functionality in remote areas, enables geotagged farmer registration, soil sample collection, and logistics tracking. The companion web-based portal provides role-based access for administrators, field enumerators, laboratory staff, and managers, ensuring data quality, transparency, and oversight. Its interactive dashboards and digital soil maps make monitoring intuitive and actionable.

The important features of MSOIL are generating the digital soil maps of soil nutrient and the soil health card module. The digital soil mapping (McBratney et al. 2003) module is powered by a machine learning algorithm that leverages the open-sourced geospatial products and ground truth data of soil nutrient status captured through MSOIL platform, which provide estimate of soil fertility status of unsampled locations. The soil health card module combines laboratory soil nutrient data or estimates from digital soil maps with farmer-specific information to generate personalized fertilizer recommendations. These advisories, delivered via mobile apps, PDFs, or vernacular chatbots on platforms like Telegram, optimize nutrient application, reduce imbalances, and improve both productivity and sustainability.

11.4.2 DIPA: Digital Platform for Precision Agro-Advisory

Complementing MSOIL, the Digital Platform for Precision Agro-Advisory (DIPA) is a decision-support system that integrates soil health information with crop modeling outputs and agronomic knowledge. Its core innovation lies in the site-specific nutrient management workflow, which customizes fertilizer recommendations based on local soil data and crop needs. DIPA employs dual-method architecture to ensure reliability and flexibility:

1. Rule-based recommendations derived from agronomic guidelines of the National Agricultural Research System (NARS). Nutrient levels are classified (low, medium, high), with adjustment rules applied, e.g., increasing fertilizer by 25% if a nutrient is low, or reducing it when levels are high.
2. Dynamic crop response functions, which model how yields respond to varying nutrient applications. Farmers can set target yields or input budgets, and the platform calculates optimized nutrient dosages and associated costs.

Merging both approaches, DIPA produces context-sensitive, optimized fertilizer plans, which are delivered through multiple channels such as mobile apps, downloadable PDFs, and AI-powered chatbots in local languages (Fig. 11.3). This multi-modal dissemination ensures last-mile connectivity and inclusivity, even for farmers with limited digital literacy.

Together, MSOIL and DIPA represent a paradigm shift in soil and crop management, moving from generic advisories to data-driven, precision agro-advisories that are responsive to farmers' contexts. By integrating soil intelligence, predictive modeling, and participatory design, these platforms lower input costs, boost productivity, and promote environmental sustainability. They mark a critical step toward climate-smart, resource-efficient agriculture, reinforcing ICRISAT's vision of empowering smallholders with actionable, science-based digital tools.

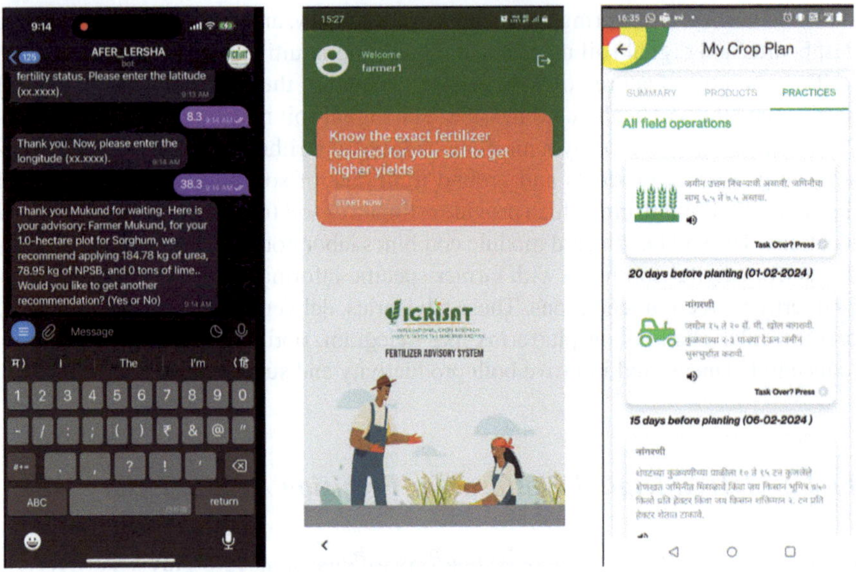

Fig. 11.3 Farmer-centric mobile applications for providing precision advisories

11.4.3 Digital Tools for Plant Protection

ICRISAT has long recognized the potential of digital innovations for improving crop protection. In its early efforts, ICRISAT partnered with the Progressive Environmental and Agricultural Technologies (PEAT) an Indo-German startup, supporting the development and field validation of its AI-powered plant health diagnostic tool—Plantix (ICRISAT 2025). Through collaborative trials with farmer networks, ICRISAT played a key role in strengthening Plantix's image-recognition models and usability for smallholder conditions. This experience not only demonstrated the power of AI-based plant protection but also highlighted the need for institutional capacity to develop in-house, scalable, and farmer-centric digital solutions.

Building on these learnings, ICRISAT has developed its own digital ecosystem for AI-driven plant health management, most notably through the Plant Health Detector (PHD) platform. PHD integrates computer vision and advanced machine learning algorithms to identify crop pests, diseases, and nutrient deficiencies with high accuracy from simple smartphone photographs. Farmers and extension workers can capture field-level images of affected crops, which the system analyzes in real time to deliver rapid diagnoses and actionable recommendations.

A central objective of PHD is to promote integrated pest management (IPM), encouraging sustainable practices that reduce the overuse of pesticides. The recommendations generated by the system are context-specific, science-based, and environmentally friendly, enabling farmers to adopt timely, targeted interventions. This leads to reduced crop losses, lower input costs, and improved yields, while safeguarding ecosystem health.

11.4.4 Digital Tools for Strengthening Climate Financing

Agriculture is both a significant source of greenhouse gas (GHG) emissions and a major opportunity for climate mitigation. Methane emissions from rice fields, nitrous oxide from fertilizers, and carbon losses from land-use practices contribute substantially to global warming. At the same time, agriculture, particularly through regenerative practices such as crop residue management, cover cropping, and minimum tillage, offers enormous potential for carbon sequestration and reducing emission intensity. Recognizing this dual role, ICRISAT has been at the forefront of developing scientific and digital frameworks that enable agriculture to transition from a climate challenge to a climate solution.

At the heart of climate financing in agriculture lies MRV that is essential for credible carbon credit programs and sustainability certifications, to ensure the impacts of regenerative farming are measured, validated, and reported with scientific rigor. Without robust MRV systems, claims of carbon sequestration or GHG reduction lack credibility, undermining both farmer incentives and investor confidence. ICRISAT is exploring a new generation of MRV approaches by combining long-term field experimentation with advanced digital platforms (Fig. 11.4), ensuring both scientific reliability and scalability across diverse semi-arid landscapes.

The first pillar of this framework is the scientific validation of regenerative practices. ICRISAT conducts extensive field studies that measure changes in soil organic carbon, track soil health dynamics, and quantify GHG emissions under different crop management systems. These observations provide a robust evidence base to calibrate

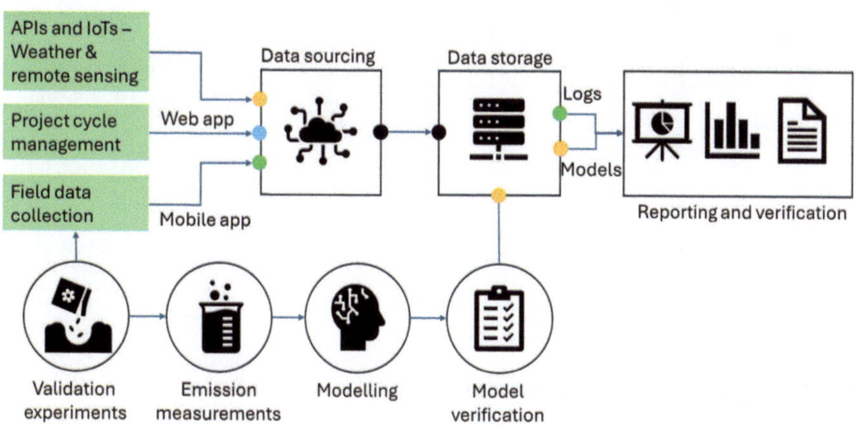

Fig. 11.4 Schematic representation of the approach to be adopted in the proposed project

and validate simulation models, enabling accurate estimation of carbon sequestration potential and emission reductions. The second pillar is the development of a digital MRV platform, which transforms the way carbon monitoring and reporting are conducted. Through mobile applications and sensor-based inputs, extension workers can record details of farm operations, residue management, and soil sampling, even in areas with limited internet connectivity. Once synchronized, this data flows seamlessly into cloud-based systems, where advanced models compute sequestration rates and emission reductions. Automated dashboards and reporting functions then translate these results into evidence-based summaries that can be directly used for verification and auditing (Fig. 11.5). By ensuring full traceability, every dataset tagged by time, location, and user identity, The platform builds transparency and trust across the carbon credit lifecycle by ensuring full traceability as every dataset tagged by time, location, and user identity. The platform is built to align with global protocols such as Verra and Gold Standard, ensuring that reported outcomes are market-ready and compliant with international standards. At the same time, the system simplifies the complexity of carbon accounting for smallholder farmers, converting technical data into actionable insights and linking them directly to potential revenue streams from climate finance.

With the blend of empirical research with digital innovation, the digital MRV tool unlocks new opportunities for farmers to participate in carbon markets and benefit financially from climate-smart agriculture. More importantly, it creates a transparent system that reassures policymakers, investors, and verification agencies that claimed climate benefits are both real and measurable. In this way, digital MRV platform evolve from being compliance mechanisms into catalysts of a new climate finance ecosystem, where science, technology, and financial flows converge to accelerate the adoption of regenerative agriculture across the semi-arid tropics and beyond.

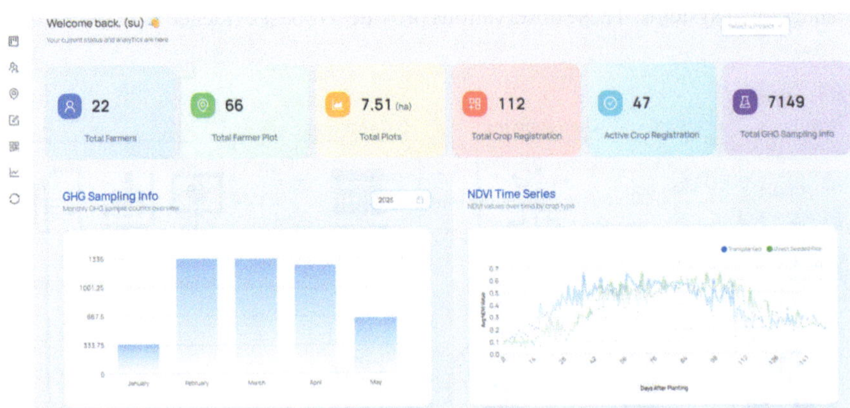

Fig. 11.5 Digital MRV dashboard showing the information on key indicators

11.5 Cultivating a Science-Driven ag-Tech Startup Culture: The Role of Public–Private Partnerships

The complexity of modern agriculture, driven by climate variability, land degradation, shrinking water resources, and volatile markets, highlights the fact that no single actor can solve these challenges in isolation. Research institutions alone cannot ensure that innovations reach farmers at scale, while the private sector on its own often struggles to keep solutions inclusive, affordable, and farmer-centric. This reality highlights the importance of public–private partnerships (PPPs) as engines of agricultural transformation. PPPs create a vital synergy: research institutions contribute scientific rigor, trusted farmer networks, and credibility, while startups and enterprises bring speed, entrepreneurial agility, and access to markets. Together, these partnerships accelerate the deployment of science-based solutions that are responsive to smallholder needs.

It was in this context that ICRISAT launched the Innovations Hub (iHub) in February 2017, marking a pivotal moment in its digital agriculture journey. Conceived as more than a traditional incubator, iHub became an open innovation platform where startups, AgriTech enterprises, and technology developers collaborate with scientists, extension agents, and farmers. Its model of engagement revolves around co-creation and validation, ensuring that new technologies are tested under real-world smallholder conditions before scaling. Startups gain access to ICRISAT's experimental plots, datasets, and farmer networks, while also receiving support through business and investor linkages. With the focus on digital-first innovations such as AI, machine learning, IoTs, remote sensing, and mobile-based platforms, iHub positions itself at the forefront of data-driven agriculture.

Among its most celebrated successes is *Plantix* (Fig. 11.6), an AI-powered mobile app developed by the Indo-German startup PEAT, which diagnoses crop diseases, pest infestations, and nutrient deficiencies through smartphone images. Supported by iHub, Plantix underwent rigorous field validation across multiple Indian states, with governments and universities contributing image datasets and extension inputs. Today, Plantix serves more than 40 million farmers in 18 languages, achieving accuracy above 93% and reducing crop losses by 10–20%. Its success exemplifies how PPPs can transform agriculture by blending scientific credibility, entrepreneurial innovation, and public-sector support.

Beyond Plantix, iHub has nurtured a diverse portfolio of startups. These include SenseAcre, offering drone—and AI-based crop health monitoring; Kalgudi Digital, which connects farmers, FPOs, and traders in a transparent digital marketplace; XMachines, developing autonomous farming robots; and PureScan AI, providing portable aflatoxin detection devices to improve food safety. Each startup adds to the larger mosaic of solutions aimed at making dryland agriculture more resilient, productive, and sustainable.

The lessons from iHub suggested that building a science-driven startup ecosystem is essential for translating agricultural research into impact at scale. Leveraging the credibility and networks of public institutions with the agility and design thinking of

Fig. 11.6 Launch of the Plantix application in Andhra Pradesh state of India (2017, left) and demonstrating the application to Prime Minister of India (2018). *Photo credit* Srikanth Rupavatharam

entrepreneurs, PPPs foster an innovation culture where smallholder farmers become active beneficiaries of co-created, context-specific solutions, rather than passive recipients of external technologies.

11.6 Summary

ICRISAT's journey in digital agriculture illustrates a paradigm shift from information-centric interventions to holistic, farmer-centric ecosystems that embed science, technology, and partnerships across the agricultural value chain. Beginning with ICT-based initiatives in the early 2000s, the institute progressively expanded into climate information services, precision soil health platforms, AI-driven plant protection, and digital tools for carbon farming and climate finance. Each phase built on the previous, moving from knowledge sharing and awareness creation toward intelligent, predictive, and context-specific decision support that addresses the complex realities of farming in the Semi-Arid Tropics.

The chapter highlighted how ICRISAT's philosophy of farmer-centric digital transformation integrates IoT, AI/ML, remote sensing, and climate services to empower smallholders with actionable insights across the entire production cycle, from crop choice and mechanization to irrigation, nutrient management, pest and disease control, yield forecasting, and post-harvest quality assessment. Digital MRV platforms have opened pathways for farmers to participate in carbon markets, while soil health tools like MSOIL and DIPA enable precision nutrient management that is both sustainable and affordable. Complementing these science-driven platforms, climate information services such as the Sowing App, Meghdoot, and ISAT have shown measurable yield improvements by providing hyperlocal advisories that strengthen resilience against weather variability.

Equally significant has been the establishment of iHub, which demonstrates how public–private partnerships and ag-tech startups can translate research into scalable, market-ready solutions. Success stories like Plantix, along with emerging innovations in drones, robotics, digital marketplaces, and food safety diagnostics, highlight how co-creation between science and entrepreneurship accelerates digital transformation. Collectively, these efforts ensure that smallholders are not passive recipients of technology but active participants and beneficiaries in a rapidly evolving digital ecosystem.

Looking forward, ICRISAT envisions the next decade of digital agriculture to be shaped by AI copilots for farmers, digital twins of dryland agriculture, and autonomous farm systems. This future will be anchored in open data, global collaborations, and farmer-centric design, ensuring inclusiveness, sustainability, and resilience. Ultimately, ICRISAT's digital agriculture strategy follows its commitment to empowering smallholders in the semi-arid tropics as leaders of a digital revolution that simultaneously addresses food security, climate challenges, and rural livelihoods.

References

Butler D (2006) The web-wide world. Nature 439(7078):776–778. https://doi.org/10.1038/439776a

Dhulipala RK, Gogumalla P, Rao KPC, Palanisamy R, Smith A, Nagaraji S, Rao SA, Vishnoi L, Singh KK, Bhan SC (2021) Meghdoot—a mobile app to access location-specific weather-based agro-advisories pan India

Fabregas R, Kremer M, Schilbach F (2019) Realizing the potential of digital development: the case of agricultural advice. Science 366(6471):eaay3038. https://doi.org/10.1126/science.aay3038

Fairbairn M, Faxon HO, Montenegro de Wit M, Bronson K, Kish Z, Ruder S-L, Ezirigwe J, Abdella S, Oguamanam C, Schnurr MA (2025) Digital agriculture will perpetuate injustice unless led from the grassroots. Nat Food 6(4):312–315. https://doi.org/10.1038/s43016-025-01137-8

ICRISAT (2025) ICRISAT and plantix celebrate a decade of collaboration. https://pressroom.icrisat.org/icrisat-and-plantix-celebrate-a-decade-of-collaboration

McBratney AB, Mendonça Santos ML, Minasny B (2003) On digital soil mapping. Geoderma 117(1–2):3–52. https://doi.org/10.1016/S0016-7061(03)00223-4

Mehrabi Z, McDowell MJ, Ricciardi V, Levers C, Martinez JD, Mehrabi N, Wittman H, Ramankutty N, Jarvis A (2021) The global divide in data-driven farming. Nat Sustain 4(2):154–160. https://doi.org/10.1038/s41893-020-00631-0

Navarro RL (2004) Reaching the unreached. Media Asia 31(3):128–132. https://doi.org/10.1080/01296612.2004.11726746

Patil M, Anantha KH, Wani SP (2016) Digital technologies for agricultural extension. In: Harnessing dividends from drylands: innovative scaling up with soil nutrients. CABI Wallingford UK, pp 99–115

Patil M, Burry S, Gogumalla P, Rozario G, Katakamshetti N, Shaik J, Choudhari P (2024) Digital platform for soil health monitoring and management: requirements specifications. International Food Policy Research Institute. https://hdl.handle.net/10568/162941

Ramaraj AP, Rao KPC, Kumar GK, Ugalechumi K, Sujatha P, Rao SA, Dhulipala RK, Whitbread AM (2023) Delivering context specific, climate informed agro-advisories at scale: a case study of iSAT, an ICT linked platform piloted with rainfed groundnut farmers in a semi-arid environment. Clim Serv 31:100403. https://doi.org/10.1016/J.CLISER.2023.100403

Rao A, Dixit S (2019) Weather based crop advisories for climate resilience: crop management advisories through mobile phones yield positive results for groundnut farmers in Kurnool Andhra Pradesh. Ecologic:1–4

Shalander Kumar KPC, Rao KK, G, Mukund Patil, ML Jat (2024) Customization, parameterization, and scaling of the iSAT: an ICT based agro advisory platform for location-specific informed decision-making. International Food Policy Research Institute. https://hdl.handle.net/10568/162942

Snapp S, Chamberlin J, Winowiecki L, Amede T, Aynekulu E, Gameda S, Herrick JE, Lal R, Marenya P, Nagarajan L, Stewart Z, Vågen T (2025) Realizing soil health for food security in Africa. Nat Sustain 8(1):3–5. https://doi.org/10.1038/s41893-024-01482-9

Chapter 12
Climate Resilient Technologies for Sustainable Food Security: Experiment and Simulation Analysis

Ahmad Faisal, Arshdeep Singh, Debasruti Bhattacharya, Poonam Biswal, Sanghamitra Swain, and Dillip Kumar Swain

Abstract Achieving food security amidst climate change requires sustainable agricultural practices that address environmental impact while improving soil and human health. Organic food production plays a pivotal role, providing higher nutrient and antioxidant content for human diet and improving soil health compared to conventional farming. Reducing reliance on excessive chemical fertilizers and pesticides is critical for mitigating health risks and preserving soil health, besides climate change mitigations. The climate-smart agriculture offers solutions such as crop rotation, resilient crop/variety selection, crop diversification, advanced water management techniques, and site-specific nutrients management are designed to enhance productivity with limited use of resources. Additionally, integrated farming systems aim to achieve zero-emission agriculture, promoting resource efficiency and resilience. On the other hand, technologies like drone surveillance and satellite-based NDVI calculations support efficient monitoring and targeted interventions, optimizing land use and other resources. The innovative approaches like hydroponics and vertical farming present sustainable options for meeting the quality food demand of urban population with minimal environmental impact. Through the integration of production systems with technologies, the agricultural sector can address production shortages, product quality, and soil health issues, creating a pathway to sustainable food security for environmental and human wellbeing.

Keywords Climate-smart agriculture · Food security · Organic food · Sustainable practices · Zero-emission farming

---

A. Faisal · A. Singh · D. Bhattacharya · P. Biswal · D. K. Swain (✉)
Agricultural and Food Engineering Department, Indian Institute of Technology Kharagpur, Kharagpur, India
e-mail: swain@agfe.iitkgp.ac.in

S. Swain
Rajendra Mishra School of Engineering Entrepreneurship, Indian Institute of Technology Kharagpur, Kharagpur, West Bengal, India

D. Kumar Swain et al. (eds.), *Climate-Smart Technologies for Sustainable Agricultural Production*, https://doi.org/10.1007/978-3-032-25666-9_12

12.1 Introduction

12.1.1 Background

Climate change is one of the most significant global challenges threatening food security and agricultural sustainability. Increasing temperatures, changing precipitation patterns, extreme weather events, and frequent droughts affect crop productivity, soil health, water availability, and ecosystem stability. Subtropical and tropical regions, including large parts of India, are especially vulnerable due to dependence on rainfed agriculture and limited resource access. The Green Revolution markedly increased global food production through extensive use of chemical fertilizers and pesticides. However, this reliance on chemical inputs has resulted in environmental degradation, human health risk, and diminishing yield gains due to soil deterioration and pollution. Excessive fertilizer and pesticide use contribute to greenhouse gas emissions such as nitrous oxide, groundwater contamination, eutrophication of water bodies, and persistence of toxic residues in food chains. These combined pressures challenge the capacity of agriculture to meet the food demands of a rapidly growing population under changing climatic conditions, necessitating urgent strategies for adaptation and sustainability.

12.1.2 Importance of Sustainable Agricultural Practices

Sustainable agricultural practices offer a pathway to address the intertwined challenges of climate change, food security, and environmental health. They emphasize rational and balanced use of inputs, conservation of soil and water resources, biodiversity preservation, and enhanced resilience of agroecosystems. Practices such as organic farming, Integrated Nutrient Management and Integrated Pest Management (INM/IPM), conservation agriculture, crop diversification, and precision farming improve nutrient-use efficiency, reduce environmental pollution, and safeguard human health. Climate-Smart Agriculture (CSA) frameworks integrate productivity enhancement with climate adaptation and mitigation, focusing on increasing yields sustainably, building resilience to climate variability, and reducing greenhouse gas emissions. Emerging technologies including drone surveillance, satellite-based vegetation indices (NDVI), remote sensing, and artificial intelligence aid in efficient resource management and timely decision-making. Additionally, innovative urban farming models like hydroponics and vertical farming offer solutions for sustainable food production in space-limited environments. Together, these sustainable approaches realign agriculture towards ecosystem health, food safety, and long-term viability in a warming world.

12.1.3 *Objectives and Scope*

This chapter aims to comprehensively analyze climate adaptive technologies and sustainable agricultural practices with potential to secure food production under climate change scenarios. It integrates experimental findings and simulation studies to evaluate challenges from excessive chemical inputs, and viable solutions including organic inputs, precision nutrient management, and integrated pest control. The chapter further explores CSA practices such as crop rotation, diversification, resilient crop varieties, conservation agriculture, and advanced water management to enhance productivity and resilience. Technological innovations like remote sensing, drone-based monitoring, ICT decision support systems, and Artificial Intelligence (AI)/Machine Learning (ML) applications in crop management are highlighted as transformative tools. The role of organic farming in improving nutritional outcomes and environmental sustainability is examined, alongside integrated and zero-emission farming systems that optimize resource use and mitigate emissions. Finally, the chapter presents emerging urban farming models like hydroponics and vertical farming as scalable options for sustainable intensification in densely populated areas. The chapter is directed towards researchers, policymakers, extension agents, and agribusiness stakeholders to guide the development and adoption of climate-resilient food systems that are environmentally responsible and health-conscious.

12.2 Chemical Input Management: Challenges and Solutions

Global crop production increased immensely after the Green Revolution, where chemical fertilisers and pesticides played a crucial role in boosting yields and improving food security (Sadaf et al. 2024). However, the intensive and excessive use of these chemical inputs has seriously threatened soil health, water quality, ecosystems, and human well-being. It is estimated that only 10–40% of applied fertilizers and pesticides are absorbed by crops, while the majority is lost to the environment, causing contamination, nutrient imbalances and pollution (Baweja et al. 2020). Excessive use of nitrogen fertilisers contributes to greenhouse gas emissions and ozone depletion, while pesticide runoff and leaching cause eutrophication of water bodies and groundwater contamination (Mitra et al. 2022). Despite the increased use of annual chemical fertilizer and pesticides, which reached approximately 188 million metric tons and 4.4 million tons, respectively by the end of year 2022, the crop yields have grown only marginally, reflecting diminishing returns from chemical inputs (Brief 2022; Vos et al. 2025).

Addressing these challenges requires a shift toward more sustainable input management strategies. Scientists and policymakers are now promoting practices such as organic inputs, integrated nutrient and pest management, site-specific nutrient application, and precision farming to minimise chemical dependence while maximising productivity. Biotechnological innovations, including nanotechnology and Genetically Modified (GM) crops, are also being explored for their potential to enhance efficiency and reduce the use of chemicals. Thus, effective chemical input management lies not in the elimination of fertilisers and pesticides, but in their rational and balanced use, guided by strategies that protect ecosystems, safeguard human health, and ensure sustainable agricultural growth.

12.2.1 Risks of Excessive Chemical Use

The Green Revolution significantly increased food production through the widespread use of chemical fertilisers and pesticides. While these inputs remain necessary for ensuring food security, their excessive and indiscriminate application has created a serious threat to the environment and human health. One of the most serious threats of excessive chemical use is the persistence of toxic pesticide residues in soil and food chains, leading to chronic health effects such as neurological disorders, cancers, endocrine disruption, and reproductive toxicity (Zhou et al. 2025). Chemical inputs, particularly nitrogen-based compounds, contribute to the emission of nitrous oxide (N_2O), a greenhouse gas 320 times more potent than CO_2 and play a crucial role in ozone layer depletion (Rizwan et al. 2024). Similarly, phosphorus and nitrogen runoff from fields accelerate eutrophication of water bodies, causing algal blooms, fish mortality, and the collapse of aquatic ecosystems (Tiwari and Pal 2022). The excessive nitrogen and phosphorus application has led to nitrate pollution of groundwater beyond the World Health Organisation (WHO) permissible limit of 50 mg/l, raising concerns about methemoglobinemia in infants (Fewtrell 2004; Nieder et al. 2018). Additionally, long-term exposure of pesticide has also been linked to cancers, congenital disorders, and compromised immune responses, posing a major public health risk (Upadhayay et al. 2020). The risk of excessive use of chemical fertilisers and pesticides on health and the environment is listed in Table 12.1.

12.2.2 Strategies for Reducing Chemical Input in Agriculture

The overuse of fertilisers and pesticides has raised critical concerns about soil degradation, water pollution, biodiversity loss, and human health hazards. To address these challenges, multiple sustainable strategies are being promoted that balance productivity with environmental safety. Organic farming and biofertilizers reduce dependency on synthetic inputs by enhancing soil fertility through natural processes. Biocontrol agents and IPM offer eco-friendly alternatives to chemical pesticides, while conservation agriculture and intercropping improve resource use efficiency and ecological balance. Other strategies, such as precision farming, site-specific nutrient management, and nanotechnology, enable efficient nutrient delivery and thus reduce losses and pollution. Additionally, recent biotechnological intervention also

Table 12.1 Adverse impacts of chemical fertilisers and pesticides on health and environment

Type of input	Examples	Human health risks	Environmental risks	References
Nitrogen fertilizers	Urea, ammonium nitrate, ammonium sulfate	• Methemoglobinemia (blue baby syndrome) • Carcinogenic nitrosamines from N transformation • Respiratory irritation	• Nitrate leaching leading to groundwater contamination • N_2O emission: potent greenhouse gas (GHG) • Soil acidification • Eutrophication of rivers and lakes	Majumdar (2003), Sarwar (2018), Nieder and Benbi (2022)
Phosphorus fertilizers	Diammonium phosphate, triple super phosphate	Heavy metal accumulation (Cd, Pb, As) causes kidney, bone, and neurological disorders	• Runoff causes eutrophication • Soil heavy metal contamination • long-term toxicity to plants and soil microbes	Gupta et al. (2014), Chen et al. (2020)
Potassium fertilizers	Muriate of potash, potassium sulfate	Reduces vitamin C and carotene in vegetables	Alters soil nutrient balance and microbial activity	Perkins-Veazie and Roberts (2002), Khan et al. (2018)
Sulfur and aluminium-based fertilisers	Ammonium sulfate, Al-salts	• Birth defects, asthma, Alzheimer's disease • Developmental and cognitive delays	• Acidification of soil and water bodies • Mobilization of toxic metals like Al^{3+}, Fe^{2+}, Mn^{2+}	Khan et al. (2018), Powlson and Dawson et al. (2022)
Pesticides—Organochlorines	Dichlorodiphenyltrichloroethane (DDT), hexachlorocyclohexane (HCH), dieldrin	Carcinogenic, endocrine disruption • Neurological disorders, Parkinson's disease	• Persistent in soil and water • Bioaccumulation in fish, birds, humans • Long half-life and remains in the food chain	Kaushik and Kaushik (2007), Jayaraj et al. (2016)

(continued)

Table 12.1 (continued)

Type of input	Examples	Human health risks	Environmental risks	References
Pesticides—Organophosphates	Chlorpyrifos, malathion, monocrotophos	Fetal malnutrition, respiratory paralysis, cancers • Obesity, diabetes, endocrine disruption	Soil microbial toxicity • Leaching and runoff to groundwater • Toxic to fish and aquatic ecosystems	Androutsopoulos et al. (2013). Sidhu et al. (2019)
Pesticides—Herbicides	Glyphosate, butachlor, atrazine	Potential carcinogenic in nature	• Bioaccumulation in fish • Herbicide-resistant weeds • Reduced microbial diversity	Choudri et al. (2020), Zhang et al. (2025)
Fungicides	Mancozeb, captan, tebuconazole	Skin/eye irritation, possible carcinogenic effects	Inhibit N-fixation and denitrification • Reduce microbial biomass and enzymatic activity • Toxic to beneficial fungi	Thomas Cusick (2025)
Heavy metal contamination from fertilisers	Cd, Pb, As, Cr, Zn (from phosphate and mixed fertilisers)	• Kidney and bone damage • Alzheimer's disease	Long-term soil contamination • Toxicity to crops and soil microbes • Entry into the food chain	Rahman and Singh (2019)

provides opportunities to develop pest-resistant and nutrient-efficient crop varieties. Together, these strategies represent a shift towards sustainable agricultural systems that maintain yields while protecting natural resources and public health (Table 12.2).

12.2.3 Regulatory Framework for Safe Pesticide Use

The safe use of pesticides requires an extensive regulatory framework to protect human health, ecosystems, and agricultural sustainability. Globally, several conventions and directives have been established, such as the Stockholm Convention on Persistent Organic Pollutants (2001), which bans or restricts highly hazardous pesticides like DDT, and the Rotterdam Convention (1998), which promotes prior informed consent in the trade of restricted chemicals (Brunelle et al. 2024). In the European Union, the Sustainable Use of Pesticides Directive (2009/128/EC) sets restrictions on training, certification, integrated pest management, and pesticide reduction targets. However, in India, pesticide regulation is governed by the Insecticides Act (1968), which controls registration, manufacture, certification and use of pesticides. However, the Act has been criticised for being outdated and insufficient to address modern challenges. To address these gaps, the Pesticides Management Bill (2020) was proposed, aiming to tighten the registration process and enhance transparency in pesticide use. Despite these efforts, the regulatory interventions in India remain weak, and the chemical pesticide industry has grown rapidly, raising concerns about indiscriminate use and residues in food chains (Kumar et al. 2025a, b, c). Thus, while strong frameworks exist globally, effective implementation, monitoring, and farmer awareness are crucial to ensure pesticide use remains within safe limits, minimising health hazards and environmental contamination.

12.3 Climate-Smart Agricultural Practices

The CSA is a transformative approach developed by the Food and Agriculture Organization (FAO) to address the intertwined challenges of food security and climate change. It focuses on three core objectives: sustainably increasing agricultural productivity and income; enhancing resilience to climate variability and extreme events; and reducing greenhouse gas (GHG) emissions or increasing carbon sinks.

12.3.1 Principles and Pillars of CSA

CSA is guided by four overarching principles. First, integrated resource management promotes the efficient use of water, nutrients, land, and energy, reducing

Table 12.2 Strategies for reducing chemical inputs in agriculture

Solution/approach	Benefits	Past research
Balanced fertilisation and site-specific nutrient management (SSMN)	Improves nutrient-use efficiency, prevents nutrient losses, reduces leaching and runoff and increase productivity	Sinha et al. (2019)—Eastern gangetic plains: Nutrient balances under SSNM in rice–wheat and rice–maize systems were positive for N and P (< 50 kg ha^{-1}), but negative for K (up to—90 kg ha^{-1}), under zero or conventional tillage practices Singh et al. (2015)—Modipuram, Meerut: Productivity gain under SSNM over farmer's practice was 30.8% in rice–wheat, 41.8% in rice–mustard, 67.3% in rice–potato, 70.1% in rice–garlic, 42.1% in rice–chickpea, and 24.1% in rice–berseem Abdulrachman et al. (2002)—Tamil Nadu: 12–36%, 8–13% nearly 100% increase in N, P and K use efficiency, respectively in rice fields
Integrated nutrient management (INM)	Enhances soil fertility, reduces chemical dependence, improves soil organic matter and increase yield	Chander et al. (2023)—Karnal, Haryana: Cost of cultivation, gross return, and net return in conventional inorganic fertilization (CIF) + farmyard manure (FYM) were 13.7, 12.8 and 11.9% higher, respectively, compared to CIF alone Mahato and Kafle (2018)—Nepal: wheat yield increased 33.3% with *Azotobacter* + 120:80:80 kg NPK/ha + 10 t FYM/ha compared to 120:80:80 NPK/ha. Saini et al. (2004)—Udaipur, Rajasthan: Chickpea yield increased 66% with 10 t FYM ha − 1 + *A. brasilense* or *Rhizobium* + *B. megaterium* (PSB) + *G. fasciculatum* (VAM) conpared to 100% recommended dose of N and P through fertilizers
Integrated pest management (IPM)	Lowers pesticide load, reduces pest resistance, protects beneficial organisms	Meena et al. (2025)—Haryana: spray of aureofungin/ bavistin+2,4-D reduced Kinnow mandarin fruit drop by 69.07% in IPM compared to farmer practice Khokhar et al. (2024)—Uttar Pradesh, Uttarakhand, Haryana: IPM practices increased beneficial microbial population such as *Trichoderma harzianum* and *Pseudomonas flurorescens* and decreased level of pathogen *Fusarium verticillioides* compared to non IPM practices. Dhawan et al. (2009)—Bathinda, Punjab: Pesticide load in cotton with proper implementation of IPM reduced 3–4 times reduction in pesticide load.

(continued)

Table 12.2 (continued)

Solution/approach	Benefits	Past research
Conservation agriculture	Conserves soil moisture, improves soil structure and biological health, reduces nutrient losses	Hossen et al. (2018)—Ghazipur and Kushtia, Bangladesh: Strip tillage significantly reduced irrigation input compared to bed tillage (12.7% in Ghazipur and 45.9% in Kushtia and conventional tillage (11.8% in Ghazipur and 52.8% in Kushtia). Strip tillage reduced fuel consumption by 50–70% compared to bed tillage and conventional tillage Lal (2015)—N-use efficiency increased by 30% while irrigation decreased by 15–20%; Jat et al. (2019)—Haryana: Conservation agriculture based zero-till direct seeded rice followed by zero tilled maize and permanent raised beds with residue practices reduced the total system water use by 151–341 mm and 212–423 mm, and resulted enhanced system water productivity by 27–57% and 39–68%, respectively compared to conventional tillage
Precision farming	Minimises wastage, reduces input costs, increases efficiency and yield	Gao and Bian (2017)—China: fertiliser use reduced 25–30% in wheat–maize system Hefty (2020)—USA: Corn yields increased by an average of 6% with VRS technologies Bora et al. (2012)—USA: Guided tractor application saved expenses by 2.4% in the seed; 2.2% in fertilizer, and 10.4% in tractor fuel
Use of nano-fertilisers and pesticides	Increases nutrient-use efficiency, lowers application frequency, and reduces pollution	Rizwan et al. (2019)—Pakistan: SiO_2 in Rice increased growth and yield Rossi et al. (2019)—USA: ZnO in coffee plant increased photosynthetic efficiency Srivastava et al. (2014)—India: FeS_2 in Chickpea, carrot, sesame, Brassica Enhanced germination, stress tolerance
Agroecological practices	Enhances ecological balance, reduces pest incidence, and improves resilience to climate change	Tripathi et al. (2021)—Karnal, Haryana: Maize + cowpea–wheat system recorded the highest productivity and profitability (land equivalent return = 1.71, sustainable value index = 0.96, benefit:cost = 1.94) with a 43.6% increase in net return and 25% N saving (37.5 kg ha^{-1}) among other wheat intercropped system Singh et al. (2009)—Intercropping maize with legumes can reduce chemical fertiliser application by up to 25–30%, while improving yield and farm income
Policy and awareness programs	Promotes the adoption of sustainable practices, ensures safe food and environment	Eco farmer certification in Japan: Farmers must reduce chemical input usage by 50% per hectare to receive a subsidy (MAFF 2022) Andhra Pradesh Community Managed Natural Farming (APCNF), India (2016): The initiative promotes natural inputs like cow dung, cow urine, and bio-stimulants instead of synthetic fertilisers and pesticides (UNEP 2018)

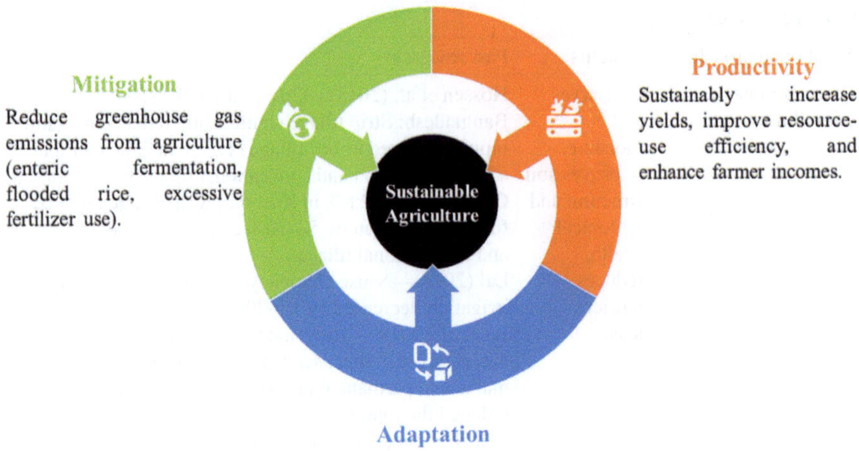

Fig. 12.1 Pillars of CSA

waste and environmental degradation. Second, climate risk assessment and vulnerability profiling enable location-specific interventions suited to diverse agro-climatic zones and socio-economic contexts. Third, participatory approaches actively involve farmers, researchers, and extension agents in co-developing adaptive solutions. Fourth, agroecological intensification harnesses natural processes—such as biological nitrogen fixation, pollination, and natural pest control—to enhance productivity while conserving resources.

CSA is built on three interconnected pillars: productivity, adaptation, and mitigation (Fig. 12.1). The productivity pillar focuses on sustainably increasing yields, improving input efficiency, and raising farmer incomes. Adaptation strengthens the ability of farming systems to cope with droughts, floods, heat stress, and erratic monsoon patterns. Mitigation aims to reduce greenhouse gas emissions from agriculture, addressing sources such as enteric fermentation, flooded rice systems, and excessive fertilizer application. Together, these pillars foster agricultural systems that are resilient, low-emission, and sustainable in the face of climate change.

12.3.2 Crop Rotation and Diversification

Crop rotation and diversification are foundational strategies in sustainable and CSA, enhancing soil health, nutrient cycling, and farm profitability while reducing vulnerability to pests, diseases, and climate shocks. Crop rotation involves growing different crops sequentially on the same field to disrupt pest cycles and balance nutrient use, whereas diversification introduces multiple crops spatially (intercropping) or temporally (sequential cropping) to spread risk and optimize resource utilization. Together, these practices strengthen ecological resilience and stabilize farm incomes under variable climatic conditions.

In subtropical India, the dominance of rice–wheat systems have led to soil fatigue, declining groundwater, and heightened climate risks (Khedwal et al. 2023). Diversified alternatives like rice–lentil and rice–mungbean systems in West Bengal and Odisha have improved soil fertility through biological nitrogen fixation and allowed for triple cropping (Kumar et al. 2025a, b, c). Similarly, rice–pulse rotations reduce weed pressure, break disease cycles, and enhance soil organic matter. In rainfed uplands of Jharkhand and Chhattisgarh, short-duration pulses like chickpea and urad are intensifying rice-fallow systems, increasing cropping intensity and income (Pande et al. 2012; Tiwari and Chaturvedi 2021). In peri-urban eastern India, women-led vegetable cultivation (okra, brinjal, spinach) after cereals has improved household nutrition and income. These diversified rotations also promote higher soil microbial activity, better nutrient cycling, and reduced GHG emissions by lowering chemical fertilizer demand. Despite challenges like poor seed access and market linkages, government schemes such as National Food Security Mission (NFSM) and Paramparagat Krishi Vikas Yojana (PKVY) are promoting rotation and diversification.

12.3.3 Climate-Resilient Crop Varieties

Climate-resilient crop varieties are a cornerstone of CSA and a critical adaptation strategy in the face of increasing climate variability. These varieties are bred or selected to withstand abiotic stresses such as drought, heat, submergence, salinity, and cold, as well as emerging biotic stresses including pests and diseases aggravated by changing climatic conditions. In India, where over 60% of agriculture is rainfed and climate risks are high (Sharma et al. 2009), the development and dissemination of such varieties have become a national priority. The development of climate-resilient varieties involves traditional breeding, marker-assisted selection, and modern genomic approaches. These varieties aim to stabilize yields under stress rather than maximize output under ideal conditions. Key traits include early maturity, deeper rooting, osmotic adjustment, and better canopy regulation—enabling crops to escape or tolerate climatic extremes. Such varieties are especially vital in smallholder systems, where farmers often lack irrigation, insurance, or resources to re-sow.

India has made notable progress in promoting climate-resilient crop varieties through programs like National Innovations in Climate Resilient Agriculture (NICRA), the All India Coordinated Research Project (AICRP) and the Krishi Vigyan Kendra (KVK) network. By 2024, Indian Council of Agricultural Research (ICAR) had released over 200 stress-tolerant varieties across cereals, pulses, oilseeds, and vegetables. Institutional mechanisms such as seed hubs, community seed banks, and public–private partnerships have played a key role in ensuring timely access to quality seeds. For example, during the delayed 2022 monsoon, Farmer Producer Organisation (FPO)-led seed hubs in Jharkhand distributed drought-tolerant rice and pulses to over 50 villages, reducing crop losses (ICAR 2024). Moreover, digital platform like mKisan provide real-time seed availability and advisory services (Nagpal and Mittal 2021). However, adoption is often constrained by low awareness, supply bottlenecks, and farmer hesitation to switch from traditional varieties. Bridging these gaps requires participatory varietal trials, strengthened extension services, and collaboration among breeders, agri-departments, and startups to scale adoption effectively.

12.3.4 Conservation Agriculture

Conservation Agriculture (CA) is a climate-smart approach built on three core principles: minimal soil disturbance (e.g., zero tillage), permanent soil cover (using residues or mulches), and diverse crop rotations. Together, these practices aim to improve soil health, enhance water and nutrient efficiency, reduce greenhouse gas emissions, and build long-term farm resilience. CA is especially relevant in subtropical regions like eastern India, where intensive tillage, residue burning, and monocropping have degraded soils and reduced input-use efficiency (Singh et al. 2025). One widely adopted practice is zero tillage in the rice–wheat system of eastern Uttar Pradesh and Bihar, which has improved yields while reducing irrigation and input costs (Laxmi et al. 2007). Similarly, the use of rice straw mulch in potato cultivation has demonstrated multiple benefits in West Bengal—enhancing soil moisture retention, boosting tuber yields, and reducing weed pressure (Sarangi et al. 2021; Biswal et al. 2022). Beyond these, CA interventions such as residue retention in pulses, diversified rice–legume rotations, and conservation tillage have also shown positive impacts on soil structure, biological activity, and carbon sequestration (Hazra et al. 2019; Kumar et al. 2019). Despite some constraints—such as limited access to machinery, competing uses of crop residues, and variable farmer awareness—CA adoption is growing through support from schemes like the Sub-Mission on Agricultural Mechanization (SMAM) and FPO-led Custom Hiring Centres, which promote resource-sharing and mechanization at the grassroots level.

12.3.5 Advanced Water Management

Advanced water management is central to CSA, aiming to boost water-use efficiency, match water supply with crop needs, and build climate resilience through precision and conservation technologies. Eastern India's agro-climatic zones are especially vulnerable due to monsoon dependence, smallholder dominance, and limited irrigation coverage. In these areas, expanding micro-irrigation and adopting smart scheduling tools are key to sustainable intensification and climate adaptation.

12.3.5.1 Micro-Irrigation Systems

Micro-irrigation is a water-efficient irrigation technology that delivers water directly to the root zone of plants in small, frequent interval. It includes systems such as drip and sprinkler irrigation, both designed to minimize water losses from evaporation, runoff, and deep percolation—issues commonly associated with traditional flood or furrow irrigation methods (Ray and Majumder 2024). These systems allow for precise, uniform, and timely application of water, making them beneficial in undulating terrains and water-scarce regions.

Drip irrigation, which uses a network of pipes, emitters, and valves to supply water directly to the root zone, is highly effective in high-value and water-sensitive crops such as vegetables, fruits, and tuber crops like potato (Trout and Kincaid 2007; Qamar et al. 2025). Studies in Odisha and West Bengal have shown that drip-irrigated potato can reduce water usage by 30–40% while increasing yields by 20–25% compared to furrow irrigation (Jata et al. 2013; Biswal et al. 2022). Biswal et al. (2025a, b, c) demonstrated that integrating drip irrigation with rice straw mulch in the rice–potato system significantly improved soil moisture retention, reduced irrigation frequency, and enhanced economic returns—highlighting the synergistic benefits of combining water-saving and soil-conserving practices. In resource-poor and topographically constrained areas like Jharkhand and Chhattisgarh, low-cost, gravity-fed drip systems have been introduced through government schemes and Non-Governmental Organizations (NGOs). The Pradhan Mantri Krishi Sinchayee Yojana (PMKSY) has been instrumental in expanding micro-irrigation coverage by offering subsidies, training, and institutional support.

Sprinkler irrigation, including mini- and micro-sprinklers, disperses water in fine droplets over a wide area, making it suitable for field crops such as mustard, pulses, and maize (Tanwar 2003; Chauhdary et al. 2023). Research by ICAR-RCER, Patna, found that mini-sprinklers improved water distribution uniformity, boosted seedling emergence, and reduced labor and energy costs in mustard and lentil cultivation. Together, drip and sprinkler systems under the micro-irrigation umbrella offer scalable, climate-resilient solutions for improving water productivity in Indian agriculture.

12.3.5.2 Smart Irrigation Scheduling

Smart irrigation scheduling is a data-driven approach that involves determining the optimal timing and amount of irrigation by integrating information on soil moisture status, weather conditions, crop stage, and evapotranspiration demand. Several approaches are currently being used or tested in India:

(a) **Soil Moisture-Based Scheduling**

Soil moisture status is one of the most direct and accurate indicators of irrigation need. Instruments like Tensiometers, Time Domain Reflectometry (TDR) sensors, and capacitance probes are increasingly used in both research and field settings.

These sensors provide real-time data on soil water availability at different depths, allowing farmers to irrigate only when moisture levels fall below a crop-specific threshold. Studies show that soil moisture-guided drip irrigation can reduce water use by 40–60% and increase water productivity, especially in shallow-rooted crops (Singh et al. 2010; Narayanamoorthy 2022).

(b) **Weather-Based Forecasting**

Short- to medium-range weather forecasts (3–7 days), particularly rainfall probability and temperature trends, are valuable for planning irrigation in advance. By aligning irrigation schedules with forecasted rainfall, farmers can avoid unnecessary watering, thereby reducing water wastage, pumping costs, and the risk of waterlogging stress. In India, agro-advisory services provided by the India Meteorological Department (IMD), ICAR institutes, and the KVK network routinely deliver location-specific irrigation guidance tailored to crop stage and soil type.

(c) **Evapotranspiration-Based Scheduling**

Evapotranspiration (ET), which combines atmospheric demand and crop transpiration, provides a comprehensive measure of water loss. ET-based scheduling enables precise irrigation by aligning water application with actual crop water requirements. It can be implemented using on-ground weather data (temperature, humidity, wind speed, solar radiation) or derived from remote sensing platforms such as the FAO's WaPOR, MODIS-ET, or OpenET. This approach has proven particularly useful in canal-command areas and peri-urban farming zones, where it helps optimize irrigation efficiency and avoid both under- and over-watering.

(d) **Automated and IoT-Enabled Systems**

More advanced setups use Internet of Things (IoT) devices to automate irrigation based on sensor data. These systems integrate real-time soil moisture, weather forecasts, and crop growth models to automatically trigger irrigation through mobile apps or embedded controllers. Studies have demonstrated that automated drip scheduling reduced water use by up to 40% with increased water productivity, especially in high-value vegetables (Kapoor et al. 2022; Kumar et al. 2024). Despite their benefits, advanced water management technologies face adoption challenges in India due to high initial costs, energy limitations in remote areas, low technical know-how, and fragmented landholdings. Overcoming these barriers requires targeted subsidies, shared access through custom hiring centers, farmer training on digital tools and scheduling, and public–private partnerships to scale IoT-based irrigation advisory services.

12.3.6 Soil Health and Carbon Sequestration

Soil health refers to the capacity of soil to function as a living system that sustains plants, animals, and humans. A healthy soil is marked by balanced nutrients, good structure, high organic matter, strong water-holding capacity, and active microbial communities. These properties underpin nutrient cycling, water infiltration, root growth, and resilience to climatic stress. Within the framework of CSA, soil health is central to both adaptation—by supporting higher productivity and stress tolerance—and mitigation—by storing carbon and reducing greenhouse gas emissions.

Carbon sequestration in soils refers to the process by which atmospheric CO_2 is absorbed by plants and stored in the soil as stable organic matter. This process not only enhances soil fertility and structure but also reduces net greenhouse gas concentrations, offering a win–win for climate and agriculture. The buildup of soil organic carbon (SOC) depends on residue return, root biomass, and organic amendments.

In subtropical India, intensive rice-based systems, frequent tillage, and residue burning have led to declining SOC levels (Yadav et al. 2019). Conservation agriculture (zero tillage and residue retention), organic inputs (like FYM, compost, and green manures), and crop diversification (especially with legumes) are scientifically validated strategies for restoring SOC. Long-term trials in eastern India have shown significant increases in SOC and soil quality under such practices (Jat et al. 2014; Samal et al. 2017). Thus, managing soils as a living carbon sink rather than just a medium for plant growth is essential for advancing CSA goals in the region.

12.3.7 Simulation-Based Climate Adaptation

Simulation-based adaptation offers a predictive, data-driven approach to design and evaluate crop and resource management strategies under both current and future climate scenarios. Crop models are especially useful in resource-constrained, climate-vulnerable regions, where targeted adaptation is essential. By integrating soil, weather, crop, and management data, simulation enables virtual experimentation across diverse agroecological zones. It supports testing of "what-if" scenarios, optimizing practices like sowing dates, variety selection, nutrients management, irrigation scheduling, etc. Crucially, simulation bridges field research and policy planning, offering evidence-based insights for improving productivity, resilience, and sustainability.

12.3.7.1 Use of Crop Simulation Models

Crop simulation models are mathematical tools that represent the physiological and phenological processes driving crop growth, development, and yield in relation to environmental and management factors. Widely applied models such as DSSAT (Decision Support System for Agrotechnology Transfer), APSIM (Agricultural Production Systems sIMulator), AquaCrop, and InfoCrop have been employed globally to assess the impacts of climate variability and adaptation strategies.

In subtropical regions, particularly in rice-based systems, DSSAT modules such as CERES-Rice and SUBSTOR-Potato have been extensively used to simulate crop performance under different climatic, soil, and management scenarios (Satapathy et al. 2014; Srivastava et al. 2019; Biswal et al. 2024). These simulations help evaluate the performance of stress-tolerant varieties, quantify yield losses due to delayed sowing or heat stress, and assess the efficiency of resource-conserving practices. For example, in subtropical India, DSSAT has been applied to determine optimal sowing windows for rice and potato under RCP 4.5 and RCP 8.5 scenarios (Rajwade et al. 2018; Biswal et al. (2025a, b, c). Results show that adjusting sowing by 10–15 days could significantly mitigate yield losses under projected mid-century heat stress. In addition, AquaCrop, a model developed by FAO, with its lower data requirement and focus on water productivity, has helped extension agencies issue better water-saving advisories to farmers under limited irrigation. The InfoCrop model, developed by ICAR-IARI, has been utilized to simulate nitrous oxide emissions from rice fields and the carbon sequestration potential of CA-based systems, making it highly relevant for integrating adaptation and mitigation perspectives.

12.3.7.2 Model Validation and Scenario Analysis

Model validation is essential for ensuring the reliability of crop simulation outputs, particularly when informing climate adaptation strategies and agricultural policy. Validation involves calibrating the model with site-specific parameters and comparing simulated outputs—such as phenological stages, biomass accumulation, and yield—with observed field data across different seasons and agro-climatic conditions. Once validated, these models can be used to assess future climate impacts through scenario analysis. Using climate projections from Global Climate Models (GCMs) and Representative Concentration Pathways (RCPs), simulations help estimate medium- and long-term effects of rising temperatures, altered rainfall, and CO_2 changes on crop performance. Increasing night temperatures could reduce rice grain filling, but adaptation strategies such as early sowing, heat-tolerant varieties, and improved irrigation efficiency can mitigate yield losses (Jagadish et al. 2015; Khan et al. 2019).

Scenario modeling also aids in designing resilient cropping systems. For instance, simulations have shown that shifting from cereal–cereal to cereal–legume rotations enhanced productivity and reduced yield variability (Ghosh et al. 2020). Similarly, in Jharkhand, modeling demonstrated that practices like line sowing and short-duration legume intercropping improve resilience and profitability under future climates (Bastia et al. 2025). However, the broader adoption of simulation tools faces barriers such as limited access to high-quality weather and soil data, lack of trained personnel, and computational constraints. To address this, efforts are underway to develop user-friendly, region-specific interfaces and conduct capacity-building programs for extension agents, FPOs, and agri-tech start-ups.

12.4 Technological Innovations for Climate Adaptation

Technological advancements are becoming central to agricultural resilience in the face of climate change. In regions where erratic rainfall, heatwaves, and declining resource bases threaten smallholder livelihoods, the integration of digital and geospatial tools is revolutionizing how crops are monitored, managed, and protected. These technologies enable real-time decision-making, precision input use, and predictive planning—crucial for building climate-smart agriculture systems. This section discusses four key innovation domains crucial for building climate-smart systems.

12.4.1 Remote Sensing and Drone-Based Monitoring

Remote sensing through satellites and unmanned aerial vehicles (UAVs) provides critical data on vegetation health, soil moisture, land use, and climatic anomalies. In the context of climate adaptation, it supports early detection of crop stress, mapping of floods and droughts, and surveillance of pest and disease outbreaks. High-resolution imagery is particularly useful for tracking canopy growth and phenology, thereby guiding decisions on sowing dates, irrigation scheduling, and nutrient application. While satellites such as Sentinel and Landsat offer regional-scale monitoring, drones provide micro-level insights into field variability. UAVs can quickly identify chlorophyll deficiencies, weed patches, or localized waterlogging, allowing interventions well before ground scouting would detect them. In India, drone-based applications are being piloted through initiatives like the Digital Agriculture Mission and state-level public–private partnerships, with KVKs and FPOs facilitating services for field mapping, yield estimation, and precision input management. Integration of remote sensing with GPS and GIS further enhances decision support by linking spatial data with farm-level advisories. Drone-generated elevation and crop maps have also been used for land leveling and irrigation planning, improving water-use efficiency and crop uniformity. With costs declining and accessibility improving, remote sensing—especially UAV-based monitoring—holds strong potential for scaling climate-resilient agriculture across diverse farming systems.

12.4.2 Satellite-Based NDVI Applications

The NDVI is a widely used satellite-based indicator that measures vegetation health by analyzing the difference between near-infrared (strongly reflected by vegetation) and red light (absorbed by vegetation). NDVI values range from -1 to $+1$, with higher values indicating denser and healthier green cover. As a non-invasive and scalable tool, NDVI plays a vital role in monitoring crop growth, biomass, and stress across time and space—making it integral to climate-resilient agriculture (Biswal et al. (2025a, b, c).

NDVI enables early detection of drought stress, nutrient deficiencies, pest outbreaks, and poor crop vigor. By providing near real-time information, it allows for more efficient water, fertilizer, and pesticide use, while reducing input waste. Satellite platforms such as MODIS, Landsat, and Sentinel-2 generate NDVI products at different resolutions, supporting both regional assessments and field-level monitoring. At the farm scale, NDVI-based maps reveal intra-field variability, enabling site-specific management and precision agriculture. With the growing availability of drones and high-resolution imagery, users are beginning to benefit from plot-level NDVI monitoring. However, cert mlmmn limitations persist, including cloud cover during the monsoon, low resolution in some sensors, and challenges in interpreting signals from mixed cropping or multilayered canopies. Nonetheless, when integrated with weather forecasts, crop models, and ground-based observations, NDVI enhances adaptive planning, yield prediction, and sustainable intensification strategies.

12.4.3 ICT Tools and Decision Support Systems

Information and Communication Technology (ICT) tools and Decision Support Systems (DSS) are central to climate-resilient agriculture as they enable timely, data-driven decisions at both farm and policy levels. These digital platforms integrate weather forecasts, soil information, crop models, and market trends to support adaptive planning, resource use efficiency, and risk mitigation in real time. In rainfed and resource-constrained regions, where decisions are closely tied to erratic rainfall and limited extension services, ICT-based advisories help bridge critical information gaps. Services such as mKisan, e-Sagu, and Kisan Suvidha deliver localized weather updates, crop advisories, and input recommendations through SMS or mobile apps, enabling farmers to adjust sowing, irrigation, and pest management decisions. Beyond advisories, DSS tools such as DSSAT, InfoCrop, and AquaCrop simulate crop growth under varying climatic and management scenarios. These models predict yield, water demand, and nutrient efficiency, and are increasingly embedded into regional planning frameworks to guide state-level agricultural strategies. GIS-based DSS platforms also support risk mapping and contingency planning. For example, the National Agricultural Drought Assessment and Monitoring System (NADAMS) and Crop Weather Outlook (CWO) provide drought and flood assessments, aiding crop insurance schemes and disaster preparedness. Despite their promise, challenges remain in ensuring data accuracy, user-friendly design, farmer training, and last-mile connectivity. However, with rapid smartphone penetration, initiatives like Digital India and PM-KISAN, and integration with satellite data and AI/ML tools, ICT and DSS platforms are set to play a transformative role in advancing climate-smart agriculture in India and beyond.

12.4.4 Artificial Intelligence and Machine Learning in Crop Management

AI and ML are transforming agriculture by enabling predictive, data-driven, and adaptive crop management strategies. These technologies process vast amounts of structured and unstructured data ranging from satellite imagery and weather forecasts to soil profiles and market trends to support timely, accurate decisions that enhance productivity and resilience. In climate-smart agriculture, AI/ML tools are particularly valuable for modeling complex interactions among climate, crop growth, and management practices. ML algorithms such as Random Forest, Support Vector Machines, and Neural Networks have been effectively used to forecast crop yields, detect diseases and pest infestations, and classify crop types from remote sensing data. These applications help farmers and extension agents anticipate risks, optimize input use, and adopt timely interventions.

AI-powered decision support systems integrate real-time weather data, soil moisture levels, and crop stage information to recommend precise irrigation schedules, fertilizer doses, or pest control measures. Startups and research institutions in India are leveraging these tools to develop advisory platforms, some of which provide voice-based or mobile app interfaces in regional languages to reach smallholders. For instance, AI models trained on historical climate and yield data are being used to identify suitable sowing windows, especially under delayed or erratic monsoon conditions (Bowden et al. 2023). In regions, where climatic risks and resource constraints are high, AI/ML applications offer a scalable solution for improving adaptive capacity. However, challenges persist regarding data availability, algorithm transparency, and accessibility for resource-poor farmers. Continued investment in open datasets, interdisciplinary research, and user-friendly interfaces will be essential to scale these technologies, ensuring AI and ML become integral to climate-resilient agriculture.

12.5 Organic Farming as Climate-Resilient Agriculture

12.5.1 Definition and Principles of Organic Agriculture

Organic agriculture is a sustainable production system that reduces the use of synthetic fertilisers, pesticides, genetically modified organisms (GMOs), and other harmful chemical inputs instead, it relies on ecological processes, biodiversity, and natural cycles to maintain soil fertility, control pests, and enhance farm productivity (Reganold and Wachter 2016). According to the International Federation of Organic Agriculture Movement (IFOAM 2008), "Organic Agriculture is a production system that sustains the health of soils, ecosystems, and people. It relies on ecological processes, biodiversity and cycles adapted to local conditions, rather than the use of inputs with adverse effects. Organic Agriculture combines tradition, innovation, and science to benefit the shared environment and promote fair relationships

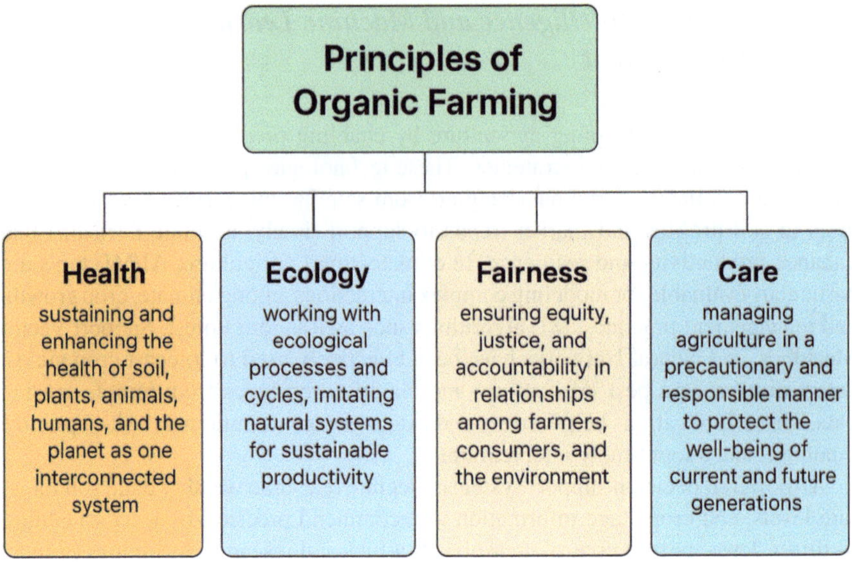

Fig. 12.2 Principles of organic farming

and good quality of life for all involved." The four principles of organic farming emphasise health, ecology, fairness, and care (Fig. 12.2).

12.5.2 Comparative Nutritional Benefits: Organic Vs Conventional Food

The nutritional value of food is influenced by multiple factors, including genotype, environmental conditions, nutrient management practices, and year of production (Hornick 1992). The question of whether organic nutrient management provides superior nutritional benefits in fruits and vegetables over chemical nutrient management continues to be a matter of scientific debate for a long time. Several studies have demonstrated that organic crops contain higher concentrations of bioactive compounds, particularly antioxidants and polyphenols, which are important for human health (Murmu et al. 2013; Liao et al. 2019; Hallmann et al. 2019). Barański et al. (2014) in a meta-analysis concluded that organic food is richer in several antioxidant compounds, including phenolic acids, flavonoids and anthocyanins. These compounds are associated with reduced risks of chronic diseases such as cardiovascular disorders and cancer. A systematic review of beneficial nutrients presents in organic crops revealed an increase of vitamin C (+28.7%), iron (+21.1%), magnesium (+29.3%) and phosphorus (+13.6%) (Rembiałkowska 2007). Using 75% of the recommended chemical fertilizer in combination with *Trichoderma*-enriched bio-organic fertilizer significantly increased soluble sugars (+24%) and vitamin C

(+57%), compared to 100% chemical fertilization (Ye et al. 2020). However, in a later meta-analysis, Barański et al. (2017) cautioned that it may still be too early to draw definitive conclusions on nutritional superiority of organic products due to variability across crops and environments.

In addition to nutrient content, organic foods are consistently reported to contain lower levels of contaminants. Johansson et al. (2014) highlighted that pesticide residues are significantly reduced in organically grown produce, leading to an increased food safety advantage. Similarly, Suciu et al. (2022) noted that cadmium concentrations, often associated with phosphorus fertilisers, tend to be lower in organic crops. Consumer perception aligns strongly with these findings, as the absence of synthetic pesticide residues is widely considered a major health benefit of organic products (Koch et al. 2017). This perception has contributed to the higher willingness of consumers to pay for organic product, despite its higher price and lower yield compared to conventionally grown food (Roy et al. 2023). Overall, while evidence indicates that organic foods tend to be richer in certain antioxidants and nutrients and contain lower levels of harmful residues, variability across studies suggests that further long-term, crop-specific research is needed. Nevertheless, from both a nutritional and food safety perspective, organic products present advantages over their conventional counterparts.

12.5.3 Contribution to Human Health

Organic farming contributes to human health both directly, through improving the nutritional quality of food, and indirectly, through reduced exposure to harmful chemicals and the promotion of environmental sustainability. Several studies have demonstrated that organic produce tends to have higher concentrations of bioactive compounds, including antioxidants, phenolic acids, and certain vitamins, which may contribute to the prevention of chronic diseases such as cancer and cardiovascular disorders (Olsson et al. 2006; Kazimierczak et al. 2014). These enhanced nutritional properties are largely attributed to the stress conditions under which crops are grown in organic systems, which stimulate secondary metabolite synthesis (Jan et al. 2021).

From a toxicological perspective, organic foods reduce consumer exposure to pesticide residues, heavy metals, and antibiotic-resistant bacteria (Mie et al. 2017). This is particularly relevant for vulnerable groups such as children, pregnant women, and immunocompromised individuals, who are more sensitive to chemical contaminants in food. For example, long-term dietary exposure to organic diets with reduced pesticide residues has been associated with decreased rate of developmental disorders, metabolic disorders, and cardiovascular diseases (Bradbury et al. 2014; Baudry et al. 2015; Baudry et al. 2018). Direct pesticide exposure of conventional farmers has been also reported to have higher risk of metabolic disorder, cardiovascular disorder, hepatic and nerve dysfunction compared to organic farmers (Huang et al. 2016; Kongtip et al. 2018).

Beyond nutritional and toxicological benefits, organic farming indirectly promotes human health by supporting ecosystem services that maintain clean air, water, and biodiversity (Maitra et al. 2020). Healthier soils and reduced environmental pollution lower the risk of contamination in food and drinking water, thereby decreasing the incidence of associated illnesses. Moreover, organic livestock management, which prohibits the use of antibiotics and synthetic growth promoters, reduces the risk of antimicrobial resistance, which is again a growing global public health concern (Brown et al. 2017). Despite these benefits, debates remain regarding the extent of health advantages linked to organic diets. Some meta-analyses report higher nutrient levels and reduced contaminants, while others highlight inconsistencies and attribute differences more to dietary diversity than farming systems (Barański et al. 2017; Rahman et al. 2024). Nonetheless, literature increasingly supports the role that organic agriculture plays an important part for an integrated approach to healthier food systems, contributing not only to individual well-being but also to long-term public health sustainability.

12.5.4 Contribution to Environment

Organic agriculture plays an important role in addressing environmental challenges associated with modern intensive farming. By prioritising ecological processes, biodiversity, and natural nutrient cycles over synthetic inputs, it reduces greenhouse gas (GHG) emissions, enhances soil health, conserves biodiversity, and strengthens resilience to climate change. One of the most significant contributions of organic farming lies in carbon sequestration. Through practices such as compost application, incorporation of crop residues, cover cropping, and reduced tillage, organic systems increase soil organic matter and carbon storage (Krauss et al. 2022). This not only mitigates climate change but also improves soil structure, water retention, and long-term fertility. Sharma et al. (2021) estimates that global adoption of organic practices could sequester up to 32% of current man-made GHG emissions. By avoiding synthetic nitrogen fertilisers in organic farming, nitrous oxide (N_2O) release is also minimised (Sahil et al. 2023). Moreover, recycling of plant and animal wastes into compost and biogas further lowers methane emissions while simultaneously providing renewable energy and organic manure for crops (Phillip et al. 2024). Moreover, organic agriculture also contributes to reducing environmental pollution. The elimination of synthetic pesticides and fertilisers decreases the risk of water contamination, nitrate leaching, and chemical residues in food and soil. The energy efficiency has also been reported to be increased in organic farming compared to conventional farming (Babu et al. 2022). Table 12.3 describes the range of environmental indicators and contribution of organic farming for each indicator accompanied with Indian and global examples.

Table 12.3 Environmental impacts of organic farming

Environmental indicator	Contribution of organic farming	Examples
Soil organic matter	Maintains and enhances soil organic carbon through compost, green manures, and crop residues	Krauss et al. (2022)—Switzerland: Reduced tillage in comparison with conventional tillage increased soil organic carbon (SOC) stocks by 20.8% in the surface layer (0–10/15 cm) and by 14.4% in the deepest soil layer (70–100 cm) Prasad et al. (2016)—India; Carbon sequestration rate from 2000 to 2010 was 62 kg/ha/year for conventional tillage and 186 kg/ha/year for minimum tillage
Nitrogen leaching	Reduces nitrate leaching by optimising nutrient cycling and avoiding synthetic fertilisers	Kanthle et al. (2016)—India; In low-carbon soil, nitrate leaching can be mitigated by biochar at a rate of 20 g/kg soil Kostensalo et al. (2024)—Finland: A reduction of 20% nitrogen leached per unit area was reported in organic farming compared to conventional farming in sand and clay soil in a 16-year field experiment
Nitrous oxide emissions	Mitigates emissions through balanced organic nutrient management and improved soil health	Sahil et al. (2023)—India; Nitrous oxide (N_2O) emissions from Indian agricultural soils increased nearly fourfold, from 73.7 Gg in 1966 to 279.3 Gg in 2017. The study reported that synthetic fertilisers contributed 57% of total emissions, followed by animal manure (15%) and crop residues (9%) Skinner et al. (2019)—Switzerland: 40.2% lower N_2O emissions per unit area of organic system compared to conventional system in a long-term organic farming study Westphal et al. (2018)—Manitoba: N_2O emissions (g N_2O-N ha^{-1}) with organic wheat were reported as half of conventional management
Ammonia emissions	Decreases emissions due to the restricted use of synthetic nitrogen inputs	Thirunagari et al. (2024)—India; From the spatially resolved emission inventory, it was observed that Uttar Pradesh has the highest NH_3 emission (706 Gg/yr), both from animal and crop farming and urea fertiliser contributing about 93.85% of the emission Sahoo et al. (2024)—India; The total NH_3 emission in India was reported to be 10.54 Tg in 2022, where synthetic fertiliser application and livestock excreta were the dominant contributors with 4.9 Tg/yr and 3.2 Tg/yr, respectively
Phosphorus losses	Minimises P losses via organic amendments and controlled nutrient release	Das et al. (2017)—India; A 7-year long-term study reported a 9.17% increase in phosphorus uptake in raised crops treated with farmyard manure + rock phosphate compared to chemical fertilisers

(continued)

Table 12.3 (continued)

Environmental indicator	Contribution of organic farming	Examples
Land use and biodiversity	Promotes diversified cropping and rotations, improving land-use efficiency over time and increases above- and below-ground biodiversity through habitat diversification and the absence of synthetic pesticides	Yang et al. (2021)—Europe; An increase of herbivores (82%), bacterivores (206%), and omnivores (135%) was reported in organic systems compared to conventional systems. Additionally, the herbivore footprints and composite in the organic system increased by 451% and 195% respectively
Energy use	Lowers energy consumption by avoiding synthetic agrochemicals and relying on renewable inputs	Babu et al. (2022)—India; A higher energy productivity (0.22 kg M J^{-1}) was reported in an integrated system of diversified cropping + livestock + poultry + piggery compared to a monocropping system (0.03 kg M J^{-1})
Greenhouse gas emissions	Reduces overall GHG emissions and enhances carbon sequestration in soils	Babu et al. (2023a, b)—India; The study demonstrated that zero tillage reduced greenhouse gas intensity (GHGI) to 0.15 kg CO_2-eq per kg production, which was 13.3 and 6.3% lower than conventional tillage and reduced tillage, respectively. Among the cropping systems, the maize–rajmash–buckwheat system recorded the lowest GHGI (0.14 kg CO_2-eq per kg production) compared to the monocropping system
Eutrophication potential	Limits nutrient runoff, thereby lowering risks of water eutrophication	Balasuriya et al. (2022)—Thailand; Marine eutrophication potential (209.6 million kg N eq) and freshwater eutrophication potential (2.0 million kg PO_4 eq) were reported in conventional nutrient management, whereas integrated nutrient management strategies such as 4R practices (Right Source, Right Rate, Right Time, and Right Place for nutrient management) combined with nitrification inhibitors reduced eutrophication potential and associated ecosystem damage
Acidification potential	Avoids soil acidification by reducing reliance on synthetic fertilisers	Zhang et al. (2020)—China; Soil acidification is induced by excessive nitrogen fertilisation, and it significantly enhanced SOC accrual and stability in Chinese croplands

12.5.5 Integrated and Zero-Emission Farming Systems

The Green Revolution substantially enhanced global grain and livestock productivity, yet ensuring food and nutritional security for a projected 9.1 billion people by 2050 remains a major challenge (Hamdan et al. 2022). In India, where nearly 80%

of farm households belong to small and marginal categories, shrinking landholdings and limited investment capacity further constrain agricultural growth (Chand et al. 2011). Under such conditions, sustainable intensification that enhances productivity while conserving natural resources is critical. Conventional reliance on high levels of chemical fertilisers and pesticides during the twentieth century substantially increased yields but at the cost of soil degradation, declining productivity, and environmental pollution (Tripathi et al. 2020). These unsustainable practices disproportionately affect smallholders, nearly 75% of whom face reduced farm viability in developing countries (Amede et al. 2023). As a result, farm profitability is declining, whereas rural poverty and distress migration are increasing (Nienkerke et al. 2023). To counter these challenges, agricultural systems must be redesigned to integrate ecological principles, minimise external input dependency, and improve economic resilience.

Integrated Farming Systems (IFS) provide a multidisciplinary framework to address these needs by linking crops, livestock, horticulture, aquaculture, and other enterprises within a single production unit (Singh and Dubey 2023). By recycling crop residues and by-products across components, IFS reduces nutrient losses, improves soil fertility, enhances productivity, and stabilises farm income. Functioning as a "system of systems," IFS ensures complementarities among enterprises, leading to higher resource-use efficiency, location-specific sustainability, and resilience against climatic and market shocks (Soni et al. 2014). Building on these principles, Zero-Emission Farming Systems (ZEFS) extend the concept further by aiming for complete nutrient and energy recycling within the farm boundary (Kumar et al. 2025a, b, c). Through innovations in renewable energy use, manure and waste management, carbon sequestration, and low-emission nutrient management strategies, ZEFS minimises greenhouse gas emissions while maintaining or improving productivity. Together, IFS and ZEFS represent ecologically intensive farming models that can strengthen smallholder livelihoods, reduce environmental externalities, and contribute significantly to long-term food and nutritional security (Tran et al. 2020).

12.5.5.1 A Zero-Emission Integrated Small-Scale Cattle Farm in the Mekong Delta, Vietnam

A successful example of zero-emission farming has been demonstrated through the integration of cattle breeding with crop cultivation in rural communities of the Mekong Delta, Vietnam (Tran et al. 2020). The system was designed to achieve environmental sustainability while ensuring economic viability and livelihood security for smallholder families (Fig. 12.3). The model established a closed-loop cycle where waste streams were transformed into valuable resources. Cattle manure was utilised in a biogas digester to generate clean energy for household cooking, thereby reducing dependence on non-renewable fuels. The slurry and wastewater from the digester, once treated with biochar, met the quality standards for irrigation and were applied to crops, enhancing soil fertility and water-use efficiency. Solid residues were

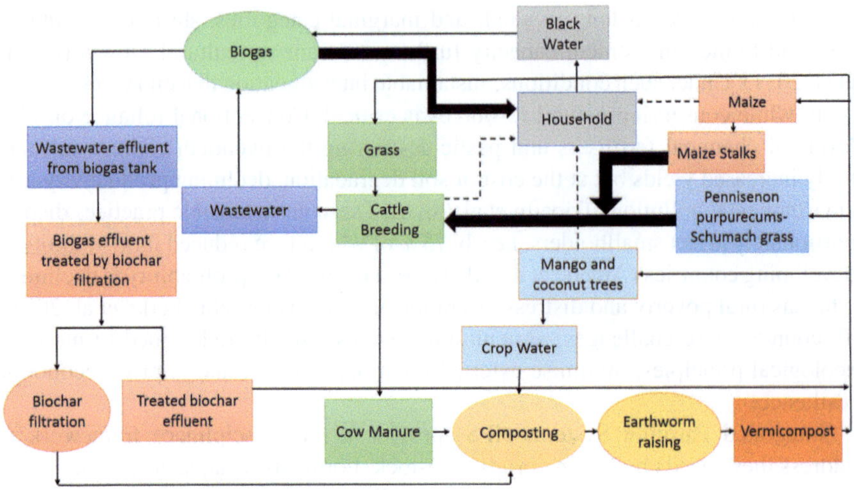

Fig. 12.3 Integrated zero emission farming model of Mekong Delta, Vietnam

processed into organic fertilisers, further improving crop yields while minimising external input requirements. In return, the crops cultivated provided sufficient feed for the herd of 30 cattle, completing the nutrient cycle. From this system, the farm family income increased by 41.55%, supported by higher agricultural productivity, renewable energy use, and reduced input costs. Environmental and sanitation conditions improved markedly, and greenhouse gas emissions were minimised through efficient nutrient and energy recycling. This case study demonstrates that zero-emission integrated farming is a practical and replicable model for rural development.

12.5.5.2 Integrated Farming System Models for Enhancing Economic and Environmental Sustainability of Smallholder Farms in Telangana, India

The study was conducted in the southern plateau and hill region of India with the objective of developing suitable IFS models capable of improving economic returns and energy efficiency with minimal GHG emissions (Karthik et al. 2024). The seven IFS models were M1: Rice – Groundnut; M2: Rice – Groundnut, Pigeonpea + Sweetcorn (1:3) – Bajra, Bt cotton + Greengram (1:2) – Maize; M3: Rice – Groundnut, Pigeonpea + Sweetcorn (1:3) – Bajra, Pigeonpea + Maize (1:3) – Sunhemp; Napier grass, Sheep (5 + 1); M4: Rice – Groundnut, Pigeonpea + Sweetcorn (1:3) – Bajra, Bt cotton + Greengram (1:2) – Maize, Pigeonpea + Maize (1:3) – Sunhemp, Poultry unit; M5: Guava, Hedge Lucerne, Napier grass, Bt cotton + Greengram (1:2) – Maize, Sheep (5 + 1); M6: Guava, Bt cotton + Greengram (1:2) – Maize, Rice – Groundnut, Poultry; M7: Rice – Groundnut, Pigeonpea + Sweetcorn (1:3) – Bajra, Pigeonpea + Maize (1:3) – Sunhemp; Napier grass,

Hedge lucerne, Poultry (100), Sheep (5 + 1). The performance of these models was compared with respect to economic efficiency, energy balance, GHG emissions, and employment generation. The results of the study indicated that models M7 and M3 substantially outperformed the other systems. Economic efficiency reached ₹342.3 day^{-1} in M7 and ₹263.7 day^{-1} in M3, compared with much lower values in M1 model. Output energy was highest in M7 (228,529 MJ), followed by M3 (183,231 MJ), while net energy gains were 258,184 MJ and 198,920 MJ, respectively. Both models also demonstrated a positive environmental effect, recording net negative GHG emissions of −2842 kg CO_2 eq. in M7 and − 2399 kg CO_2 eq. in M3. Furthermore, these two models also ensured sustained rural employment by generating 112.5 and 110.5 man-days per year in M7 and M3, respectively. Thus, from this study, M7 and M3 models contribute to the inclusion of many components such as fodder crops, livestock and horticultural crops, which enhanced the diversification and resource recycling. These results highlight the potential of well-designed IFS models to serve as climate-resilient and economically viable options for smallholder farmers in Telangana, India simultaneously improving profitability, reducing environmental footprints, and creating year-round employment opportunities.

12.6 Emerging Urban Farming Models

As projected by the UN, India will add 416 million new urban dwellers by 2050, highlighting the urgent need to supplement traditional agricultural models with innovative urban farming systems. These models offer a solution for creating, distributing, and marketing food within metropolitan boundaries, enhancing food security, reducing the urban carbon footprint, and reconnecting city dwellers with their food sources (Fei et al. 2025). Traditional agriculture is heavily dependent on long supply chains vulnerable to disruptions from climate events, transportation costs, and spoilage (Rahman et al. 2022). Inefficiencies in this chain lead to significant post-harvest losses, ranging from 4.65 to 15.88% in India, translating into substantial economic and nutritional deficits (CIPHET 2022). To counter these challenges, urban farming utilizes advanced technologies to convert underutilized urban spaces into highly productive agricultural hubs (Kabir et al. 2023a). This approach facilitates a resilient, efficient, and sustainable urban food system by bringing food production closer to consumers. Two of the most prominent models are hydroponics and vertical farming.

12.6.1 Hydroponics

Hydroponics is the technique of growing plants without soil, using nutrient-rich aqueous solutions in a controlled environment (Ragaveena et al. 2021). The method supplies essential nutrients directly to the plant roots, a critical advantage in urban areas where land is scarce and soil may be contaminated. By enabling year-round production independent of external weather conditions, hydroponics emerges as a promising solution to enhance urban food security.

12.6.1.1 Principles and Design

Hydroponics directly delivers a balanced nutrient solution to the roots to accelerate nutrient uptake and promote efficient plant growth. Inert media such as perlite, rockwool, or clay balls provide physical support without altering the nutrient solution's composition. Standard designs include the wick system, ebb and flow (EBBF), deep water culture (DWC), and the nutrient film technique (NFT), each employing a unique method to deliver oxygen and nutrients (Ranade and Chaudhari n.d.).For example, a study on lettuce grown in a DWC system demonstrated maturity in approximately 41 days, a 15-day reduction compared to protected soil-based cultivation (Majid et al. 2021).

12.6.1.2 Benefits in Resource-Limited Environments

Hydroponics offers substantial and multifaceted benefits in resource-constrained urban environments (Lakhiar et al. 2025). Water efficiency is its most critical advantage; studies confirm that hydroponic systems use 80–90% less water than traditional agriculture (Lakhiar et al. 2025). A study by (Majid et al. 2021) calculated the benefit-cost ratio for a hydroponic DWC system at over 2.1, indicating high profitability despite greater initial setup costs. Land productivity is also significantly enhanced. As a soilless system, hydroponics can be implemented in unconventional spaces and is a key enabling technology for vertical farming, where multiple layers of crops increase yield per square meter (Kabir et al. 2023b). The absence of soil also eliminates the risk of soil-borne diseases and weeds, reducing or eradicating the need for pesticides and herbicides (Visen et al. 2024). Finally, the optimized growing conditions in hydroponics promote faster growth cycles compared to conventional agriculture (Bouarroudj et al. 2025).

12.6.2 Vertical Farming

Vertical farming increases crop yield per unit of land by growing crops in stacked vertical layers. This is typically achieved by stacking horizontal hydroponic systems or cultivating crops on vertical surfaces (Beacham et al. 2019). The technique aims to maximize crop yield within a minimal footprint, presenting an ideal solution for dense urban areas (Van Gerrewey et al. 2022). Vertical farms can utilize abandoned buildings, repurposed skyscrapers, insulated warehouses, or modular shipping containers, turning the built environment into agricultural production hubs (D'Ostuni et al. 2024).

12.6.2.1 Urban Agriculture Potential

Integrating vertical farming into urban settlements bridges the geographic and logistical gap between food production and consumption (Thwaites et al. 2025). This proximity enables the delivery of fresh, nutrient-dense produce to local communities, restaurants, and grocery stores within hours of harvesting. The results are lower transportation costs, reduced post-harvest losses, and superior flavor and quality (Martin and Bustamante 2021). Moreover, these systems provide a reliable source of year-round produce, even in regions with erratic weather and high climatic variability (Lwasa and Dubbeling 2015).

12.6.2.2 Energy and Water Efficiency

While vertical farming models outperform traditional agriculture in productivity, they raise critical concerns regarding resource efficiency and long-term sustainability (Kaiser et al. 2024). Because these farms are primarily based on hydroponic or aeroponic systems, they are highly water-efficient, using up to 95% less water than conventional farming (Pomoni et al. 2023).

Despite this advantage, the sustainability of vertical farming is challenged by its high energy consumption. These systems require artificial lighting—typically from LEDs—and sophisticated climate control systems to manage temperature, humidity, and airflow, all of which demand continuous energy. This energy demand contributes significantly to operational costs and the farm's overall carbon footprint (Cai et al. 2025). However, continuous innovations are addressing these challenges. Key strategies include developing highly efficient, crop-specific LED lighting; co-locating farms with data centers to utilize waste heat; and integrating renewable energy sources like rooftop solar panels to power operations (Abdelhamid et al. 2025; Pereira and Gomes 2025; Shi et al. 2025). These advancements pave the way for vertical farms to become cornerstones of sustainable, resilient, and secure urban food systems.

12.7 Conclusion and Recommendations

The dual challenges of climate change and rising global populations demand a transformative shift toward sustainable and climate-resilient agricultural systems. While the Green Revolution enhanced food production through intensive use of fertilizers and pesticides, it also caused environmental degradation, soil health decline, and human health risks from chemical residues. Moreover, yields have stagnated despite higher input use, underscoring the unsustainability of current practices. The future of food security lies in approaches that integrate productivity with ecological stewardship and public health. Sustainable input management—such as site-specific nutrient application and integrated pest management—reduces environmental pollution while maintaining yields. The CSA practices, including crop diversification, resilient varieties, conservation agriculture, and precision water management, strengthen farm-level adaptation to climate variability and mitigate greenhouse gas emissions. Technological innovations, such as drone-based remote sensing, NDVI monitoring, and AI-powered decision support systems, enable data-driven management, helping farmers optimize inputs and minimize risks.

Organic farming has emerged as a key pillar of sustainability, offering improved nutritional value, reduced pesticide residues, and environmental benefits through enhanced soil health and carbon sequestration. Integrating organic principles with zero-emission and circular farming models can boost smallholder incomes, close nutrient loops, and lower ecological footprints. Urban agriculture innovations such as hydroponics and vertical farming provide fresh food in cities while reducing supply chain vulnerabilities, though their energy needs require further innovation. Accelerating this transition requires robust policies, investments, and partnerships. Policymakers must promote balanced input use, climate-resilient seeds, and wider CSA adoption through incentives and extension. Expanding digital and AI technologies, precision irrigation, and monitoring infrastructure will enable smarter decisions and higher efficiency. Support for organic and zero-emission farming through certification, markets, and training can improve livelihoods and ecosystem health. Urban agriculture also needs enabling policies and awareness campaigns. Overall, resilient and sustainable food systems demand ecological farming, digital innovations, strong institutions, and inclusive partnerships. Such integration offers the best pathway to safeguard food security, natural resources, and human health for current and future generations.

References

Abdelhamid MA, Abdelkader TK, Sayed HAA, Zhang Z, Zhao X, Atia MF (2025) Design and evaluation of a solar powered smart irrigation system for sustainable urban agriculture. Sci Rep 15(1):11761. https://doi.org/10.1038/s41598-025-94251-3

Abdulrachman, S., Gines, H. C., Nagarajan, R., Satawathananont, S., Son, T. T., Tan, P. S., & Wang, G. H. (2002). Variation in the performance of site-specific nutrient management among different environments with irrigated rice in Asia

Amede T, Konde AA, Muhinda JJ, Bigirwa G (2023) Sustainable farming in practice: building resilient and profitable smallholder agricultural systems in sub-Saharan Africa. Sustainability 15(7):5731

Androutsopoulos VP, Hernandez AF, Liesivuori J, Tsatsakis AM (2013) A mechanistic overview of health associated effects of low levels of organochlorine and organophosphorous pesticides. Toxicology 307:89–94

Babu S, Das A, Singh R, Mohapatra KP, Kumar S, Rathore SS, Yadav SK, Yadav P, Ansari MA, Panwar AS, Wani OA, Singh VK (2023a) Designing an energy efficient, economically feasible, and environmentally robust integrated farming system model for sustainable food production in the Indian Himalayas. Sustain Food Technology 1(1):126–142

Babu S, Singh R, Avasthe R, Rathore SS, Kumar S, Das A, Layek J, Sharma V, Wani OA, Singh VK (2023b) Conservation tillage and diversified cropping enhance system productivity and eco-efficiency and reduce greenhouse gas intensity in organic farming. Front Sustain Food Syst 7:1114617

Balasuriya BTG, Ghose A, Gheewala SH, Prapaspongsa T (2022) Assessment of eutrophication potential from fertiliser application in agricultural systems in Thailand. Sci Total Environ 833:154993

Barański M, Rempelos L, Iversen PO, Leifert C (2017) Effects of organic food consumption on human health; the jury is still out! Food Nutr Res 61(1):1287333

Barański M, Średnicka-Tober D, Volakakis N, Seal C, Sanderson R, Stewart GB, Benbrook C, Biavati B, Markellou E, Giotis C, Gromadzka-Ostrowska J, Leifert C (2014) Higher antioxidant and lower cadmium concentrations and lower incidence of pesticide residues in organically grown crops: a systematic literature review and meta-analyses. Br J Nutr 112(5):794–811

Bastia DK, Behera SK, Tripathy S, Nanda A, Rath S, Srivastava RK (2025) Building resilience: climate adaptation practices in Indian Rainfed farming. In: Mitigation and adaptation strategies against climate change in natural systems. Springer Nature Switzerland, Cham, pp 109–147

Baudry J, Lelong H, Adriouch S, Julia C, Allès B, Hercberg S, Touvier M, Lairon D, Galan P, Kesse-Guyot E (2018) Association between organic food consumption and metabolic syndrome: cross-sectional results from the NutriNet-santé study. Eur J Nutr 57(7):2477–2488

Baudry J, Méjean C, Péneau S, Galan P, Hercberg S, Lairon D, Kesse-Guyot E (2015) Health and dietary traits of organic food consumers: results from the NutriNet-santé study. Br J Nutr 114(12):2064–2073

Baweja P, Kumar S, Kumar G (2020) Fertilizers and pesticides: their impact on soil health and environment. In: Soil health. Springer International Publishing, Cham, pp 265–285

Beacham AM, Vickers LH, Monaghan JM (2019) Vertical farming: A summary of approaches to growing skywards. J Hortic Sci Biotechnol 94(3):277–283. https://doi.org/10.1080/14620316.2019.1574214

Biswal P, Faisal A, Swain DK, Bhowmick GD, Mohan G (2025a) NDVI is the best parameter for yield prediction at the peak vegetative stage of potato (*Solanum tuberosum* L.). Climate smart Agriculture 2(2):100053

Biswal P, Faisal A, Swain DK, Mohan G (2025b) Climate-resilient agricultural strategies for sustainable rice and potato production: a multi-scale review. Discover Plants 2(1):1–25

Biswal P, Faisal A, Tripathy S, Swain DK, Jha MK, Mohan G (2025c) Influence of drip irrigation and straw mulching on economic feasibility and soil fertility of rice-potato system in subtropical India. Irrig Sci 43(3):363–376

Biswal P, Swain DK, Jha MK (2022) Straw mulch with limited drip irrigation influenced soil microclimate in improving tuber yield and water productivity of potato in subtropical India. Soil Tillage Res 223:105484

Biswal P, Swain DK, Jha MK (2024) Understanding the significance of weather and soil microclimate for improvement of potato yield using SUBSTOR and statistical models. Potato Res 67(4):1339–1353

Bora GC, Nowatzki JF, Roberts DC (2012) Energy savings by adopting precision agriculture in rural USA. Energy Sustain Soc 2(1):22

Bouarroudj K, Babaa F, Touil A (2025) IoT-based monitoring and control for optimized plant growth in smart greenhouses using soil and hydroponic systems. Internet of Things 33:101710. https://doi.org/10.1016/j.iot.2025.101710

Bowden C, Foster T, Parkes B (2023) Identifying links between monsoon variability and rice production in India through machine learning. Sci Rep 13(1):2446

Bradbury KE, Balkwill A, Spencer EA, Roddam AW, Reeves GK, Green J, Key TJ, Beral V, Pirie K (2014) Organic food consumption and the incidence of cancer in a large prospective study of women in the United Kingdom. Br J Cancer 110(9):2321–2326

Brief FA (2022) Pesticides use, pesticides trade and pesticides indicators. FAO, Rome, Italy

Brown K, Uwiera RR, Kalmokoff ML, Brooks SP, Inglis GD (2017) Antimicrobial growth promoter use in livestock: a requirement to understand their modes of action to develop effective alternatives. Int J Antimicrob Agents 49(1):12–24

Brunelle T, Chakir R, Carpentier A, Dorin B, Goll D, Guilpart N, Maggi F, Makowski D, Nesme T, Roosen J, Tang FH (2024) Reducing chemical inputs in agriculture requires a system change. Commun Earth Environ 5(1):369

Cai W, Bu K, Zha L, Zhang J, Lai D, Bao H (2025) Energy consumption of plant factory with artificial light: challenges and opportunities. Renew Sust Energ Rev 210:115235. https://doi.org/10.1016/j.rser.2024.115235

Chand R, Prasanna PL, Singh A (2011) Farm size and productivity: understanding the strengths of smallholders and improving their livelihoods. Econ Polit Wkly:5–11

Chander S, Tripathi SC, Venkatesh K, Meena RP, Kumar N, Chhokar RS, Kumar N, Samota SR, Sharma D, Singh G (2023) Sustaining rice-wheat system through integrated nutrient management via FYM or *Sesbania aculeata* in India on long term basis. Int J Plant Prod 17(4):775–793

Chauhdary JN, Li H, Jiang Y, Pan X, Hussain Z, Javaid M, Rizwan M (2023) Advances in sprinkler irrigation: a review in the context of precision irrigation for crop production. Agronomy 14(1):47

Chen XX, Liu YM, Zhao QY, Cao WQ, Chen XP, Zou CQ (2020) Health risk assessment associated with heavy metal accumulation in wheat after long-term phosphorus fertilizer application. Environ Pollut 262:114348

Choudri BS, Charabi Y, Al-Nasiri N, Al-Awadhi T (2020) Pesticides and herbicides. Water Environ Res 92(10):1425–1432

CIPHET (2022) Assessment of quantitative harvest and post-harvest losses of major agricultural produce in India. ICAR-Central Institute of Post-Harvest Engineering and Technology

D'Ostuni M, Zou T, Sermarini A, Zaffi L (2024) Integrating greenhouses into buildings: a renewed paradigm for circular architecture and urban regeneration. Sustainability 16(23):10685. https://doi.org/10.3390/su162310685

Das A, Patel DP, Kumar M, Ramkrushna GI, Mukherjee A, Layek J, Ngachan SV, Buragohain J (2017) Impact of seven years of organic farming on soil and produce quality and crop yields in eastern Himalayas, India. Agric Ecosyst Environ 236:142–153

Dhawan, A. K., Singh, S., & Kumar, S. (2009). Integrated pest management (IPM) helps reduce pesticide load in cotton

Fei S, Wu R, Liu H, Yang F, Wang N (2025) Technological innovations in urban and peri-urban agriculture: pathways to sustainable food systems in metropolises. Horticulturae 11(2):212. https://doi.org/10.3390/horticulturae11020212

Fewtrell L (2004) Drinking-water nitrate, methemoglobinemia, and global burden of disease: a discussion. Environ Health Perspect 112(14):1371–1374

Gao W, Bian X (2017) Evaluation of the agronomic impacts on yield-scaled N_2O emission from wheat and maize fields in China. Sustainability 9(7):1201

Ghosh PK, Hazra KK, Venkatesh MS, Praharaj CS, Kumar N, Nath CP, Singh U, Singh SS (2020) Grain legume inclusion in cereal–cereal rotation increased base crop productivity in the long run. Exp Agric 56(1):142–158

Gupta DK, Chatterjee S, Datta S, Veer V, Walther C (2014) Role of phosphate fertilizers in heavy metal uptake and detoxification of toxic metals. Chemosphere 108:134–144

Hallmann E, Rozpara E, Słowianek M, Leszczyńska J (2019) The effect of organic and conventional farm management on the allergenic potency and bioactive compounds status of apricots (*Prunus armeniaca* L.). Food Chem 279:171–178

Hamdan MF, Mohd Noor SN, Abd-Aziz N, Pua TL, Tan BC (2022) Green revolution to gene revolution: technological advances in agriculture to feed the world. Plants 11(10):1297

Hazra KK, Nath CP, Ghosh PK, Swain DK (2019) Inclusion of legumes in rice–wheat cropping system for enhancing carbon sequestration. In: Carbon management in tropical and sub-tropical terrestrial systems. Springer, Singapore, pp 23–36

Hefty D (2020) Variable rate variety planting in wheat and soybeans. AgPhD

Hornick SB (1992) Factors affecting the nutritional quality of crops. Am J Altern Agric 7(1–2):63–68

Hossen, M. A., Hossain, M. M., Haque, M. E., & Bell, R. W. (2018). Transplanting into non-puddled soils with a small-scale mechanical transplanter reduced fuel, labour and irrigation water requirements for rice (*Oryza sativa* L.) establishment and increased yield. Field Crop Res, 225, 141-151.4

Huang X, Zhang C, Hu R, Li Y, Yin Y, Chen Z, Cai J, Cui F (2016) Association between occupational exposures to pesticides with heterogeneous chemical structures and farmer health in China. Sci Rep 6(1):25190

IFOAM (2008) Definition of organic agriculture. IFOAM. https://www.ifoam.bio/why-organic/organic-landmarks/definition-organic

Indian Council of Agricultural Research, (n.d.), Natural resource management (PDF). https://icar.org.in/sites/default/files/inline-files/nrm.pdf

Indian Council of Agricultural Research (2024) Promising climate resilient technologies for Jharkhand [PDF]. ICAR. https://www.icar-crida.res.in/assets_c/img/Books/New%20PCRT%20Books/Jharkhand%20PCRT%20book.pdf

Jagadish SVK, Murty MVR, Quick WP (2015) Rice responses to rising temperatures—Challenges, perspectives and future directions. Plant Cell Environ 38(9):1686–1698

Jan R, Asaf S, Numan M, Lubna, Kim KM (2021) Plant secondary metabolite biosynthesis and transcriptional regulation in response to biotic and abiotic stress conditions. Agronomy 11(5):968

Jat RK, Sapkota TB, Singh RG, Jat M, Kumar M, Gupta RK (2014) Seven years of conservation agriculture in a rice—Wheat rotation of eastern Gangetic Plains of South Asia: yield trends and economic profitability. Field Crop Res 164:199–210. https://doi.org/10.1016/j.fcr.2014.04.015

Jat RK, Singh RG, Kumar M, Jat ML, Parihar CM, Bijarniya D, Sutaliya JM, Jat MK, Parihar MD, Kakraliya SK, Gupta RK (2019) Ten years of conservation agriculture in a rice–maize rotation of eastern Gangetic Plains of India: yield trends, water productivity and economic profitability. Field Crop Res 232:1–10

Jata SK, Nedunchezhiyan M, Sahoo TR, Sahoo V (2013) Fertigation in high value tuber crops—A review. Odisha Review, May 2013. Information & Public Relations Department, Government of Odisha, pp 69–78. https://magazines.odisha.gov.in/Orissareview/2013/may/engpdf/69-78.pdf

Jayaraj R, Megha P, Sreedev P (2016) Organochlorine pesticides, their toxic effects on living organisms and their fate in the environment. Interdiscip Toxicol 9(3–4):90–100

Johansson E, Hussain A, Kuktaite R, Andersson SC, Olsson ME (2014) Contribution of organically grown crops to human health. Int J Environ Res Public Health 11(4):3870–3893

Kabir MSN, Reza MN, Chowdhury M, Ali M, Samsuzzaman, Ali MR, Lee KY, Chung S-O (2023a) Technological trends and engineering issues on vertical farms: A review. Horticulturae 9(11):1229. https://doi.org/10.3390/horticulturae9111229

Kabir MSN, Reza MN, Chowdhury M, Ali M, Samsuzzaman, Ali MR, Lee KY, Chung S-O (2023b) Technological trends and engineering issues on vertical farms: A review. Horticulturae 9(11):1229. https://doi.org/10.3390/horticulturae9111229

Kaiser E, Kusuma P, Vialet-Chabrand S, Folta K, Liu Y, Poorter H, Woning N, Shrestha S, Ciarreta A, van Brenk J, Karpe M, Ji Y, David S, Zepeda C, Zhu X-G, Huntenburg K, Verdonk JC, Woltering E, Gauthier PPG, Courbier S, Taylor G, Marcelis LFM (2024) Vertical farming goes dynamic: optimizing resource use efficiency, product quality, and energy costs. Fron Sci 2. https://doi.org/10.3389/fsci.2024.1411259

Kanthle AK, Lenka NK, Lenka S, Tedia K (2016) Biochar impact on nitrate leaching as influenced by native soil organic carbon in an Inceptisol of Central India. Soil Tillage Res 157:65–72

Kapoor R, Kumar A, Sandal SK, Sharma A, Raina R, Thakur KS (2022) Water and nutrient economy in vegetable crops through drip fertigation and mulching techniques: a review. J Plant Nutr 45(15):2389–2403

Karthik R, Ramana MV, Kumari CP, Prakash TR, Goverdhan M, Naik DS, Kumar NV, Chandra MS, Bhatt R, Elhindi KM, Mattar MA (2024) Suitable integrated farming system models in terms of energetics, greenhouse gas emissions and employment generation for the small and marginal farmers. Sustainability 16(23):10189

Kaushik, P., & Kaushik, G. (2007). An assessment of structure and toxicity correlation in organochlorine pesticides. J Hazard Mater, 143(1–2), 102–111

Kazimierczak R, Hallmann E, Lipowski J, Drela N, Kowalik A, Püssa T, Matt D, Luik A, Gozdowski D, Rembiałkowska E (2014) Beetroot (*Beta vulgaris* L.) and naturally fermented beetroot juices from organic and conventional production: metabolomics, antioxidant levels and anticancer activity. J Sci Food Agric 94(13):2618–2629

Khan MN, Mobin M, Abbas ZK, Alamri SA (2018) Fertilizers and their contaminants in soils, surface and groundwater. In: Encyclopedia of the anthropocene, vol 5, pp 225–240

Khan S, Anwar S, Ashraf MY, Khaliq B, Sun M, Hussain S, Gao ZQ, Noor H, Alam S (2019) Mechanisms and adaptation strategies to improve heat tolerance in rice. A review. Plants 8(11):508

Khedwal RS, Chaudhary A, Sindhu VK, Yadav DB, Kumar N, Chhokar RS, Poonia TM, Kumar Y, Dahiya S (2023) Challenges and technological interventions in rice–wheat system for resilient food–water–energy-environment nexus in north-western indo-gangetic plains: a review. Cereal Res Commun 51(4):785–807

Khokhar MK, Kumar R, Kumar A, Sehgal M, Singh SP, Meena PN, Singh N, Acharya LK, Birah A, Singh K, Bana RS, Choudhary M (2024) Impact of IPM practices on microbial population and disease development in transplanted and direct-seeded rice. Front Microbiol 15:1388754

Koch S, Epp A, Lohmann M, Böl GF (2017) Pesticide residues in food: attitudes, beliefs, and misconceptions among conventional and organic consumers. J Food Prot 80(12):2083–2089

Kongtip P, Nankongnab N, Tipayamongkholgul M, Bunngamchairat A, Yimsabai J, Pataitiemthong A, Woskie S (2018) A cross-sectional investigation of cardiovascular and metabolic biomarkers among conventional and organic farmers in Thailand. Int J Environ Res Public Health 15(11):2590

Kostensalo J, Lemola R, Salo T, Ukonmaanaho L, Turtola E, Saarinen M (2024) A site-specific prediction model for nitrogen leaching in conventional and organic farming. J Environ Manag 349:119388

Krauss M, Wiesmeier M, Don A, Cuperus F, Gattinger A, Gruber S, Haagsma WK, Peigné J, Palazzoli MC, Schulz F, Van Der Heijden MGA, Steffens M (2022) Reduced tillage in organic farming affects soil organic carbon stocks in temperate Europe. Soil Tillage Res 216:105262

Kumar J, Patel N, Sahoo PK, Sudhishri S, Yadav R, Vibhute SD, Singh AK (2024) Development of sensor and decision support system (DSS)-based automated irrigation system for enhancing water productivity of tomato crop: development of sensor and DSS-based irrigation system. J Soil Salinity Water Qual 16(2):307–316

Kumar N, Nath CP, Hazra KK, Das K, Venkatesh MS, Singh MK, Singh SS, Praharaj CS, Singh NP (2019) Impact of zero-till residue management and crop diversification with legumes on soil aggregation and carbon sequestration. Soil Tillage Res 189:158–167

Kumar R, Das A, Mondal S, Upadhyay PK, Bhatt BP, Mishra JS, Singh AK, Choudhary JS, Kumar S, Sundaram PK, Biswas AK, Chaudhari SK (2025a) Greening rice-fallow areas: integrating pulses and oilseeds for sustainable cropping in eastern India. Front Sustain Food Syst 9:1502759

Kumar S, Bhattacharya D, Banerjee M (2025b, April) Key factors affecting the adoption of biopesticide using logit and probit model: A study from Bihar, India. In: 2025 IEEE conference on technologies for sustainability (SusTech). IEEE, pp 1–7

Kumar S, Meena RS, Ghosh S (2025c) Agriculture toward net zero emissions: an overview. In: Agriculture toward net zero emissions, pp 1–9

Lakhiar IA, Yan H, Syed TN, Zhang C, Shaikh SA, Rakibuzzaman M, Vistro RB (2025) Soilless agricultural systems: opportunities, challenges, and applications for enhancing horticultural resilience to climate change and urbanization. Horticulturae 11(6):568. https://doi.org/10.3390/horticulturae11060568

Lal R (2015) Sequestering carbon and increasing productivity by conservation agriculture. J Soil Water Conserv 70(3):55A–62A

Laxmi, V., Erenstein, O., & Gupta, R. K. (2007). Impact of zero tillage in India's rice-wheat systems

Liao J, Xu Q, Xu H, Huang D (2019) Natural farming improves soil quality and alters microbial diversity in a cabbage field in Japan. Sustainability 11(11):3131

Lwasa S, Dubbeling M (2015) Urban agriculture and climate change. In: Cities and agriculture. Routledge

Mahato S, Kafle A (2018) Comparative study of Azotobacter with or without other fertilizers on growth and yield of wheat in Western hills of Nepal. Ann Agrarian Sci 16(3):250–256

Maitra S, Shankar T, Gaikwad DJ, Palai JB, Sagar L, Gaikwad DJ (2020) Organic agriculture, ecosystem services and sustainability: a review. Int J Mod Agric 9(4):370–378

Majid M, Khan JN, Ahmad Shah QM, Masoodi KZ, Afroza B, Parvaze S (2021) Evaluation of hydroponic systems for lettuce (*Lactuca sativa L.*, var. *Longifolia*) and comparison with protected soil-based cultivation. Agric Water Manag 245:106572. https://doi.org/10.1016/j.agwat.2020.106572

Majumdar D (2003) The blue baby syndrome: nitrate poisoning in humans. Resonance 8(10):20–30

Martin M, Bustamante MJ (2021) Growing-service systems: new business models for modular urban-vertical farming. Front Sustain Food Syst 5. https://doi.org/10.3389/fsufs.2021.787281

Meena PN, Raghavendra D, Singh S, Kumar N, Khokhar MK, Chander S, Lal MK, Tiwari RK, Kumar R (2025) Integrated pest management techniques in a Kinnow mandarin (*Citrus reticulata Blanco*) orchard with an emphasis on yield improvement. Heliyon 11(4):e42574

Mie A, Andersen HR, Gunnarsson S, Kahl J, Kesse-Guyot E, Rembiałkowska E, Quaglio G, Grandjean P (2017) Human health implications of organic food and organic agriculture: a comprehensive review. Environ Health 16(1):111

Ministry of Agriculture, Forestry and Fisheries (MAFF) (2022) Organic JAS. Ministry of Agriculture, Forestry and Fisheries (MAFF), Tokyo, Japan

Mitra B, Chowdhury AR, Dey P, Hazra KK, Sinha AK, Hossain A, Meena RS (2022) Use of agrochemicals in agriculture: alarming issues and solutions. In: Input use efficiency for food and environmental security. Springer, Singapore, pp 85–122

Murmu K, Ghosh BC, Swain DK (2013) Yield and quality of tomato grown under organic and conventional nutrient management. Arch Agron Soil Sci 59(10):1311–1321

Nagpal R, Mittal M (2021) Digital empowerment of Indian agriculture. In: Citizen Empowerment through Digital Transformation in Government, p 23

Narayanamoorthy A (2022) Economic impact of drip irrigation in India: an empirical analysis with farm level data. In: The irrigation future of India: development, resource and policy. Springer, Cham, pp 329–360

Nieder R, Benbi DK (2022) Reactive nitrogen compounds and their influence on human health: an overview. Rev Environ Health 37(2):229–246

Nieder R, Benbi DK, Reichl FX (2018) Reactive water-soluble forms of nitrogen and phosphorus and their impacts on environment and human health. In: Soil components and human health. Springer, Dordrecht, pp 223–255

Nienkerke IM, Thorat A, Patt A (2023) From distress migration to selective migration: transformative effects of agricultural development on seasonal migration. World Dev Perspect 29:100483

Olsson ME, Andersson CS, Oredsson S, Berglund RH, Gustavsson KE (2006) Antioxidant levels and inhibition of cancer cell proliferation in vitro by extracts from organically and conventionally cultivated strawberries. J Agric Food Chem 54(4):1248–1255

Pande, S., Sharma, M., & Ghosh, R. (2012). Role of pulses in sustaining agricultural productivity in the rainfed rice-fallow lands of India in changing climatic scenario

Pereira J, Gomes MG (2025) Lighting strategies in vertical urban farming for the enhancement of plant productivity and energy consumption. Appl Energy 377:124669. https://doi.org/10.1016/j.apenergy.2024.124669

Perkins-Veazie P, Roberts WARREN (2002) Can potassium application affect the mineral and antioxidant content of horticultural crops. In: Proceedings of the symposium on fertilizing crops for functional food, vol 2, p 1

Phillip A, Bhatt SM, Sharma N, Poudel B (2024) Enhanced biogas production using anaerobic co-digestion of animal waste and food waste: a review. J Sci Res 30:761–781

Pomoni DI, Koukou MK, Vrachopoulos MG, Vasiliadis L (2023) A review of hydroponics and conventional agriculture based on energy and water consumption, environmental impact, and land use. Energies 16(4):1690. https://doi.org/10.3390/en16041690

Powlson DS, Dawson CJ (2022) Use of ammonium sulphate as a Sulphur fertilizer: implications for ammonia volatilization. Soil Use Manag 38(1):622–634

Prasad JVNS, Rao CS, Srinivas K, Jyothi CN, Venkateswarlu B, Ramachandrappa BK, Dhanapal GN, Ravichandra K, Mishra PK (2016) Effect of ten years of reduced tillage and recycling of organic matter on crop yields, soil organic carbon and its fractions in Alfisols of semi arid tropics of southern India. Soil Tillage Res 156:131–139

Qamar R, Garcia y Garcia A, Shoaib M, Ahmad S (2025) Irrigation management in vegetable crops. In: Sustainable and innovative vegetable production in times of climate change. Springer, Singapore, pp 269–288

Ragaveena S, Shirly Edward A, Surendran U (2021) Smart controlled environment agriculture methods: a holistic review. Rev Environ Sci Biotechnol 20(4):887–913. https://doi.org/10.1007/s11157-021-09591-z

Rahman A, Baharlouei P, Koh EHY, Pirvu DG, Rehmani R, Arcos M, Puri S (2024) A comprehensive analysis of organic food: evaluating nutritional value and impact on human health. Foods 13(2):208

Rahman MM, Nguyen R, Lu L (2022) Multi-level impacts of climate change and supply disruption events on a potato supply chain: an agent-based modeling approach. Agric Syst 201:103469. https://doi.org/10.1016/j.agsy.2022.103469

Rahman Z, Singh VP (2019) The relative impact of toxic heavy metals (THMs) (arsenic (As), cadmium (Cd), chromium (Cr)(VI), mercury (Hg), and lead (Pb)) on the total environment: an overview. Environ Monit Assess 191(7):419

Rajwade YA, Swain DK, Tiwari KN (2018) Effect of irrigation method on adaptation capacity of rice to climate change in subtropical India. Int J Plant Prod 12(3):203–217

Ranade Y, Chaudhari N (n.d.) Integrating hydroponics into tropical agricultural systems: implications for food security. J Hortic Sci Biotechnol 0(0):1–8. https://doi.org/10.1080/14620316.2025.2505167

Ray S, Majumder S (2024) Water management in agriculture: innovations for efficient irrigation. Mod Agron:169–185

Reganold JP, Wachter JM (2016) Organic agriculture in the twenty-first century. Nat Plants 2(2):1–8

Rembiałkowska E (2007) Quality of plant products from organic agriculture. J Sci Food Agric 87(15):2757–2762

Rizwan M, Ali S, ur Rehman MZ, Malik S, Adrees M, Qayyum MF, Alamri SA, Alyemeni MN, Ahmad P (2019) Effect of foliar applications of silicon and titanium dioxide nanoparticles on growth, oxidative stress, and cadmium accumulation by rice (*Oryza sativa*). Acta Physiol Plant 41(3):35

Rizwan M, Tanveer H, Ali MH, Sanaullah M, Wakeel A (2024) Role of reactive nitrogen species in changing climate and future concerns of environmental sustainability. Environ Sci Pollut Res 31(39):51147–51163

Rossi L, Fedenia LN, Sharifan H, Ma X, Lombardini L (2019) Effects of foliar application of zinc sulfate and zinc nanoparticles in coffee (*Coffea arabica* L.) plants. Plant Physiol Biochem 135:160–166

Roy A, Ghosh A, Vashisht D (2023) The consumer perception and purchasing attitude towards organic food: a critical review. Nutr Food Sci 53(3):578–599

Sadaf T, Sehar S, Hassan S (2024) Agricultural revolutions and food security. In: Food security in the developing world, pp 95–122

Sahil FM, Narayanan M, Ilampooranan I (2023) Nitrous oxide emissions dynamics in Indian agricultural landscapes: revised emission rates and new insights. Agric Ecosyst Environ 356:108639

Sahoo P, Sahu SK, Mangaraj P, Mishra A, Beig G, Gunthe SS (2024) Reporting of gridded ammonia emission and assessment of hotspots across India: a comprehensive study of 24 anthropogenic sources. J Hazard Mater 479:135557

Saini VK, Bhandari SC, Tarafdar JC (2004) Comparison of crop yield, soil microbial C, N and P, N-fixation, nodulation and mycorrhizal infection in inoculated and non-inoculated sorghum and chickpea crops. Field Crop Res 89(1):39–47

Samal S, Rao K, Poonia S, Kumar R, Mishra J, Prakash V, Mondal S, Dwivedi S, Bhatt B, Naik SK, Choubey AK, Kumar V, Malik R, Mc Donald A (2017) Evaluation of long-term conservation agriculture and crop intensification in rice-wheat rotation of Indo-Gangetic Plains of South Asia: carbon dynamics and productivity. Eur J Agron 90:198–208. https://doi.org/10.1016/j.eja.2017.08.006

Sarangi SK, Maji B, Sharma PC, Digar S, Mahanta KK, Burman D, Mandal UK, Mandal S, Mainuddin M (2021) Potato (*Solanum tuberosum* L.) cultivation by zero tillage and paddy straw mulching in the saline soils of the Ganges Delta. Potato Res 64(2):277–305

Sarwar M (2018) Harmful effects of intensive nitrogen fertilizer usage to humans and managing of risks. Spec J Agric Sci 4(3–2018):9–15

Satapathy SS, Swain DK, Herath S (2014) Field experiments and simulation to evaluate rice cultivar adaptation to elevated carbon dioxide and temperature in sub-tropical India. Eur J Agron 54:21–33

Sharma BR, Rao K, Vittal K, Ramakrishna Y, Amarasinghe U (2009) Estimating the potential of rainfed agriculture in India: prospects for water productivity improvements. Agric Water Manag 97(1):23–30. https://doi.org/10.1016/j.agwat.2009.08.002

Sharma SK, Ravisankar N, Chander M, Jat G, Choudhary R, Kumar Fagodiya R (2021) Organic agriculture under changing climate scenario. Indian J Agron 66:S82–S94

Shi X, Shi C, Tablada A, Guan X, Cui M, Rong Y, Zhang Q, Xie X (2025) A review of research progress in vertical farming on façades: design, technology, and benefits. Sustainability 17(3):921. https://doi.org/10.3390/su17030921

Sidhu GK, Singh S, Kumar V, Dhanjal DS, Datta S, Singh J (2019) Toxicity, monitoring and biodegradation of organophosphate pesticides: a review. Crit Rev Environ Sci Technol 49(13):1135–1187

Singh, C. J., Thind, H. S., Manchanda, J. S., & Kansal, B. D. (2009). Effect of coal fly ash on crop yield and soil health under cotton-wheat cropping sequence

Singh PK, Dubey A (2023) Integrated farming system. In: Test book of resource conservation practices, pp 99–113

Singh R, Kundu DK, Bandyopadhyay KK (2010) Enhancing agricultural productivity through enhanced water use efficiency. J Agric Phys 10(2):1–15

Singh VK, Shukla AK, Dwivedi BS, Singh MP, Majumdar K, Kumar V, Mishra RP, Rani M, Singh SK (2015) Site-specific nutrient management under rice-based cropping systems in Indo-Gangetic Plains: yield, profit and apparent nutrient balance. Agric Res 4(4):365–377

Sinha AK, Ghosh A, Dhar T, Bhattacharya PM, Mitra B, Rakesh S, Paneru P, Shrestha SR, Manandhar S, Beura K, Dutta S, Menzies NW (2019) Trends in key soil parameters under conservation agriculture-based sustainable intensification farming practices in the Eastern Ganga Alluvial Plains. Soil Research 57(8):883–893

Skinner C, Gattinger A, Krauss M, Krause HM, Mayer J, Van Der Heijden MG, Mäder P (2019) The impact of long-term organic farming on soil-derived greenhouse gas emissions. Sci Rep 9(1):1702

Soni RP, Katoch M, Ladohia R (2014) Integrated farming systems-a review. IOSR J Agric Veterinary Sci 7(10):36–42

Srivastava G, Das A, Kusurkar TS, Roy M, Airan S, Sharma RK, Singh SK, Sarkar S, Das M (2014) Iron pyrite, a potential photovoltaic material, increases plant biomass upon seed pretreatment. Mater Express 4(1):23–31

Srivastava RK, Talla A, Swain DK, Panda RK (2019) Quantitative approaches in adaptation strategies to cope with increased temperatures following climate change in potato crop. Potato Res 62(2):175–191

Suciu NA, De Vivo R, Rizzati N, Capri E (2022) Cd content in phosphate fertilizer: which potential risk for the environment and human health? Curr Opin Environ Sci Health 30:100392

Tanwar BS (2003) Saline water management for irrigation. International Commission on irrigation and drainage, New Delhi, India, p 140

Thirunagari BK, Kancheti M, Kumar R, Kota SH (2024) Managing synthetic N-fertilizer emissions in India: insights from field surveys across 102 districts. J Environ Manag 366:121909

Thomas Cusick H III, H. (2025). Fungicides. Farm Toxicology: A Primer for Rural Healthcare Practitioners, 85–91

Thwaites HJ, Suh J, Coote GM, Nursey-Bray M, Cavagnaro TR, Salomon MJ (2025) Urban food forests: seeing the fruit for the trees—A systematic quantitative literature review and emerging research gaps. J Clean Prod 501:145358. https://doi.org/10.1016/j.jclepro.2025.145358

Tiwari AK, Chaturvedi SK (2021) Pulses revolution in India through rice-fallows management. In: Scaling-up solutions for farmers. Springer, Cham, pp 229–263

Tiwari AK, Pal DB (2022) Nutrients contamination and eutrophication in the river ecosystem. In: Ecological significance of river ecosystems. Elsevier, pp 203–216

Tran QB, Nguyen TPT, Le TN, Schnitzer H, Braunegg G, Le S, Hoang CT, Nguyen XC, Nguyen VH, Peng W, Kim SY, Van Le Q (2020) Integrated farming system producing zero emissions and sustainable livelihood for small-scale cattle farms: case study in the Mekong Delta, Vietnam. Environ Pollut 265:114853

Tripathi S, Srivastava P, Devi RS, Bhadouria R (2020) Influence of synthetic fertilizers and pesticides on soil health and soil microbiology. In: Agrochemicals detection, treatment and remediation. Butterworth-Heinemann, pp 25–54

Tripathi SC, Venkatesh K, Meena RP, Chander S, Singh GP (2021) Sustainable intensification of maize and wheat cropping system through pulse intercropping. Sci Rep 11(1):18805

Trout TJ, Kincaid DC (2007) On-farm system design and operation and land management. In: Irrigation of agricultural crops, vol 30, pp 133–179

UNEP-WCMC, I. U. C. N (2018). NGS (2018). Protected planet report, 70

Upadhayay J, Rana M, Juyal V, Bisht SS, Joshi R (2020) Impact of pesticide exposure and associated health effects. In: Pesticides in crop production: physiological and biochemical action, pp 69–88

Van Gerrewey T, Boon N, Geelen D (2022) Vertical farming: the only way is up? Agronomy 12(1):2. https://doi.org/10.3390/agronomy12010002

Visen SS, Thakur D, Atluri S (2024) Sustainable Management of Soilborne Diseases of horticultural crops. In: Thakur M, Belwal T (eds) Advances in postharvest and analytical technology of horticulture crops. Springer, pp 349–363. https://doi.org/10.1007/978-981-97-7247-6_19

Vos R, Glauber J, Hebebrand C, Rice B (2025) Global shocks to fertilizer markets: impacts on prices, demand and farm profitability. Food Policy 133:102790

Westphal M, Tenuta M, Entz MH (2018) Nitrous oxide emissions with organic crop production depends on fall soil moisture. Agric Ecosyst Environ 254:41–49

Yadav GS, Babu S, Das A, Bhowmik SN, Datta M, Singh R (2019) Soil carbon dynamics and productivity of rice–rice system under conservation tillage in submerged and unsubmerged ecologies of Eastern Indian Himalaya. Carbon Management 10(1):51–62

Yang B, Banerjee S, Herzog C, Ramírez AC, Dahlin P, van der Heijden MG (2021) Impact of land use type and organic farming on the abundance, diversity, community composition and functional properties of soil nematode communities in vegetable farming. Agric Ecosyst Environ 318:107488

Ye L, Zhao X, Bao E, Li J, Zou Z, Cao K (2020) Bio-organic fertilizer with reduced rates of chemical fertilization improves soil fertility and enhances tomato yield and quality. Sci Rep 10(1):177

Zhang C, Zhang W, Cao W, Guo X, Zeng D, Chen Z, Li X (2025) Unraveling the hidden pathways: bioaccumulation and metabolism of atrazine, acetochlor, and metolachlor in aquatic organisms. Environ Res 279:121852

Zhou W, Li M, Achal V (2025) A comprehensive review on environmental and human health impacts of chemical pesticide usage. Emerging Contaminants 11(1):100410

Chapter 13
From Farmland Abandonment to Rural Revitalization: Evidence from Toyama Prefecture and Other Comparative Case Studies in Japan

Mohan Geetha, Norio Horie, Farida Ayu, Kikuko Shoyama, Indrek Melts, and Naoya Wada

Abstract Agricultural land abandonment has emerged as a critical issue in rural areas of Japan, driven by an aging and declining population, labor shortages, and declining agricultural profitability. Toyama Prefecture, despite its historical strength in a rice-based agricultural system, has experienced a growing area of uncultivated agricultural land, with significant socio-economic and environmental consequences. This chapter examines the drivers, spatial patterns, and impacts of agricultural land abandonment in Toyama, with a focus on the role of local business organizations and community initiatives in land revitalization. Using an integrated methodology that combines geographic information system analysis, vegetation survey data, key informant interviews, and comparative case studies from Tajima in Hyogo Prefecture and Tsukuba in Ibaraki Prefecture, the study identifies abandonment hotspots and evaluates alternative land-use pathways. The findings demonstrate that local community-led cooperative farming, livestock-based land use, corporate land leasing for biomass production, and specialty crop cultivation can effectively transform abandoned agricultural land into productive and multifunctional landscapes. Comparative case studies highlight the importance of local adaptation, institutional support, and market

M. Geetha · N. Horie · F. Ayu · N. Wada
Graduate School of Sustainability Studies, University of Toyama, Toyama, Japan

M. Geetha (✉) · N. Horie · N. Wada
Global Research Centre for Advanced Sustainability Science (GRASS), University of Toyama, Toyama, Japan
e-mail: geetha@eco.u-toyama.ac.jp

K. Shoyama
Department of Regional and Comprehensive Agriculture, College of Agriculture, Ibaraki University, Ibaraki, Japan

I. Melts
Independent Consultant, Tallinn, Estonia

D. Kumar Swain et al. (eds.), *Climate-Smart Technologies for Sustainable Agricultural Production*, https://doi.org/10.1007/978-3-032-25666-9_13

integration in ensuring successful land transitions. The chapter argues that sustainable rural revitalization requires multi-stakeholder collaboration and policy frameworks that balance economic viability, environmental sustainability, and community welfare, offering transferable insights for other aging and depopulating rural regions all over Japan.

Keywords Land abandonment · Paddy field reuse · Corporate land leasing · Community-based agriculture · Biomass production · Novel crops · Sustainable landscape management

13.1 Introduction

Agricultural land abandonment is an increasingly pressing issue in Japan, including in Toyama Prefecture, where a significant share of agricultural land is no longer actively cultivated for food production (Su et al. 2018). Japanese agricultural landscapes have long been maintained through intensive rice-based agriculture, sophisticated irrigation networks, and robust community-based land management systems. Nowadays, however, the continuity of these socio-ecological systems can no longer be taken for granted (Takeuchi et al. 2016). In Toyama, historically characterized by fertile lowlands, well-developed water infrastructure, and a strong tradition of paddy farming, the expansion of abandoned agricultural land reflects not a temporary fluctuation in land use but a deeper structural transformation of rural society in Japan.

The consequences of agricultural land abandonment extend beyond declines in agricultural productivity. Underutilized, uncultivated agricultural land reshapes rural economies, alters the landscapes, damages cultural heritage, and weakens the collective ability to manage land and water. Demographic change plays a key role in this process. Japan's aging population and persistently low birth rates have reduced the rural labor force and disrupted the exchange of intergenerational local agricultural knowledge and practices (Rieher 2024). As the average age of agricultural workers continues to rise, the physical demands of hard labor are becoming increasingly difficult to meet, particularly in fragmented or marginal fields that require high labor inputs (Roberts and Fujita 2023).

Simultaneously, agricultural land abandonment is reinforced by broader socio-economic pressures. The declining profitability of traditional crops (e.g. rice), combined with rising input costs and economic vulnerability, is shaping decisions to scale back cultivation or exit farming altogether. These dynamics have accelerated the abandonment of active land use and contributed to the spatial expansion of abandoned agricultural land. While land abandonment is often understood as a consequence of demographic and economic decline, it also generates feedback effects that can intensify rural vulnerability. Unmanaged agricultural land is prone to ecological succession, which may lead to the spread of weeds and pests and to the deterioration of irrigation and drainage systems. A traditional agricultural landscape that previously supported biodiversity, flood regulation, and coordinated land management can become fragmented and nonresilient to global change. In addition, visible land abandonment can undermine local residents' perception of the agricultural viability, which in turn discourages younger people from living in or returning to rural areas and reinforces the population bottleneck.

At the same time, land abandonment can also represent a resource for rural renewal. Under appropriate institutional and economic conditions, such land may provide space for experimentation with alternative land-use models, innovative production systems, and collaborative revitalization strategies that engage multiple stakeholders. In this context, business organizations have become increasingly important stakeholders. While traditional revitalization approaches have focused mainly on individual farmers, cooperatives, or local governments, private-sector businesses can mobilize financial capital, technological know-how, and supply-chain integration, while offering long-term operational stability.

In Toyama, such initiatives include corporate leasing of abandoned agricultural land for fast-growing, short-rotation biomass production systems, experimentation with high-value or specialty crops, and hybrid models that combine business investment with community-led cooperative farming and livestock-sharing arrangements. These emerging forms of engagement suggest new pathways for reconnecting abandoned land with local economies and environmental management. This chapter, therefore, examines the revitalization of abandoned agricultural land through the involvement of business organizations in Toyama Prefecture. It analyzes how these initiatives influence economic opportunities, environmental outcomes, and community resilience, and considers their implications for the future governance of Japan's rural landscape.

13.2 Literature Review: Key Observations on Drivers, Impacts, and Land Reuse Pathways

13.2.1 Structural Drivers: Demographic Decline, Labor Shortages, and Lack of Successors

A central finding across the scholarly literature is that agricultural land abandonment in Japan is fundamentally driven by demographic and labor-related transformations. Aging populations, rural depopulation, labor shortages, and the absence of young successors are repeatedly identified as primary determinants of farmland abandonment in rural Japan (Su et al. 2018; Su et al. 2020; Shoji et al. 2020; Takuya 2023). These factors are mutually reinforcing and collectively accelerate land neglect. Depopulation reduces the availability of labor for both routine and seasonal agricultural activities, while population aging increases the proportion of farmland managed by elderly farmers with declining physical capacity. The lack of successors further undermines long-term land stewardship; even when older farmers remain active, retirement, illness, or death can result in abrupt land abandonment if effective mechanisms for land transfer or consolidation are absent. These demographic pressures are closely intertwined with economic conditions. Cultivation decisions are increasingly shaped by cost–benefit considerations, particularly in contexts where labor inputs outweigh expected returns. As rice profitability declines, maintaining small and marginal fields becomes economically unviable for households facing labor constraints. This decision-making dynamic has been widely documented in rural Japan and is especially pertinent to Toyama Prefecture, where agriculture remains predominantly paddy-based and labor-intensive.

Empirical population change data from Toyama Prefecture strongly reinforce these arguments. Long-term population trends from 1976 to 2024 reveal a clear demographic transition from population growth to stagnation and, ultimately, to sustained and accelerating population decline (Fig. 13.1). Although Toyama experienced positive population growth until the mid-1980s, the steady decline in both absolute population change and growth rates during this period already signaled weakening demographic momentum. From the late 1990s onward, population change became consistently negative, and after the late 2000s, population loss accelerated sharply, reaching approximately −1.0% by 2023–2024. This prolonged decline reflects not only numerical population loss but also a deeper structural transformation in the population profile.

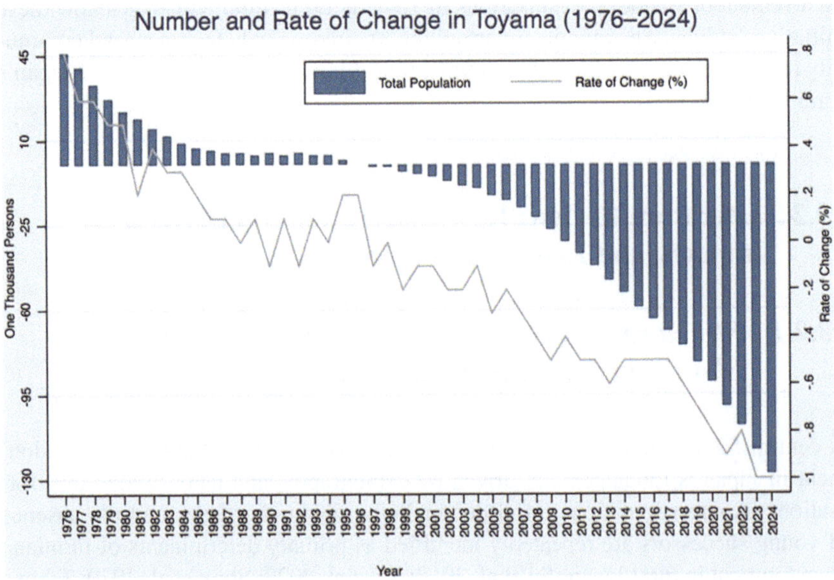

Fig. 13.1 Changes in population size and rate of population change in Toyama Prefecture (1976–2024)

These demographic shifts have direct implications for agricultural labor availability and land management capacity. As rural populations contract and age, the pool of individuals capable of sustaining labor-intensive farming practices—particularly paddy rice cultivation—declines substantially. In Toyama's agricultural landscape, where cultivation depends on regular labor input, irrigation management, and seasonal coordination, the erosion of labor capacity translates into practical constraints on continued farming. Consequently, farmers increasingly prioritize easily manageable plots while allowing marginal, distant, or less productive fields to fall out of cultivation.

Taken together, Toyama's population trends provide concrete empirical evidence that demographic decline and labor shortages are not merely background conditions but active drivers of farmland abandonment. The convergence of long-term depopulation, accelerated aging, and declining agricultural profitability has created a structural environment in which traditional family-based land management systems are increasingly difficult to sustain. This demographic reality forms a critical foundation for understanding the expansion of abandoned farmland in Toyama Prefecture and underscores the need for alternative land-use strategies, collective arrangements, and external actors, such as business organizations, to support rural land revitalization.

13.2.2 Geographic and Soil Constraints: Hilly Regions and Productivity Loss

Beyond demographic considerations, geographical and soil-related constraints significantly influence the spatial patterns of land abandonment. In mountainous and hilly areas, the abandonment of agricultural land is amplified by steep slopes, fragmented fields, limited accessibility, and soil conditions that raise production costs while reducing yields (Sasaki et al. 2021; Kitano 2021). These characteristics restrict mechanization, increase labor requirements, and thereby reduce the economic viability of cultivation.

The consequences extend beyond local agricultural households. As marginal lands are progressively abandoned, aggregate agricultural productivity declines with implications for food security and domestic supply stability (Liu and Zhou 2021; Liu et al. 2021). Although Toyama Prefecture is characterized by productive lowlands, it also encompasses peripheral and rural areas that share challenges related to accessibility, field conditions, and demographic vulnerability. Consequently, findings from studies such as Shoyama et al. (2021) and Su et al. (2018) on hilly regions remain relevant to understanding the uneven distribution of land abandonment in Toyama Prefecture. These insights highlight the importance of geographically differentiated policy measures rather than uniform land-management strategies.

13.2.3 Environmental Impacts of Abandonment: Biodiversity Loss, Carbon Emissions, and Soil Degradation

The literature further emphasizes that agricultural land abandonment generates substantial environmental consequences. Abandoned agricultural land may adversely affect agroecosystems through reduced biodiversity, increased carbon emissions, and soil degradation (Uchida and Ushimaru 2014; Lécuyer et al. 2021; Irie and Kawahara 2023; Katayama et al. 2024; Wang et al. 2025). Traditional agricultural landscapes in Japan function as semi-managed ecosystems that support biological diversity, regulate hydrological processes, and sustain culturally valued landscapes.

When traditional agricultural management ceases, the dominance of weeds and ecological succession may diminish biological diversity. In parallel, soil and land-cover deterioration can lead to erosion, while changes in land cover can alter greenhouse gas emissions. In the context of Toyama, these environmental implications are particularly relevant. Paddy-based agricultural systems rely on collective water management and shared infrastructure. Therefore, land abandonment can trigger cascading effects, including the collapse of irrigation networks, increased flood risk, and the loss of ecological services. Land abandonment is not only an agricultural challenge but also a broader environmental and landscape management issue.

13.2.4 Scale of Abandonment and the Land Use Debate: Bioenergy and Afforestation Versus Food Security

An additional observation concerns the rapid acceleration in the scale of land abandonment. Research indicates that the area of abandoned agricultural land has increased rapidly, with reported annual growth rates exceeding 67% per decade (Haga et al. 2019). This trend has stimulated growing interest in alternative land use strategies, including bioenergy production, afforestation, and other forms of land conversion aimed at generating new economic and ecological value from underutilized land (Mitsui et al., 2010; Timilsina 2012; Saez de Bikuña et al. 2017; Kgathi et al. 2017; Huang et al. 2024; Russo et al. 2025; Ménard et al. 2025; Song et al. 2025). However, these approaches have also intensified debates, raised concerns about food security. The repurposing of agricultural land usage for non-food biomass production may diminish domestic food production capacity and increase dependence on imports—a particularly sensitive issue in Japan, where food self-sufficiency has historically been a significant national policy concern. Accordingly, land reuse strategies must be evaluated from a multifunctional perspective: they should aim to balance agricultural productivity, ecological restoration, climate change mitigation potential, and community well-being. In Toyama Prefecture, this debate transitions from theoretical discourse to practical application, exemplified by local initiatives such as biomass-oriented willow cultivation, experimentation with novel crops, and discussions of integrated land-use solutions, including solar energy sharing.

13.2.5 *Implications for this Study: Why Business Organizations Matter in the Literature*

Across these themes above, a significant gap becomes apparent in the existing literature. While numerous studies (Li and Li 2017; Su et al., 2018; Shoyama et al. 2021) document the drivers and impacts of agricultural land abandonment, relatively few analyze the role of business organizations as operational counterparts in land reuse and rural revitalization (Lankauskienė et al. 2024, Liu et al. 2026 (in press)). Enterprises can mitigate resource constraints in aging, depopulating rural communities by providing capital investment, technical expertise, and access to a wider market. Moreover, businesses can interlink local land management initiatives to broader sustainability agendas, including biomass supply chains, integration of renewable energy, and the development and branding of specialty crops and their products. Consequently, this chapter advances the existing body of literature by focusing directly on business-led land reuse and its community impacts in Toyama Prefecture, while acknowledging community initiatives as vital mechanisms for ensuring long-term resilience. Together, these perspectives offer a more comprehensive understanding of how abandoned agricultural land can be reintegrated into multifunctional rural landscapes.

13.3 Importance of the Study, Research Aim, and Objectives

The research examines the drivers and consequences of agricultural land abandonment in Toyama Prefecture, with the aim of identifying sustainable solutions through stakeholder engagement and revitalization strategies tailored to abandoned agricultural land. The significance of this study lies in its comprehensive and holistic approach. Rather than viewing farmland abandonment solely through demographic or economic perspectives, the research conceptualizes it as a spatially heterogeneous process shaped by multiple interacting factors, including population change, market pressures, physical land constraints, institutional frameworks, and stakeholder capacity. Toyama Prefecture represents a particularly relevant case for examining the regional dynamics of farmland abandonment in Japan. While the prefecture has a long history of productive, rice-based agriculture, it now includes both highly productive agricultural areas and rural zones experiencing rapid population aging and decline. This internal diversity provides an important opportunity to analyze why farmland abandonment occurs even in regions with a historically strong agricultural foundation. In addition, emerging business initiatives in Toyama—such as land leasing for biomass production, biorefinery systems, and experimentation with novel crops—offer a practical context for assessing the role of business actors in local rural revitalization. Demographic evidence underscores the scale of this transformation. As shown in Fig. 13.2, Toyama's population composition has shifted markedly over the

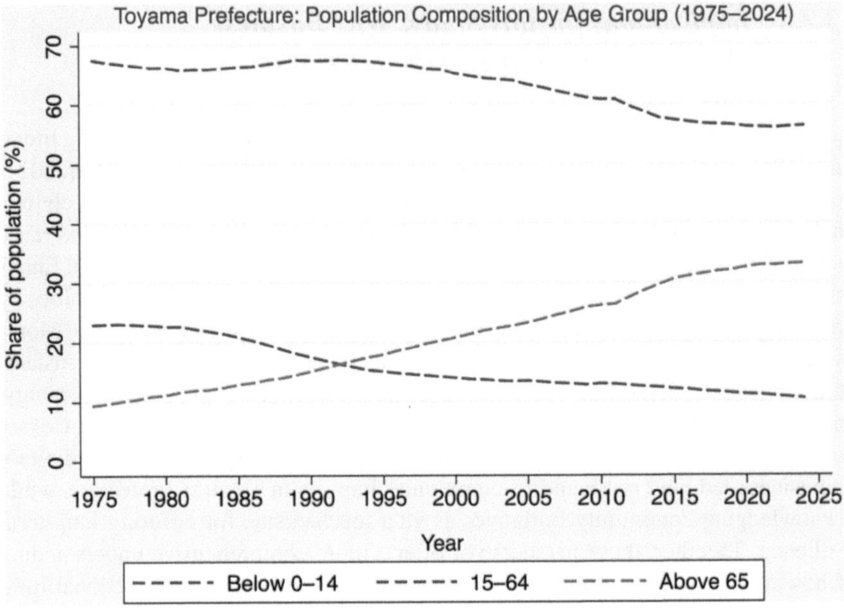

Fig. 13.2 Changes in population composition by age group in Toyama Prefecture (1975–2024)

past five decades, characterized by a continuous decline in the proportions of children and working-age residents and a rapid increase in the proportion of older adults. The share of residents aged 65 and above has risen from less than 10% in the mid-1970s to more than one-third of the population by 2024, while the working-age population has contracted substantially. This demographic restructuring has direct implications for agricultural labor availability and land management capacity—particularly in labor-intensive paddy farming systems—and provides essential context for understanding the expansion of abandoned farmland across the prefecture.

The study has three main objectives: (1) to examine the spatial distribution and extent of abandoned agricultural land in Toyama Prefecture, highlighting geographic patterns and areas with most prevalent land abandonment; (2) to analyse socio-economic, demographic, and environmental factors fostering agricultural land abandonment, with a particular focus on population aging, labor shortages, agricultural profitability, and land conditions; (3) to explore potential pathways for rural revitalization through business engagement and community participation, utilizing evidence from Toyama Prefecture and comparable case studies within Japan to generate policy-relevant and transferable insights. These objectives collectively support the broader goal of developing realistic, locally based approaches to sustainable land management that improve the livelihoods and resilience of rural communities.

13.4 Methodology

To accomplish the research objectives, this study employs an integrated methodology that combines spatial analysis, field engagement, and comparative learning. This approach moves beyond general description to generate practical insights relevant to both academic understanding and policy development. By integrating GIS-based analysis with qualitative field data from key informant interviews (KII) and comparative case studies, the methodology captures both the spatial patterns and the socio-institutional dynamics underlying farmland abandonment and revitalization.

13.4.1 GIS-Based Identification of Abandoned Farmland

The initial methodological component employs Geographic Information Systems (GIS) to delineate areas of abandoned agricultural land within Toyama Prefecture. Digitized land use datasets, together with multi-year grid-mesh data, were used to map the spatial concentration of abandonment and to analyze its temporal expansion. This spatial approach is critical for two reasons. First, farmland abandonment is not uniformly distributed across the landscape but occurs in clusters shaped by demographic structure, land quality, and accessibility. Second, effective policy interventions require geographic targeting, as identifying areas with the highest rates of abandonment facilitates resource prioritization and strategic planning. To situate Toyama within a broader analytical context, the study also incorporates comparative case study locations from other regions of Japan. Figure 13.3 presents the geographic locations of the primary study area (Toyama Prefecture) and the comparative case-study sites in Tajima (Hyogo Prefecture) and Tsukuba (Ibaraki Prefecture). This map provides spatial context for the multi-scalar design of the research and clarifies how insights from these regions inform, rather than substitute for, the Toyama-focused analysis.

A key element of this spatial analysis draws on data on abandoned agricultural land derived from vegetation surveys conducted by Japan's Ministry of the Environment and made available through the Biodiversity Center of Japan (Ministry of the Environment, Japan, n.d.a; n.d.b). Newly identified abandoned agricultural land appears in the sixth and seventh national surveys (1999–2023), with spatial data provided at a scale of 1:25,000, whereas earlier surveys (1973–1998) did not record abandoned land in Toyama. Within this dataset, abandoned agricultural land is classified as a subset of cultivated land and identified using two overlay procedures: (i) cultivated land combined with weedy communities in abandoned paddy fields, and (ii) cultivated land combined with weedy communities along roadsides and within crop fields. This classification approach captures the ecological signatures associated with abandonment and supports the interpretation of farmland abandonment as both a land-use transition and an environmental process.

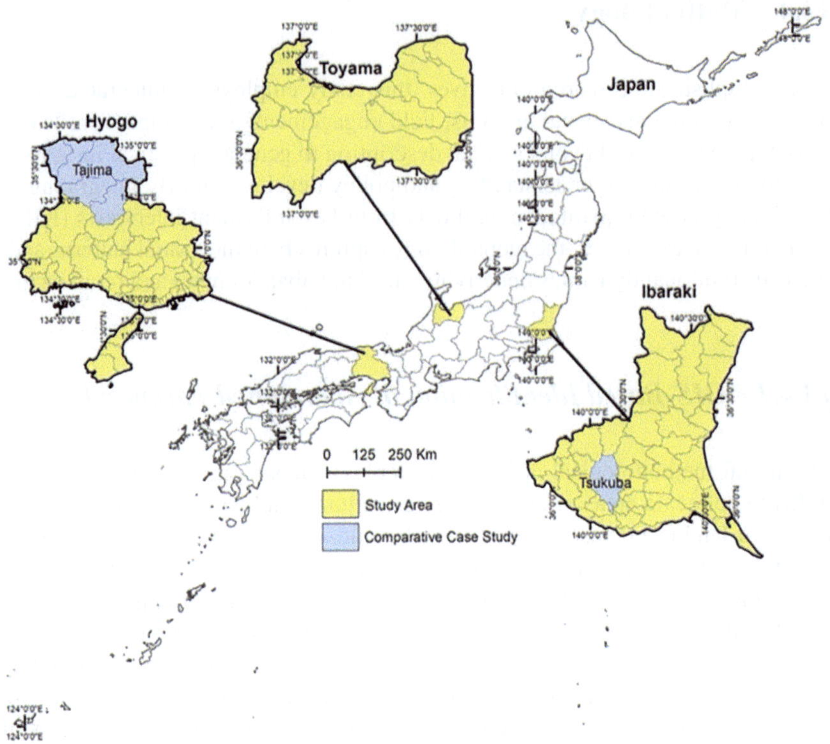

Fig. 13.3 Location of the study area (Toyama Prefecture) and comparative case study sites in Tajima (Hyogo Prefecture) and Tsukuba (Ibaraki Prefecture), Japan

13.4.2 Key Informant Interviews and Stakeholder Engagement

The second component comprises key informant interviews (KII) with farmers, corporate representatives, and officials. This qualitative methodology is vital because the reasons for abandonment are not entirely discernible from spatial datasets alone. Interviews elucidate how farmers assess profitability, labor constraints, and land conditions, and how communities assess the viability of cooperative solutions. Corporate interviews provide insights into why companies invest in rural land, how leasing arrangements function, and the perceived benefits and risks of such investments. Officials contribute perspectives on governance structures, policy instruments, and institutional limitations in managing fragmented land and facilitating revitalization.

13.4.3 Comparative Case Study: Tajima and Tsukuba

The third component draws on land-use models for livestock grazing in Tajima (Hyogo Prefecture) and for vineyards in Tsukuba (Ibaraki Prefecture). Comparative cases enable us to assess whether successful revitalization strategies in other prefectures are adaptable to abandoned agricultural land in Toyama. Tajima provides a model for converting marginal and underutilized rice fields into pasture through cattle grazing and leasing systems, particularly in contexts of demographic decline. Tsukuba offers a paradigm of converting abandoned land into high-value vineyards, illustrating how novel crop production systems together with agritourism can generate new economic value (Liu et al. 2026). These comparisons enhance the analytical framework by presenting alternative transition pathways and elucidating the conditions necessary for successful adaptation at the rural community level.

13.4.4 Integration of Methods

The strength of this methodology lies in integration. GIS identifies where abandonment occurs and how it spreads; interviews explain why stakeholders abandon or reuse land and what constraints they face; comparative cases provide concrete models that expand the menu of feasible solutions. Together, these methods support a comprehensive analysis of business-led revitalization and its impacts on Toyama's rural communities.

13.5 Results and Key Findings: Spatial Patterns, Stakeholder Insights, and Emerging Land Use Pathways in Toyama

13.5.1 Identification of Key Drivers and the Local Logic of Land Abandonment

The implementation phase reinforced the evidence that agricultural land abandonment in Toyama is influenced by a convergence of demographic pressures and the decline of agricultural economics. Key informant interviews, spatial mapping, and the literature review (Oishi 2021; Shoyama et al. 2021; Mameno and Kubo 2022; Imai et al. 2023; Reiher 2024) collectively identify consistent drivers: aging populations, declining birth rates, labor shortages—particularly the absence of young successors—and decreasing profitability of rice cultivation. These pressures are not experienced abstractly; rather, they manifest in everyday farm decisions, such as whether to maintain marginal parcels in production, invest in machinery, or reorganize household labor around seasonal agricultural activities. In Toyama, rice cultivation continues to hold cultural and economic significance; however, stakeholders repeatedly noted that declining rice prices and income volatility are primary drivers of labor withdrawal from farming and of fields becoming fallow. When agricultural land is fragmented, and the number of active household members is limited, farmers tend to prioritize parcels that are easier to manage, located closer to their homes, and more productive. Less accessible plots-particularly those affected by poor drainage, soil conditions, or inadequate field consolidation-are more susceptible to abandonment. This phenomenon helps to explain why abandonment within the same municipality varies spatially and is strongly correlated with demographic structure and practical cultivation constraints.

The literature review further emphasized the growing interest in the valorization of vacant land through innovative or multifunctional approaches, including bioenergy, agroforestry, and community-led land stewardship initiatives. In Toyama, these strategies transcend theoretical discussion: the implementation phase recorded a transition from "abandonment as decline" to "abandonment as potential space" for novel agricultural paradigms and land utilizations. Specifically, stakeholders demonstrated a growing receptiveness to (i) specialty crops, (ii) livestock-centered land management, and (iii) corporate-community collaborations via leasing arrangements.

13.5.2 Spatial Distribution and Temporal Trends: What GIS and Vegetation Data Reveal

Utilizing GIS and vegetation survey data, the research mapped the spatial concentration and trends of abandoned agricultural land across Toyama Prefecture. A key finding is the pronounced clustering of abandoned land in municipalities characterized by high proportions of elderly residents and declining local populations. Asahi-machi and Nyuzen-machi emerge as distinct hotspots, supporting the hypothesis that demographic vulnerability and limited agricultural profitability exacerbate land disuse (Fig. 13.5). A secondary spatial insight reveals that the relationship between industrial development and agricultural land abandonment is complex. Contrary to initial assumptions that industrial activity would inherently reduce abandonment by providing complementary employment opportunities or infrastructural benefits, the results demonstrate a more intricate pattern. Industrial zones such as Takaoka, Namerikawa, and Imizu exhibit relatively effective integration between industrial and agricultural land uses, with lower abandonment rates despite substantial factory areas. Conversely, Toyama City, despite its industrial presence, displays high levels of agricultural land abandonment. This indicates that industrial development alone does not mitigate land abandonment when factors such as an aging population, low agricultural profitability, and limited incentives for land management remain predominant (Fig. 13.4). A third temporal insight indicates that abandonment has increased

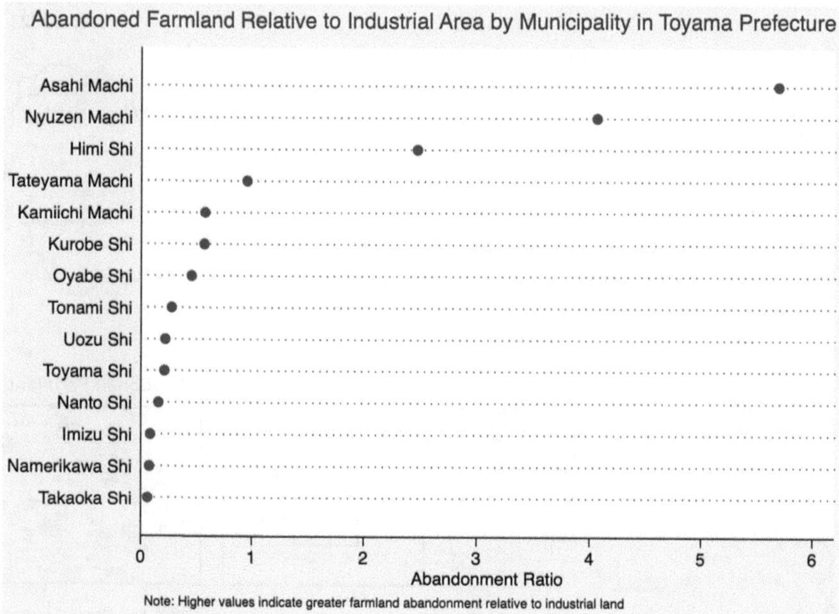

Fig. 13.4 Abandoned farmland relative to industrial area by municipality in toyama prefecture

markedly since the sixth vegetation survey (1999), with the most significant changes observed between the sixth and seventh surveys (1999–2023). This suggests that agricultural land abandonment in Toyama is a relatively recent phenomenon that has been rapidly intensifying, rather than an enduring issue. These findings underscore the importance of strategic interventions: once land abandonment is established, restoration efforts are likely to incur substantially higher costs due to weed invasion, soil degradation, and the deterioration of agricultural infrastructure.

13.5.3 Data Source and Classification Logic: Why the Vegetation Survey Matters

A significant methodological contribution of this research lies in the use of national vegetation survey data initiated by Japan's Ministry of the Environment and made available through the Biodiversity Center of Japan (Ministry of the Environment, Japan, n.d.a; n.d.b). The category of abandoned agricultural land is delineated through land-classification overlays within cultivated-land classes, providing a standardized method for identifying abandonment at a 1:25,000 scale. In Toyama, abandoned agricultural land was not detected in the second to fifth surveys (1973–1998), but became observable in the sixth and seventh surveys (1999–2023) (Ministry of the

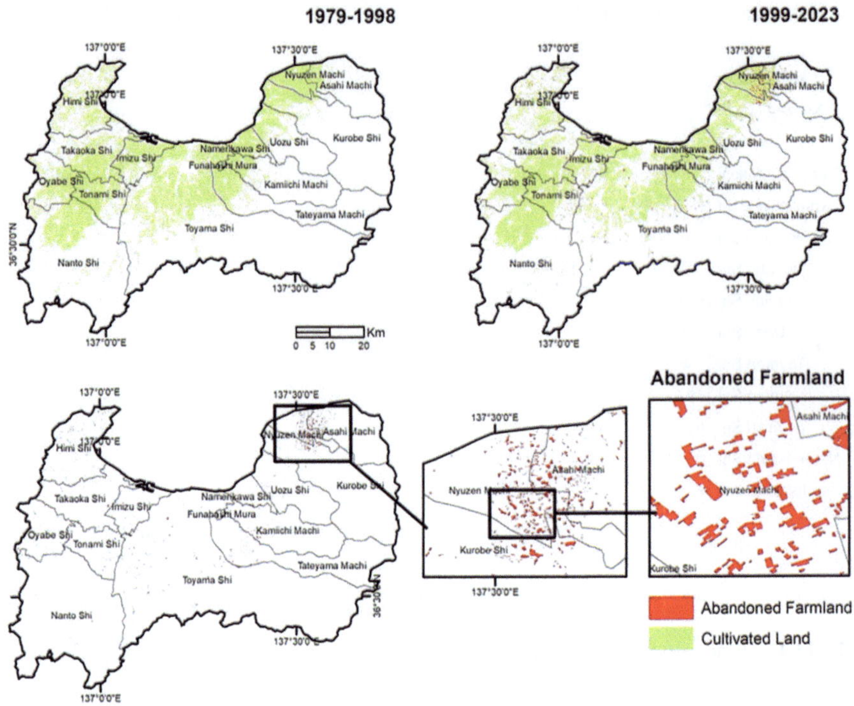

Fig. 13.5 Spatial variation in abandoned and cultivated farmland across districts of Toyama prefecture

Environment, Japan, n.d.b). Notably, the "abandoned farmland" category is considered a subset of cultivated land and is delineated using two overlays: cultivated land-weed communities in abandoned paddy fields, and cultivated land-weed communities along roadsides combined with weed communities in crop fields. This classification proves valuable as it captures the ecological signature of abandonment, namely, weed community dynamics, rather than relying solely on administrative reports. It facilitates linking land use changes to ecological processes, thereby providing a clearer understanding of how abandonment progresses into environmental risks, including soil degradation, vegetation succession, and the loss of managed landscapes.

13.6 Stakeholder Interview Insights: Community Adaptation, Corporate Innovation, and Expert Pathways

13.6.1 Farmers and Local Officials: "We Need Labor, Income, and a Shared System"

Key informant interviews (KIIs) revealed shared concerns among farmers and municipal officials regarding worsening land quality, ongoing labor shortages, mainly due to youth outmigration, and declining rice prices. These structural pressures have made small-scale farming increasingly unsustainable, especially in fragmented landscapes with limited mechanization. In response, local adaptation strategies are emerging, notably cooperative land management, which was repeatedly highlighted as a practical approach. This allows farmers to consolidate plots, share equipment, coordinate planting schedules, and pool labor, reducing individual risk and improving efficiency. Livestock-sharing systems, including joint ownership and leasing, were also identified as ways to distribute financial burdens and stabilize income. Additionally, stakeholders emphasized the importance of crop diversification, shifting toward high-value specialty crops, recognizing that traditional rice systems alone cannot support rural livelihoods. These strategies rely heavily on social cohesion, trust, shared norms, and strong local leadership. A significant adaptation involves expanding cattle farming in suitable areas, particularly on land no longer fit for crops. Inspired by Tajima's livestock model, farmers expressed interest in converting abandoned fields into managed pastures, including cattle rental and leasing. As illustrated in Fig. 13.6, grass cultivation for livestock demonstrates how marginal arable land can be transformed into productive pastures. This approach offers a practical response to demographic challenges, maintains agricultural land use, and reduces the risks of abandonment.

Together, these community-led efforts reflect a broader shift in the view of land abandonment, from a decline in agriculture to flexible land-use strategies that align with demographic changes and ecological needs.

13.6.2 Corporate Representatives: Leasing Models, Biomass, and Specialty Crops

Corporate involvement has emerged as a tangible and scalable mechanism for revitalizing agricultural land. Interviews and case documentation have underscored that business organizations can provide three practical advantages: (i) capital for land preparation and infrastructure development, (ii) technical expertise in crop systems,

Fig. 13.6 Grass cultivation for livestock rearing. *Source* Field photograph taken during fieldwork in 2024, Hyogo Prefecture, Japan

supply chains, and processing, and (iii) the capacity to commit to medium-and long-term land management beyond household-level limitations. In Toyama, a private-sector enterprise exemplifies a corporate leasing model for abandoned agricultural land. Through leasing arrangements with local communities, underutilized land has been repurposed for fast-growing, short-rotation crops that serve as biomass feedstock for biorefinery processes. This approach shows how abandoned farmland can be integrated into a regional low-carbon economy, transforming it into a productive, economically viable resource. In addition, such initiatives often promote alternative crops, including hazelnuts and soba, thereby supporting local livelihood diversification alongside commercial objectives. This integrated model, combining biomass production with specialty crop cultivation, illustrates a hybrid pathway for land revitalization that aligns environmental sustainability with rural economic resilience. In Tsukuba (Ibaraki), Tsukuba Vineyards exemplifies another business-led pathway: converting abandoned crop fields into high-value vineyards. The significance of this model extends beyond economic transformation to encompass the broader rural development chain it can stimulate, encompassing branding, agritourism, processing, and new entrepreneurial ventures (Liu et al. 2026). These initiatives show that alternative land-use systems can generate new economic opportunities and boost the identity and attractiveness of rural areas. Additionally, evidence from the implementation phase indicates a diffusion effect, where successful models become more visible and credible. This encourages neighbouring farmers and stakeholders to consider similar land-use changes, especially adopting specialty crops. The "demonstration and learning" dynamic emphasizes that private-sector initiatives serve not just as operational actors but also as catalysts for wider innovation across rural landscapes.

13.6.3 Expert Perspectives: Synecoculture and Solar Sharing as Multifunctional Land Use

Expert interviews broadened the revitalization framework from traditional agricultural restructuring to include multifunctional land-use innovation. Synecoculture, a biodiversity-focused and regenerative farming system, was identified as especially suitable for fragmented or degraded lands. By encouraging diverse plant communities and reducing monoculture practices, synecoculture boosts ecological resilience while maintaining productivity (Funabashi 2016). Figure 13.7 shows that experimental synecoculture plots illustrate how mixed vegetation can restore soil health and biodiversity on marginal lands. Instead of focusing solely on short-term yields, this model prioritizes ecological regeneration as the basis for long-term productivity and landscape sustainability.

Solar sharing is a complementary approach. Installing solar panels over cultivated land allows farmers to produce renewable energy income while continuing agricultural activities below. This dual-use approach reduces the economic pressures driving land abandonment and diversifies rural income sources. Together, synecoculture and solar sharing transform abandoned or underused land into multifunctional assets within integrated food–energy systems, aligning local revitalization efforts with national sustainability goals.

Fig. 13.7 Synecoculture experimental plot. *Source* Co-authors field experimental photograph taken during a field visit in 2024, Ibaraki Prefecture, Japan

13.7 Comparative Case Studies and Transferable Lessons for Toyama

13.7.1 Tajima (Hyogo): Cattle Grazing and Leasing as a Response to Marginal Land and Demographic Decline

Tajima's case is particularly pertinent to Toyama due to shared structural conditions, including a declining population, fewer farmers, aging communities, and the need for collective systems. From the perspective of farmers, Tajima's model encompasses traditional feeding systems, the productive utilization of abandoned land for livestock, rental of cattle from other farmers, and the expansion of cattle farming as a strategy to mitigate land abandonment, particularly in regions with suboptimal arable conditions and unfavourable rice prices. The principal contribution of the Tajima model lies in its redefinition of land value: even when land is unsuitable for crop cultivation, it retains potential value as pasture or within a livestock system. Grazing also maintains open habitats and enhances biodiversity (Nakahama et al. 2024). The leasing or cattle-rental components serve to distribute risk and lower entry barriers for farmers who lack the capacity to expand their herds independently. For Toyama, the transferable lesson is not merely to "introduce cattle," but rather to "establish enabling institutions"-including leasing arrangements, cooperative labor structures, and market access-to ensure that livestock-based revitalization is both socially and financially sustainable.

13.7.2 Tsukuba (Ibaraki): Vineyard Conversion as a High-Value Crop and Rural Branding Strategy

The Tsukuba vineyard model illustrates how abandoned farmland can be repurposed into a valuable agricultural and branding resource. By transforming neglected crop fields into vineyards, Tsukuba Vineyards created integrated value chains that connect grape growing, wine making, niche marketing, and agritourism. As presented in Fig. 13.8, previously underused cropland has been systematically restructured into vineyard plots, demonstrating the productive reintegration of abandoned land into the regional economy. This change goes beyond simple crop replacement; it represents a strategic shift toward value-added agriculture aligned with market needs and regional identity. Vineyards act not only as production sites but also as symbolic landscapes that promote community involvement and draw external visitors, enhancing rural branding and regional uniqueness. For Toyama, a key lesson is the importance of linking specialty crop production to processing, tourism, and regional storytelling. However, successful replication requires careful adaptation to local climate and soil conditions, technical training, land consolidation, and institutional capacity,

Fig. 13.8 Grape cultivation in abandoned cropland. *Source* Grape's farming field photograph taken during a 2024 field visit in Tsukuba, Japan

especially when involving corporate or cooperative actors to ensure stable market access.

13.8 Synthesis: Impacts on Rural Communities and Policy Implications for Toyama

13.8.1 Community Welfare and Rural Resilience: How Revitalization Changes Local Futures

The expanded evidence from the implementation phase indicates that agricultural land revitalization is not merely a technical land-use strategy but a substantive rural welfare intervention. In areas where abandoned agricultural land is widespread, communities encounter cascading effects: landscape degradation, reduced economic circulation, declining community pride, and weakened local governance capacity. Revitalization initiatives—whether community-led cooperative farming or corporate land leasing—can restore local confidence, stimulate economic networks, and create inclusive participation opportunities. Importantly, these initiatives enable elderly residents to remain engaged through less physically demanding tasks while creating opportunities for younger generations through specialty-crop production and entrepreneurial activities.

Spatial analysis further enhances the welfare significance or revitalization by enabling targeted policy actions (Matsui et al. 2014; Kobayashi et al. 2020). The identification of land-abandonment hotspots, such as Asahi Machi and Nyuzen Machi, enables policymakers to prioritize measures and interventions in municipalities where demographic vulnerability and land abandonment intersect. Simultaneously, mixed findings regarding industrial zones suggest that effective rural revitalization must integrate land management strategies with labor organizations and systems, and with market access, rather than assuming spillover benefits from industrial development and growth alone.

13.8.2 Corporate–Community Partnerships: Conditions for "Good" Private-Sector Involvement

Corporate involvement can be advantageous to rural revitalization when it is aligned with community priorities and structured to yield tangible local benefits. Evidence from Toyama indicates that private-sector land leasing can advance both commercial objectives, such as biomass production, and community diversification through crops such as hazelnuts and soba, as well as through adaptive land-use practices. These arrangements show how underutilized agricultural land can be reintegrated into productive systems, strengthening local economic resilience. However, careful planning of corporate engagement is essential to ensure fairness and sustainability. Without proper safeguards, such involvement might prioritize short-term resource extraction over long-term community resilience. The success of corporate participation relies on institutional frameworks that balance economic efficiency with social and environmental concerns.

Effective mechanisms include transparent leasing, prioritizing local employment, working with cooperatives, and participatory governance involving the community in land-use decisions. These strategies help build stakeholder trust and support sustainable revitalization. Evidence from other regions, such as vineyard land-use changes in Tsukuba, suggests that private-sector initiatives can foster rural innovation by demonstrating viable land-use alternatives, reducing perceived risks, and encouraging broader local adoption. In Toyama, policy frameworks could involve 'pilot projects' that combine private investment with community participation, promoting knowledge sharing and adaptable, community-driven revitalization pathways.

13.8.3 Policy Implications: A Multi-Pathway Strategy for Toyama

The findings of this study indicate that addressing agricultural land abandonment in Toyama Prefecture requires a multi-pathway policy approach rather than reliance on a single strategy, intervention. Strengthening community-driven initiatives-such as cooperative farming arrangements, shared agricultural machinery systems, and livestock-sharing schemes—is essential for overcoming labor shortages and enabling collective land management in aging rural areas. At the same time, policies should facilitate corporate participation through transparent land-leasing mechanisms, streamlined administrative procedures, and incentive structures that encourage local employment and tangible community benefits.

Promoting multifunctional land-use models, including bioenergy production (e.g., fast-growing and short-rotation plant biomass), agroforestry, synecoculture, and solar energy sharing, can enhance land-use efficiency while aligning agricultural revitalization efforts with broader climate-mitigation and sustainability goals. GIS-based targeting should be systematically employed to prioritize investment in municipalities where agricultural land abandonment overlaps with high elderly population density and economic vulnerability, thereby improving policy efficiency and impact.

Importantly, the results also demonstrate that industrial development alone does not necessarily prevent agricultural land abandonment, as evidenced by the case of Toyama City. This underscores the need for policies that integrate industrial strategies with agricultural and rural revitalization planning. Altogether, these implications highlight the importance of a diversified policy portfolio that combines livestock conversion on marginal land, specialty crop development in areas with market potential, corporate land leasing for diversification and bioenergy, and technology-enabled dual-use systems, such as solar sharing. Such an integrated approach is crucial to building long-term rural resilience and sustaining community welfare in Toyama Prefecture.

Acknowledgement The author expresses sincere gratitude to all key informants who generously contributed their time, experiences, and perspectives during interviews and field interactions. This includes farmers, corporate representatives, researchers, and local officials, whose insights were vital to this study. The author also extends special thanks to the representatives of the Agricultural Promotion Division, Tajima Bureau; the Livestock Park Division, Shinonsen Town, Hyogo Prefecture; and the Head of Tsukuba Vineyard, Ibaraki Prefecture, for their valuable support and cooperation during fieldwork. In line with confidentiality considerations, the identities of private-sector organizations involved in the Toyama case are not disclosed. Additionally, the author gratefully acknowledges the financial support provided by the President's Fund of the University of Toyama, which facilitated the research activities, fieldwork, and analysis integral to the completion of this book chapter.

References

Funabashi M (ed) (2016) Synecoculture manual 2016 version (English version). Research and Education Material of UniTwin UNESCO Complex Systems Digital Campus, e-Laboratory: Open Systems Exploration for Ecosystems Leveraging, No. 2. ISSN 2432-3950.

Haga C, Inoue T, Hotta W, Shibata R, Hashimoto S, Kurokawa H, Machimura T, Matsui T, Morimoto J, Shibata H (2019) Simulation of natural capital and ecosystem services in a watershed in northern Japan focusing on the future underuse of nature: linking forest landscape models and social scenarios. Sustain Sci 14(1):89–106

Huang W, Hashimoto S, Yoshida T, Saito O, Meraj G (2024) Understanding Japan's land-use dynamics between 1987 and 2050 using land accounting and scenario analysis. Sustain Sci 19(5):1561–1577

Imai N, Otokawa H, Okamoto A, Yamazaki K, Tamura T, Sakagami T, Ishizaka S, Shinojima H (2023) Abandonment of cropland and seminatural grassland in a mountainous traditional agricultural landscape in Japan. Sustainability 15:7742

Irie N, Kawahara N (2023) Social impact scoping using statistical methods: the case of a novel design of abandoned farmland policy. Sustainability 15(4):2929

Katayama N, Fujita T, Ueta M, Morelli F, Amano T (2024) Effects of human depopulation and warming climate on bird populations in Japan. Conserv Biol 38(2):e14175

Kgathi DL, Mmopelwa G, Chanda R, Kashe K, Murray-Hudson M (2017) A review of the sustainability of Jatropha cultivation projects for biodiesel production in southern Africa: implications for energy policy in Botswana. Agric Ecosyst Environ 246:314–324

Kitano S (2021) Estimation of determinants of farmland abandonment and its data problems. Land 10(6):596

Kobayashi Y, Higa M, Higashiyama K, Nakamura F (2020) Drivers of land use changes in societies with decreasing populations: a comparison of the factors affecting farmland abandonment in a food production area in Japan. PLoS One 15:e0235846

Lankauskienė R, Gedminaitė-Raudonė Ž, Micka R (2024) Business model innovation for sustainable multifunctional land management in abandoned rural areas: a case study. Land 13:791

Lécuyer L, Alard D, Calla S, Coolsaet B, Fickel T, Heinsoo K, Henle K, Herzon I, Hodgson I, Quétier F, McCracken D, McMahon BJ, Melts I, Sands D, Skrimizea E, Watt A, White R, Young J (2021) Conflicts between agriculture and biodiversity conservation in Europe: looking to the future by learning from the past. Adv Ecol Res 65:3–56

Li S, Li X (2017) Global understanding of farmland abandonment: a review and prospects. J Geogr Sci 27:1123–1150

Liu X, Xu Y, Engel BA, Sun S, Zhao X, Wu P, Wang Y (2021) The impact of urbanization and aging on food security in developing countries: the view from Northwest China. J Clean Prod 292:126067

Liu Y, Zhou Y (2021) Reflections on China's food security and land use policy under rapid urbanization. Land Use Policy 109:105699

Liu Y, Hu S, Chen N, Ran Z (2026, in press). Multifunctional rural areal system transformation and rural revitalization. Agric Environ Sustain 100007. https://doi.org/10.1016/j.aes.2026.100007

Mameno K, Kubo T (2022) Socio-economic drivers of irrigated paddy land abandonment and agro-ecosystem degradation: evidence from Japanese agricultural census data. PLoS One 17:e0266997

Matsui T, Ugata T, Machimura T (2014) Development of factor analyzing and predicting model of abandoned agricultural land with machine learning algorithms. J Jpn Soc Civ Eng Ser G 70:131–139

Ménard I, Thiffault E, Magnan M, Kurz WA, Hébert F, Boucher J-F (2025) Improving carbon storage and greenhouse gas emissions avoidance through harvested wood products use. Trees For People 19:100757

Ministry of the Environment, Japan. (n.d.a). Second–fifth national vegetation survey vegetation map (1:50,000 scale) [GIS dataset]. Biodiversity Center of Japan. Retrieved Feb 2025, from http://gis.biodic.go.jp/webgis/sc-025.html?kind=vg
Ministry of the Environment, Japan. (n.d.b). Sixth–seventh national vegetation survey vegetation map (1:25,000 scale) [GIS dataset]. Biodiversity Center of Japan. Retrieved Feb 2025, from http://gis.biodic.go.jp/webgis/sc-025.html?kind=vg67
Mitsui Y, Seto S, Nishio M, Minato K, Ishizawa K, Satoh S (2010) Willow clones with high biomass yield in short rotation coppice in the southern region of Tohoku district (Japan). Biomass Bioenergy 34(4):467–473
Nakahama N, Hamano T, Fujimoto M, Eto A (2024) Utilization of abandoned land as cattle grazing restores butterfly and flowering plant diversities in Japan. Ecol Res 39(5):730–743
Oishi N (2021) Skilled or unskilled?: the reconfiguration of migration policies in Japan. J Ethnic Migr Stud 47(10):2252–2269. https://doi.org/10.1080/1369183X.2020.1731984
Reiher C (2024) International labor migration and regional revitalization: divergent perceptions and support of foreign workers in Japan's rural communities. Contemp Jpn:1–19
Roberts GS, Fujita N (2023) Low-skilled migrant labor schemes in Japan's agriculture: voices from the field. Soc Sci Jpn J 20(1):1–19
Russo C, Cirillo V, Pollaro N, Terribile F, Chiodini A, Maggio A (2025) The global energy challenge: second-generation feedstocks on marginal lands for a sustainable biofuel production. Chem Biol Technol Agric 12(1):10
Saez de Bikuña K, Hauschild MZ, Pilegaard K, Ibrom A (2017) Environmental performance of gasified willow from different lands including land-use changes. GCB Bioenergy 9(4):756–769
Sasaki K, Hotes S, Ichinose T, Doko T, Wolters V (2021) Hotspots of agricultural ecosystem services and farmland biodiversity overlap with areas at risk of land abandonment in Japan. Land 10:1031
Shoji G, Yoshida K, Yokoyama S, Thompson EC (2020) Transition of farmland use in a Japanese mountainside settlement: an analysis of the residents' career histories. Geogr Rev Jpn Ser B 93(1):15–26
Shoyama K, Nishi M, Hashimoto S, Saito O (2021) Outcome-based assessment of the payment for mountain agriculture: a community-based approach to countering land abandonment in Japan. Environ Manag 68(3):353–365
Song H, Li X, Xin L, Wang X (2025) Improving mechanization conditions or encouraging non-grain crop production? Strategies for mitigating farmland abandonment in China's mountainous areas. Land Use Policy 149:107421
Su G, Okahashi H, Chen L (2018) Spatial pattern of farmland abandonment in Japan: identification and determinants. Sustainability 10(10):3676
Su K, Yang Q, Zhang Z, Bi G (2020) Investigation of differential fallow patterns and technical measures for cultivated land in China. Trans Chin Soc Agric Eng 36(9):283–291
Takeuchi K, Ichikawa K, Elmqvist T (2016) Satoyama landscape as social–ecological system: historical changes and future perspective. Curr Opin Environ Sustain 19:30–39
Takuya G (2023) Agglomeration of companies entering into agriculture and regional issues: a case study of Hokuto City, Yamanashi prefecture. Hum Geogr/Jinbun Chiri 75(2)
Timilsina GR (2012) Biofuels: the food versus fuel debate. CABI Rev:1–8
Uchida K, Ushimaru A (2014) Biodiversity declines due to abandonment and intensification of agricultural lands: patterns and mechanisms. Ecol Monogr 84:637–658
Wang M, Luo W, Wang Y, Zeng G, Chen J, Zhang L, Cai X, Cheng A, Wang S (2025) Changes in soil profile temperature, moisture content, and CO_2 concentration during the initial stages of natural vegetation restoration of abandoned farmland in karst regions. Ecol Eng 216:107646

[illegible]

Chapter 14
ITC's Climate Smart Agriculture Programme-Model and Outcomes

K. Binoy Menon

Abstract Agriculture in India is exposed to climate variability and long-term climatic change posing significant implications for rural livelihoods, food security, and natural resources. Addressing these risks requires approaches that combines scientific prioritisation frameworks with community-level implementation. This chapter provides a detailed overview of the Climate Smart Village (CSV) initiative implemented by ITC Limited in partnership with national and international institutions. ITC's CSV programme promotes climate-resilient agricultural systems which integrates adaptive farming practices, sustainable natural resource management, livelihood diversification, and institutional support for smallholder farmers across its catchment villages in India. The programme promotes practices like conservation agriculture, weather advisories, crop diversification, and water-efficient irrigation to combat frequent droughts, floods, and extreme weather. Through collaborations and leveraging partnerships with institutions like CGIAR, State Universities, ICAR, and Krishi Vigyan Kendras, the programme facilitates knowledge dissemination and farmer capacity building through demonstrations and farmer learning platforms like Farmer Field Schools. The CSV model integrates comprehensive water management with rainwater harvesting, biodiversity conservation, and soil health improvement policies to achieve carbon-positive and water-positive villages. Livelihood diversification, including livestock management and agroforestry, provides income security and enhances climate resilience. Impact assessment data indicate substantial improvements in yield resilience and sustainability metrics, with high adoption rates of climate-smart practices across the project villages. Overall, the ITC's initiative on climate smart practices demonstrates a scalable approach for strengthening climate-resilient agricultural systems in rural India.

Keywords Climate smart agriculture · Climate smart village · Livelihood diversification · Resilience · Water management

K. Binoy Menon (✉)
ITC Limited, Guntur, Andhra Pradesh, India
e-mail: k.binoy@itc.in

D. Kumar Swain et al. (eds.), *Climate-Smart Technologies for Sustainable Agricultural Production*, https://doi.org/10.1007/978-3-032-25666-9_14

14.1 Introduction

Agriculture constitutes a critical component of India's economy and provides livelihoods to a large segment of the population. Despite its importance, the sector remains highly susceptible to climate-induced stresses such as erratic monsoon patterns, extended dry spells, extreme rainfall events, and increasing temperatures (BISA–ITC 2021). In addition, rainfall variability, together with soil degradation and declining natural resource bases, has significantly heightened the vulnerability of agricultural production systems and rural livelihoods (FAO 2012; Campbell et al. 2014). These pressures contribute to yield instability, threaten food security and heightened vulnerability rural livelihoods among farmers (with less than 2-hectare land) with limited adaptive capacity. A significant proportion of Indian agriculture remains dependent on rainfall, which increases the sensitivity of farming systems to climatic variability. Crops such as wheat, soybean, rice and pulses are particularly vulnerable to fluctuations in rainfall and temperature, creating uncertainty in agricultural production systems (BISA–ITC 2021). Climate Smart Agriculture (CSA) has emerged as a strategic approach aimed at enhancing agricultural productivity while simultaneously improving resilience to climate change and reducing greenhouse gas emissions from agricultural systems (FAO 2012). Despite its promise, CSA adoption at scale remains limited due to fragmented institutional support, high initial costs, and knowledge gaps among farmers (McCarthy et al. 2018). In response to these challenges, ITC Limited has implemented the Climate Smart Village (CSV) programme to operationalise the principles of climate-resilient agriculture at the village level. The programme integrates scientific tools such as geographic information systems (GIS), remote sensing, and climate risk assessments with participatory planning processes involving farmers and other local stakeholders. This approach enables the design and implementation of interventions that are customized to local agro-ecological conditions as well as the socio-economic characteristics of farming communities (Khatri-Chhetri et al. 2019). The CSV framework was initially developed through collaboration between the CGIAR Research Program on Climate Change, Agriculture and Food Security (CCAFS) and ITC Limited. Over time, the model has expanded across 19 states in India through partnerships with research institutions, government agencies, and civil society organizations (CSOs) (ITC 2023). The model was later integrated with ITC-NITI[1] Aayog partnership Model village framework to give a holistic approach for creating rural climate resilient livelihoods.

Through its extensive engagement with rural communities, ITC has implemented large-scale initiatives aimed at improving the sustainability of agricultural production and value chain. Recognising the growing impacts of climate change on agricultural ecosystems, ITC promotes a sustainable agricultural production system that is resilient to climate variability and change, which is increasingly critical in the Indian context (ITC 2025). Through the adoption and promotion of climate-smart

[1] NITI Aayog (National Institution For Transforming India) is Govt of India's Premier Policy think thank established in 1st January 2015.

agricultural practices, the company aims to address the environmental and socio-economic implications of climate change across its agri-value chains. Beyond the introduction of technological solutions, the programme emphasises the development of an enabling ecosystem for farmers. This includes strengthening access to extension services, facilitating linkages with input and output markets, supporting farmer institutions, and improving access to relevant government support schemes. Through initiatives such as the Social Investments Programme, ITC Mission Sunehra Kal (MSK), along with interventions led by its Agri Business Division, ITC seeks to enhance farmer livelihoods while simultaneously strengthening resilience to climate-related risks (ITC 2025).

14.2 Climate Change Risks in Indian Agriculture

Climate change is increasingly altering global weather patterns and is expected to have significant implications for agricultural production systems, thereby affecting food security and the livelihoods of millions of people (IPCC 2014). Changes in temperature regimes, variability in rainfall distribution, and the growing frequency of extreme climatic events are already influencing crop productivity across several regions of India. Studies indicate that agricultural systems in the Indo-Gangetic Plains are particularly sensitive to climatic variability, especially in states such as Bihar, Uttar Pradesh, and West Bengal, where fluctuations in weather patterns have led to considerable variability in crop yields (Rao et al. 2016). Future projections suggest an increase in the occurrence of droughts as well as greater intensity of flood events in multiple parts of the country, posing additional risks to agricultural sustainability and rural livelihoods (Pathak et al. 2014).

14.2.1 National Overview of Climate Variability

Climatic trends indicate that the global average surface air temperature has risen by approximately 0.4 °C over the past century. Although the overall trend in Indian monsoon rainfall has remained relatively stable, considerable regional variation has been observed. Climatic observations reveal regional variations in monsoon rainfall patterns across India. While regions such as North Andhra Pradesh, the west coast, and north-western India have witnessed increases of about 10–12% above the long-term normal over the past century, declining rainfall trends of approximately 6–8% below the normal have been observed in parts of Madhya Pradesh, Gujarat, north-eastern India, and Kerala. Climate projections generated using the Hadley Centre

Regional Climate Model (HadRM2)[2] suggest there will be a likely increase in both minimum and maximum temperatures in most regions of India. Climate projections indicate a substantial rise in temperature across India by the mid-twenty-first century. Maximum temperatures in regions located south of 25°N are expected to increase by approximately 2–4 °C by the 2050s, while northern parts of the country may experience temperature increases exceeding 4 °C. Similarly, minimum temperatures are projected to rise by nearly 4 °C nationwide, with relatively higher increases anticipated in southern regions (GoI 2004). The growing vulnerability of India to climate-related hazards such as floods, droughts, and sea-level rise emphasises the importance of adopting proactive strategies to maintain economic stability, protect infrastructure, and ensure the security of livelihoods (Economic Survey 2025). Studies assessing climate change impacts on crop productivity suggest potential yield reductions for major crops under future climate scenarios. Studies conducted by the Indian Agricultural Research Institute and other research institutions suggest that climate change could significantly affect crop productivity in India. Under the RCP 4.5 scenario, and in the absence of adaptation measures, projected yield reductions by 2040 may reach approximately 9% for wheat, 18% for maize, 12% for irrigated rice, 12% for mustard, and 13% for potato when compared with the average yields recorded during 2000–2007, even after accounting for potential CO_2 fertilisation effects (Naresh Kumar et al. 2013, 2014a, b, 2015, 2019, 2020). Climate change is also expected to affect the livestock sector, with projections indicating substantial declines in milk production under warming scenarios—estimated at about 1.6 million tonnes by 2020 and exceeding 15 million tonnes by 2050 (NPCC Report 2012; Upadhyay et al. 2013).

14.2.2 ITC CSV Model

The Climate Smart Village (CSV) programme represents a community-centred approach for translating the principles of Climate Smart Agriculture (CSA) into village-level practice. ITC launched this initiative in 2016 in collaboration with the CGIAR Research Program on Climate Change, Agriculture and Food Security (CCAFS) as a knowledge partner. The initiative focused on identifying climate risks affecting crop production and promoting adaptation practices ensuring both technical robustness and socio-economic relevance. The concept of Climate Smart Agriculture (CSA) responds to these challenges by promoting agricultural practices that enhance productivity, strengthen adaptive capacity, and reduce greenhouse gas emissions (ITC 2024). The CSV methodology incorporates climate risk assessments focusing on variability in rainfall patterns, number of rainy days, prolonged dry spells, drought conditions, and waterlogging resulting from intense rainfall events. Based on these parameters, villages are classified into four distinct typologies according to

[2] HadRM2 is a second generation regional climate model of the Hadley Centre for Climate Prediction and Research, UK. It is a high resolution climate model that covers a limited area of 5000 km × 5000 km.

their levels of agricultural productivity and resilience: High Yield–High Resilience (HY–HR), High Yield–Low Resilience (HY–LR), Low Yield–High Resilience (LY–HR), and Low Yield–Low Resilience (LY–LR). Resilience levels were evaluated by analysing the Normalised Difference Vegetation Index (NDVI) of key crops, including soybean in kharif season and wheat in rabi season, alongside historical rainfall records. NDVI values were used as proxies for estimating crop productivity and were validated against observed yield data. A strong correlation between NDVI-derived productivity estimates and actual crop yields was observed in the project districts. Variability in NDVI values over a fifteen-year period was measured using the coefficient of variation for both kharif and rabi seasons. Similarly, rainfall variability was estimated using historical rainfall datasets derived from gridded information provided by the Indian Meteorological Department (IMD). A resilience index for each village was calculated by comparing variability in NDVI values with rainfall variability. Lower index values indicate higher resilience of agricultural production systems. (Unpublished report, ITC-CCAFS).[3]

Based on this quadrant-based classification, different intervention strategies are recommended. Villages falling under the HY–LR category require interventions that strengthen resilience to climatic shocks, whereas LY–HR villages require productivity-enhancing technologies. LY–LR villages require integrated interventions that simultaneously improve productivity and resilience. A key analytical tool used within this framework is the Climate Smart Adoption Prioritisation (CSAP) toolkit, which applies a multi-criteria decision-making approach to identify and prioritise CSA interventions (Khatri-Chhetri et al. 2019). The toolkit evaluates potential interventions across several dimensions including:

1. **CSA Performance Index (CSA-PI)**—assessing impacts on productivity, income, resilience, and emissions reduction.
2. **Implementation Feasibility (CSA-IF)**—evaluating technical requirements, costs, gender inclusiveness, and policy compatibility.
3. **Adoption Barriers**—identifying constraints such as limited access to finance, labour shortages, equipment availability, and water resources.
4. **Incentive Mechanisms**—examining enabling measures such as subsidies, credit facilities, capacity building, and market access.
5. **Institutional Roles**—specifying the roles and responsibilities of key stakeholders, including government agencies, private sector entities, cooperatives, and non-governmental organisations.

Stakeholder consultations form a critical component of the CSV process. Farmers, extension personnel, NGOs, and representatives from the private sector participate in participatory workshops where they collectively assess and rank potential interventions using defined indicators. This collaborative process ensures that proposed

[3] Increasing Adaptive Capacity of Farmers to Climate Change thru Climate Smart Villages in India, Final Report ITC-CCAFS Climate Smart Village Project Implementation, Impact and Way forward, Borlaug Institute for South Asia (BISA) CIGAR Research Program on Climate Change, Agriculture and Food Security (CCAFS) CIMMYT, New Delhi, India, March 2019.

technologies reflect both scientific evidence and field realities, while also identifying practical barriers to adoption.

For example, in Maharashtra—one of the states considered highly vulnerable to climate risks—ITC organised participatory workshops in 2017 involving more than 75 stakeholders. Through this consultative process, a total of 57 climate-smart agricultural practices were identified. The findings indicated a strong preference for water management interventions such as micro-irrigation, rainwater harvesting, and agroforestry systems to address water scarcity challenges. Stakeholders also emphasised the importance of crop and livestock insurance as mechanisms for risk mitigation. However, certain infrastructure-intensive technologies such as drainage systems were considered less feasible due to machinery and cost constraints. The Climate Smart Village (CSV) framework was also implemented in Madhya Pradesh, ensuring alignment with the priorities outlined in the State Action Plan on Climate Change (SAPCC) and existing agro-climatic zoning frameworks. This alignment enabled the development of district-specific portfolios of CSA technologies. The process involved combining scientific climate projections under CMIP6 scenarios with farmer knowledge to ensure that recommended interventions were both technically robust and locally relevant (BISA–ITC 2021). Table 14.1 illustrates the classification of districts in Madhya Pradesh based on vulnerability indicators such as drought frequency and climatic variability.

In addition to drought vulnerability, projections suggest that Madhya Pradesh may experience an increase of approximately 13% in average annual rainfall by 2050. However, this increase is expected to vary significantly across regions, with western and southern districts likely to benefit from higher rainfall while eastern districts may experience declining precipitation levels (BISA–ITC 2021).

The prioritisation process ensures that interventions are custom-made to socio-economic profiles of farmers and local agro-ecological condition. Stakeholder consultations conducted in multiple states consistently highlighted the importance of certain CSA interventions: Water-smart interventions (micro-irrigation, water harvesting), Nutrient management (integrated nutrient management, composting), Resilient seeds (drought/heat tolerant varieties), Agroforestry for income diversification.

The ITC–CCAFS implementation model is structured around a 'hub and spoke' approach. The 'hub', comprising farmer groups, serves as the site for generating evidence through the application of prioritised packages of practices tailored to specific crops, along with targeted interventions addressing soil health, water management, and biodiversity. These evidences are documented and further various

Table 14.1 District-level vulnerability classification in Madhya Pradesh

Vulnerability level	Number of districts	Representative districts
High	9	Sheopur, Morena, Bhind
Moderate	18	Raisen, Sagar, Rewa
Low	10	Indore, Ujjain, Dewas

extension tools like on-site demonstrations plots, flipcharts, capacity building trainings are designed to disseminate to the 'spokes. Field-level climate-smart interventions were implemented by engaging farmers under three categories based on their capacity, resource availability, and ability to adopt climate-smart practices:

a. **Super-Champion Farmers**: These farmers possess the capacity to implement a wide portfolio of climate-smart technologies and agricultural practices. They are proactive, innovative, and often serve as local leaders within the farming community. Due to their higher level of adoption and influence, they play a critical role in demonstrating successful practices and motivating other farmers to adopt climate-resilient agricultural systems.
b. **Champion Farmers**: Champion farmers have moderate capacity and resources, enabling them to implement a limited set of climate-smart technologies and practices compared to Super-Champion farmers. They act as early adopters and contribute to the diffusion of climate-smart practices within their local farming networks.
c. **CSA Farmers**: CSA farmers typically have limited resources and capacities. As a result, they adopt only a small set of climate-smart agricultural technologies and practices that are low-cost and suitable to their circumstances. Despite these constraints, their participation is important in ensuring broader adoption and inclusivity in climate-resilient agricultural development.

During the initial phase of the ITC–CCAFS partnership (2016–2018), programme interventions were carried out across ITC's operational regions in the states of Maharashtra, Madhya Pradesh, and Rajasthan. The primary focus during this period was on introducing a range of Climate Smart Agriculture (CSA) technologies and practices, implemented in alignment with ITC's Mission Sunehra Kal (MSK) programme. Beyond individual technology demonstrations, ITC has also worked towards transforming entire villages into Climate Smart Villages (CSVs). This has been achieved by integrating the CSA framework with the Model Village approach developed through the ITC–NITI Aayog partnership. The integrated model seeks to engage all farmers within a village and promote a comprehensive transition toward climate-resilient rural livelihoods.

The approach addresses several key components. First, climate-smart crop management practices were introduced through localised climate risk assessments conducted at the village cluster level. Based on these assessments, a package of adaptation measures and recommended practices is developed for the major crops grown in the village. Second, natural resource management interventions focus on improving access to irrigation and ensuring the sustainable management of resources essential for agriculture, including water, soil health, and biodiversity. Third, the model emphasises livelihood diversification in order to reduce farmers' dependence on crop production alone. This includes promoting additional income opportunities that are relatively resilient to climatic variability, such as on-farm tree planting with fruit, plantation, or timber species. In addition, households are encouraged to adopt at least one livestock-based livelihood option, such as dairy farming, goat rearing, sheep rearing, piggery, poultry, or fisheries. Another important element involves reducing production costs and strengthening farmers' resilience to climate-related risks. The programme ensures this objective by supporting farmers to become members of local institutions, particularly Farmer Producer Organisations (FPOs) and Agri Business Centres (ABCs), thereby strengthening collective action. These institutions facilitate collective input procurement, access to farm machinery through shared services, credit availability, and dissemination of crop advisories and weather information. Farmers are also supported in accessing relevant government programmes that provide benefits such as crop insurance, institutional credit, soil health cards, pension schemes, and other livelihood support measures. A village is considered to have achieved the status of a Climate Smart Village when a majority of households are actively participating in these interventions and when natural resource management practices are implemented across both agricultural lands and village habitations. (ITC 2024).

14.3 Implementation Outcomes and Impacts

Climate Smart Village (CSV) programme was conceptualised and implemented by ITC in 2016 the initial phase of the initiative concentrated on identifying climate-related risks affecting crop production and developing appropriate adaptation strategies for farming communities. Over time, the CSV framework has expanded beyond crop-focused interventions to incorporate broader elements such as livelihood diversification, sustainable management of natural resources, and strengthening institutional support systems for farmers (ITC 2024). The programme has been implemented across northern states, including Madhya Pradesh, Rajasthan, Punjab, Bihar, Uttar Pradesh and Maharashtra, and has more recently been extended to southern state of the country. Among the climate-smart practices promoted through the programme are direct-seeded rice cultivation, the broad-bed furrow technique for soybean farming, and zero-tillage methods for wheat cultivation. ITC's Climate Smart Agriculture initiatives have expanded significantly in recent years, currently covering approximately 3.17 million acres and engaging about 1.2 million farmers across 19 states. These efforts contribute toward the company's Sustainability 2.0 commitment of reaching four million acres by 2030. In parallel, ITC has introduced ITCMAARS (Meta Market for Advanced Agricultural and Rural Services), a crop-agnostic "phygital" platform designed to provide farmers with integrated agricultural services. The platform is now operational in ten states and supports more than 2.1 million farmers through a network of approximately 2050 Farmer Producer Organisations (FPOs) (ITC 2025).

To monitor the programme and development of villages and its transitioning toward climate-smart systems, baseline assessments are conducted to measure existing crop yields, household incomes, greenhouse gas emissions from farming practices, and the overall capacity of villages to withstand climate-related shocks. Based on these indicators, villages are classified within a Yield–Resilience framework that groups them into four categories: Low Yield–Low Resilience (LY–LR),

High Yield–Low Resilience (HY–LR), Low Yield–High Resilience (LY–HR), and High Yield–High Resilience (HY–HR).

The long-term objective of the Climate Smart Village (CSV) initiative is to support the progression of villages toward the High Yield–High Resilience typology, characterised by farming systems that combine enhanced productivity with greater resilience to climate variability. An assessment conducted implemented across 89 villages in the southern states of Telangana, Andhra Pradesh, and Tamil Nadu. Demonstrated significant improvements following the introduction of CSV interventions. The number of villages classified as High Yield–High Resilience increased from six to twenty-one, representing an increase of about 350% At the same time, the number of villages falling into the Low Yield–Low Resilience category declined from forty to nineteen, a reduction of approximately 50%.

These improvements were driven by a combination of climate risk assessments, adoption of improved crop management practices, linkages with insurance schemes, access to shared agricultural machinery through farmer groups, and promotion of water-efficient technologies such as micro-irrigation. Farmers were also connected to climate information services that provided timely weather forecasts and advisory messages to guide agricultural decision-making. As a result of these integrated interventions, the CSV programme has reported yield increases of up to 38 percent while also achieving reductions in greenhouse gas emissions of as much as 66 percent. (ITC 2024).

14.4 Conclusions

The integration of robust scientific frameworks, such as the CSAP toolkit and climate risk mapping, with ITC's large-scale field implementation offers a replicable model for CSA scaling integrated with another element of ITC holistic CSV model. The dual emphasis on technological suitability and institutional feasibility ensures that interventions are both adoptable and sustainable.

The CSV approach represents a scalable model for embedding climate resilience in India's agricultural systems. For wider replication, three priorities emerge:

1 .**Policy Convergence**—Aligning CSV interventions with national schemes such as MGNREGS, NRLM, PMKSY and PMFBY.
2 .**Digital Integration**—Leveraging AI-driven advisories and weather-linked services for precision adaptation.
3 .**Institutional Scaling**—Strengthening farmer collectives and local governance structures to sustain interventions beyond project timelines.

References

BISA–ITC (2021) Adaptation strategies for managing climatic risks in Madhya Pradesh agriculture. Borlaug Institute for South Asia (BISA) and ITC Limited

Campbell BM, Thornton PK, Zougmoré RB, Van Asten P, Lipper L (2014) Sustainable intensification: what is its role in climate smart agriculture? Curr Opin Environ Sustain 8:39–43. https://doi.org/10.1016/j.cosust.2014.07.002

The Economic Survey (2025). Available on https://www.indiabudget.gov.in/economicsurvey/doc/eschapter/echap10.pdf

FAO (2012) The state of food and agriculture. FAO Retrieved from. http://www.fao.org/docrep/017/i3028e/i3028e.pdf

GoI 2004 India's initial national communication to the United Nations framework convention on climate change ministry of environment and forests. Government of India

IPCC (2014) Summary for policymakers. Climate change 2014: impacts, adaptation, and vulnerability. part a: global and sectoral aspects. In: Contribution of working group into the fifth assessment report of the intergovernmental panel on climate change. Cambridge University Press, Cambridge, United Kingdom and New York, NY, USA, pp 1–32

Khatri-Chhetri A, Pant A, Aggarwal PK, Vasireddy VV, Yadav A (2019) Stakeholders prioritization of climate-smart agriculture interventions: evaluation of a framework. Agric Syst 174:23–31. https://doi.org/10.1016/j.agsy.2019.03.002

ITC Limited (2025) Shaping the future—purpose, progress and partnership, sustainability report 2025

ITC Limited (2023) Shaping the future—innovative, inclusive and impactful, sustainability report 2023

ITC Limited (2024) Shaping the future—evolve, nurture and sustain, sustainability report 2024

McCarthy N, Lipper L, Zilberman D (2018) Economics of climate-smart agriculture: an overview. In: Lipper (ed), Climate-smart agriculture: building resilience to climate change. Food and Agriculture Organization, Rome, Italy

Naresh Kumar S, Aggarwal PK, Swarooparani DN, Saxena R, Chauhan N, Jain S (2014a) Vulnerability of wheat production to climate change in India. Clim Res 59:173–187. Available on https://doi.org/10.3354/cr01212

Naresh Kumar S, Govindakrishnan PM, Swarooparani DN, Nitin C, Surabhi J, Aggarwal PK (2015) Assessment of impact of climate change on potato and potential adaptation gains in the Indo-Gangetic Plains of India. Int J Plant Prod 9(1):151–170

Naresh Kumar S, Aggarwal PK, Saxena R, Rani S, Jain S, Chauhan N (2013) An assessment of regional vulnerability of rice to climate change in India. Clim Chang 118(3–4):683–699

Naresh Kumar S, Aggarwal PK, Kumar U, Surabhi DNJ, Rani S, Chauhan N, Saxena R (2014b). Vulnerability of Indian mustard (Brassica juncea (L.) Czernj. Cosson) to climate variability and future adaptation strategies. Mitig Adapt Strat Glob Chang 21(3):403–420. https://doi.org/10.1007/s11027-014-9606-Z

Naresh Kumar S, Chander S, Sinha P, Padaria RN (2020) Climate change impact, vulnerability and adaptation: agriculture, a report for India's third national communication to UNFCCC

Naresh Kumar S (2015) Carbon sequestration and beyond by plantations: exploring scope for added income. In: Dhayani SK, Newaj R, Alam B, Dev I (eds) Agroforestry: present status and way forward. Biotech Books, Pub., pp 193–205

Naresh Kumar S (2019) Climate change and horticultural crops. In: Chadha KL (ed) Handbook of horticulture, Vol I and II. ICAR Publication. vol I. pp 83–92

Pathak H, Pramanik P, Khanna M, Kumar A (2014) Climate change and water availability in Indian agriculture impacts and adaptation. Indian J Agric Sci 84:671–679

Rao CR, Raju BMK, Rao AS, Rao KV, Rao VUM, Ramachandran K, Rao CS (2016) A district level assessment of vulnerability of Indian agriculture to climate change. Curr Sci 1939–1946

NPCC Report (2012) Climate change and Indian agriculture: salient achievements from ICAR network project. Naresh Kumar S, Singh AK, Aggarwal PK, Rao VUM, Venkateswarlu B. IARI Pub., p 32

Upadhyay RC, Hooda OK, Aggarwal A et al (2013) Indian livestock production has resilience for climate change. In: Singh SV, Upadhyay RC, Sirohi S, Singh AK (eds) Climate resilient livestock and production system. National Dairy Research Institute, Karnal, Haryana 132001 (India), pp 286

Chapter 15
Synthesis, Policy Recommendations, and Future Directions

Mohan Geetha, Dillip Kumar Swain, Poonam Biswal, Ahmad Faisal, and Naoya Wada

Abstract This concluding chapter synthesizes essential insights derived from the interdisciplinary analyses presented throughout the volume, framing climate-smart agriculture as a systemic transformation rather than merely isolated technological solutions. Employing empirical data from India and Japan, it highlights the interconnectedness of soil carbon restoration, watershed hydrology, digital innovation, institutional reform, and multifunctional land use in promoting sustainable agricultural resilience. The chapter emphasizes that sustainable agriculture depends on ecological renewal, coherent policy frameworks, inclusive digital governance, and adaptive strategies to demographic changes. It advocates broader policy initiatives focused on soil management, incentive adjustments, watershed planning, and data transparency, and proposes future research avenues on long-term carbon dynamics, AI policy, the food–energy–water nexus, and demographic–climate interactions. By situating climate-smart agriculture within the broader context of systems governance, this approach affirms the necessity of integrating science, policy, and community engagement to effectively address agricultural challenges amid intensifying climate threats.

Keywords Climate-smart agriculture · Soil carbon governance · Institutional reform · Multifunctional landscapes · Systems transformation

M. Geetha (✉) · N. Wada
Graduate School of Sustainability Studies, University of Toyama, Toyama, Japan
e-mail: geetha@eco.u-toyama.ac.jp

M. Geetha · D. K. Swain · N. Wada
Global Research Centre for Advanced Sustainability Science (GRASS), University of Toyama, Toyama, Japan

D. K. Swain · P. Biswal · A. Faisal
Agricultural and Food Engineering Department, Indian Institute of Technology Kharagpur, Kharagpur, India

D. Kumar Swain et al. (eds.), *Climate-Smart Technologies for Sustainable Agricultural Production*, https://doi.org/10.1007/978-3-032-25666-9_15

15.1 Reconstructing Agricultural Resilience under Structural Climate Stress

The compilation of scholarly work in this volume clearly demonstrates that climate-smart agriculture signifies not merely a technological enhancement but a fundamental structural shift in the organization of agricultural systems. Climate-smart agriculture, as defined by FAO (2012), integrates productivity enhancement, adaptation to climate change, and mitigation objectives. The empirical contributions spanning soil science, hydrology, digital innovation, economic policy, land governance, and demographic transformation collectively affirm that agricultural resilience is only achievable through the integration of ecological regeneration, institutional redesign, and socio-economic adaptation. Across India and Japan, two nations with contrasting demographic and agro-ecological trajectories, agricultural systems face different manifestations of structural stress. In India, the vulnerability of smallholder production to rising temperatures, rainfall variability, and extreme events poses direct risks to the stability of staple crops (Naresh Kumar et al. 2013, 2014, 2015), consistent with global assessments of climate change impacts on agriculture (IPCC, 2014). Yield simulations for rice and wheat reveal measurable declines under projected climate scenarios, underscoring the urgency of adaptive redesign. Meanwhile, soil erosion and land degradation processes further compound vulnerability, diminishing long-term productive capacity (Mandal et al. 2021). In Japan, climatic stress intersects with demographic contraction. Rapid rural aging, labor scarcity, and farmland abandonment introduce structural discontinuities in landscape management (Su et al. 2018; Shoyama et al. 2021). The result is not merely yield volatility but systemic fragmentation of agricultural continuity. These two contexts differ in their immediate drivers yet converge in their implications: resilience cannot be sustained through incremental adjustments to twentieth-century production paradigms. A deeper restructuring is imperative.

Taken together, the collective evidence indicates that resilience must be cultivated across five interconnected domains: soil ecological regeneration, hydrological intelligence, digital governance, incentive realignment, and landscape multifunctionality. Each domain individually provides partial benefits; collectively, they constitute the framework for climate-smart transformation.

15.2 Soil Carbon as Strategic Ecological Infrastructure

Among the most consistent observations throughout the volume is the pivotal role of soil organic carbon (SOC) in connecting productivity, climate mitigation, and adaptive stability. Advancements in Digital Soil Mapping, hyperspectral sensing, and machine learning modeling have significantly enhanced the accuracy of monitoring SOC stocks. These technological developments address historical limitations associated with laboratory-based measurement systems and facilitate scalable carbon accounting frameworks suitable for national climate strategies. Nonetheless, the implications extend beyond measurement alone. Empirical evidence from conservation agriculture systems indicates that practices such as minimal tillage, residue retention, and diversified crop rotations lead to incremental yet measurable carbon sequestration (Kassam et al. 2009). Carbon and energy footprint analyses reveal that emission reductions are achieved not merely through input substitution but also via structural modifications to soil–crop interactions. Organic farming systems further diminish dependence on synthetic nitrogen inputs, thereby reducing emission intensity and strengthening biological nutrient cycling. Crucially, soil carbon functions as ecological infrastructure; increased SOC enhances soil aggregation, elevates water-holding capacity, buffers crops against rainfall variability, and stabilizes yields under thermal stress. Consequently, soil carbon restoration serves as both a mitigation measure and an adaptation strategy.

Nevertheless, enhancing SOC is not automatic. Transition periods frequently entail short-term yield uncertainties or increased management complexity. Consequently, the policy challenge resides in aligning private incentives with the overarching public ecological advantages. Implementing payment for ecosystem services, transitional subsidies, and integrating SOC metrics into climate finance instruments are essential steps to internalize long-term benefits. Absent institutional acknowledgment of soil as a climate asset, the regenerative potential delineated throughout these chapters cannot be realized at scale.

15.3 Hydrological Intelligence and Watershed Governance

Climate variability is increasingly manifested through hydrological instability, including unpredictable monsoon patterns, extreme rainfall events, and groundwater depletion. These phenomena necessitate watershed-based responses rather than solely farm-level technological interventions. The integration of Land Resource Inventory data with hydrological models, such as SWAT, provides vital spatial insights essential for infrastructure siting and water management (Velayutham et al. 1999; Arnold et al. 2012). Case studies involving rubber dams and micro-watersheds demonstrate that decentralized water storage enhances cropping intensity, reduces migration, and increases rural incomes when guided by scientific models and participatory governance (ICAR-IIWM 2023). The overarching implication is that water-use efficiency technologies, such as micro-irrigation, direct-seeded rice, and broad-bed furrow systems, must be customized to local hydrological conditions. Without basin-level assessments, efficiency improvements risk unintentionally intensifying water extraction pressures. Consequently, climate-smart agriculture requires institutionalized watershed governance that incorporates meteorological forecasts, groundwater monitoring, sediment analysis, and active farmer participation. Hydrological data should be integrated into seasonal decision-making processes rather than regarded as secondary advisory information.

15.4 Digitalization and the Institutionalization of Climate Intelligence

Digital innovation represents a fundamental theme within this volume. AI-driven advisory platforms, NDVI-based yield monitoring, plant health diagnostics, and digital market integration collectively enhance adaptive capacity. Platforms that integrate physical extension services with cloud-based analytics exemplify how climate intelligence can be operationalized at the farm level. Nevertheless, digital transformation introduces complexities in governance. Concerns related to data ownership, algorithmic bias, platform concentration, and disparities in connectivity pose structural risks. In the absence of regulatory oversight and inclusive design, digital systems may inadvertently exacerbate rather than mitigate inequality. The evidence throughout this volume indicates that digital agriculture should be regarded as a form of public infrastructure. Essential requirements include interoperable open data standards, transparent algorithmic protocols, safeguards for farmers" data rights, and the implementation of digital literacy initiatives. Participatory governance models are recommended to guide system design and deployment, ensuring contextual relevance and promoting equitable access. Digitalization enhances resilience only when integrated within accountable institutional frameworks.

15.5 Institutional Incentive Realignment and Structural Policy Reform

A fundamental insight derived from this volume is that technological innovation alone cannot effectuate structural transformation unless incentive frameworks are recalibrated. Agricultural systems are influenced equally by price signals, subsidy regimes, and trade commitments as by agronomic practices. The historical significance of Minimum Support Price (MSP) systems in India exemplifies this dynamic convincingly. While procurement-based policies have stabilized cereal production and safeguarded farmer incomes, they have also contributed to rigidity in cropping patterns, groundwater overextraction, and regional specialization in water-intensive crops. Fiscal simulations and policy analyses indicate that price deficiency payments and decoupled income transfers may be more adaptable under WTO rules. Nevertheless, such transitions must be implemented gradually, meticulously designed, and supported by institutional safeguards to prevent smallholders" exposure to market volatility. Climate-smart transformation necessitates diversification towards legumes, pulses, oilseeds, and agroforestry systems that are both water-efficient and nutritionally strategic. However, diversification cannot be realized in environments where economic signals penalize crop switching.

The policy implication does not involve the abrupt dismantling of existing systems but rather their strategic evolution. Climate objectives must be integrated into the agricultural support framework. Subsidy reform should incorporate environmental conditionality, rewarding practices that enhance soil carbon sequestration, reduce emissions intensity, and improve water-use efficiency. Insurance programs must integrate climate risk modeling alongside reliance on historical yield baselines. Public procurement systems can progressively include climate-resilient crops, thereby aligning food security initiatives with sustainability objectives. The broader evidence chapters suggest that achieving climate-smart agriculture requires aligning institutions. When crop insurance schemes, irrigation missions, rural livelihood programs, and climate mitigation strategies function independently, policy fragmentation undermines systemic resilience. Conversely, coordinated governance among various ministries strengthens adaptation pathways. Therefore, climate-smart transformation is not solely an agricultural agenda but a cross-sector governance initiative.

15.6 Demographic Transition and the Reconfiguration of Rural Landscapes

Demography is presented in this volume as a fundamental variable of equal significance to climate. In India, factors such as smallholder fragmentation, rural migration, and generational transitions significantly influence labor availability and investment capacity. Similarly, in Japan, population aging and rural depopulation have led to widespread abandonment of farmland, particularly in mountainous prefectures such as Toyama (Su et al. 2018; Shoyama et al. 2021). These demographic shifts reshape the feasibility of various agricultural strategies. The collective findings demonstrate that multifunctionality provides a common framework applicable to various demographic contexts. In aging rural landscapes, labor-efficient systems, such as livestock grazing on marginal paddy fields, serve to restore biodiversity while alleviating management burdens (Nakahama et al. 2024). Biomass systems employing fast-growing willows integrate renewable energy production with land revitalization (Mitsui et al. 2010). Vineyard conversion and branding of specialty crops provide avenues for economic diversification that are compatible with reduced labor demands. In regions with high population density, multifunctionality manifests as crop–livestock integration, the expansion of agroforestry, and the diversification of income streams within limited landholdings. Solar sharing systems exemplify multifunctional design by enabling renewable energy generation without displacing food production. The overarching lesson is that climate-smart agriculture must focus on optimizing ecosystem service bundles rather than solely maximizing the yields of individual crops. Food production, biodiversity, carbon sequestration, renewable energy, and rural employment should be regarded as interconnected outputs of landscape systems. Accordingly, policy should shift focus from hectare-based productivity metrics toward multifunctional performance indicators that incorporate ecological and socio-economic co-benefits.

15.7 Soil–Water–Energy–Data Nexus: Toward Integrated Systems Governance

A significant synthesis emerging from the volume pertains to the imperative of nexus-based governance. Soil carbon restoration interacts with hydrological stability, while water storage infrastructure impacts cropping intensity. The integration of renewable energy influences land allocation, and digital monitoring systems affect the efficiency of input application. These interdependencies necessitate systems thinking rather than sectoral management. For instance, micro-irrigation technologies reduce water consumption per unit area but may increase total water withdrawal if cropping intensity expands without watershed regulation. Likewise, carbon sequestration programs may engender inequities if smallholders lack access to monitoring technologies or aggregation platforms. Consequently, climate-smart transformation necessitates governance frameworks that integrate soil metrics, hydrological models, digital platforms, and energy planning into coherent structures. National climate strategies must incorporate agricultural carbon accounting alongside watershed governance and renewable energy transitions. Institutional silos must be dismantled in favor of cross-ministerial coordination. The contributions within this volume demonstrate that when soil science, hydrology, digital analytics, and economic modeling are synergistically combined, policy formulation becomes evidence-based rather than reactive.

15.8 Policy Recommendations

A forward-looking policy framework should be rooted in structural coherence and ecological realism. It begins with institutionalizing soil carbon governance, whereby national agricultural policies include measurable SOC targets in subsidy schemes and link conservation agriculture and organic transitions to climate finance instruments. Carbon credit mechanisms must feature robust measurement, reporting, and verification (MRV) systems, with smallholders participating through cooperative aggregation. Watershed-scale planning should be mandatory for irrigation expansion, guided by hydrological modeling rather than political considerations, and supported by investments in check dams, rubber dams, and rainwater-harvesting infrastructure. Groundwater governance needs transparent monitoring and community involvement to prevent depletion from intensification. Digital agriculture policy must emphasize equity and transparency by establishing farmer data ownership, algorithmic transparency, and preventing monopolies in advisory platforms through accountable public–private partnerships. Price support systems should be gradually adjusted to align incentives with ecological sustainability, encouraging climate-resilient crops and diversified farming, while income protection should be decoupled from environmentally damaging practices. Land-use policies in aging societies should promote cooperative leasing, community management, and multifunctional land conversion, with corporate involvement under transparent governance that safeguards local ecological and cultural assets. Lastly, institutional convergence can be achieved through integrated, climate-smart planning units that coordinate policies across agriculture, water, energy, and rural development, all aligned with shared sustainability goals.

15.9 Future Directions

The emerging research frontier from this synthesis delineates several strategic priorities. Long-term soil carbon dynamics necessitate a comprehensive investigation. SOC saturation thresholds, interactions with nitrogen cycles, and resilience under extreme climatic variability must be quantified across various agro-ecological zones (Lal 2015). Monitoring technologies should be calibrated through long-term field trials to enhance predictive reliability. The modeling of the integrated food–energy–water nexus represents another frontier. Scenario simulations combining SWAT-based hydrology, carbon accounting, demographic projections, and land-use planning could inform regional adaptation strategies. The governance of artificial intelligence in agriculture requires systematic research. Algorithmic biases, the representativeness of training data, and barriers related to digital literacy must be addressed to prevent structural inequities. Participatory AI design models warrant further exploration. Models examining demographic–climate interactions are also essential. Aging populations, labor scarcity, and migration patterns influence the adoption of climate-smart practices. Socio-economic modeling that integrates demographic variables with climate projections would strengthen policy targeting. Ensuring access to carbon markets for smallholders remains a persistent challenge. Transaction costs, MRV complexity, and aggregation mechanisms must be analyzed to facilitate inclusive participation. Finally, the valuation of multifunctional landscapes necessitates interdisciplinary methodologies capable of quantifying ecosystem service bundles, cultural heritage preservation, renewable energy co-production, and biodiversity enhancement within unified economic frameworks.

15.10 Structural Transformation as Imperative

The cumulative evidence presented throughout this volume affirms that climate-smart agriculture constitutes a fundamental transformation rather than merely an incremental adaptation. It entails the reorganization of agricultural systems within ecological constraints and demographic realities. Resilience is derived from ecological regeneration of soil systems, hydrological intelligence at the watershed scale, inclusive digital infrastructure, realigned economic incentives, and multifunctional land-use planning. India and Japan exemplify different pathways towards this transformation; however, the underlying principles remain consistent: integration, coherence, and institutional accountability. Agricultural systems can no longer operate in isolation through technological substitution alone. They must evolve as adaptive socio-ecological systems governed by science-informed policy and community participation. The transition from vulnerability to resilience is attainable, but only when soil restoration, water governance, digital inclusion, energy integration, and policy reform function as mutually reinforcing components of a unified framework. Consequently, climate-smart agriculture is not a discretionary innovation; rather, it is the foundational structure for sustainable agricultural futures in the twenty-first century.

References

Arnold JG et al (2012) SWAT: model use, calibration, and validation. Trans ASABE 55(4):1491–1508

FAO (2012) The state of food and agriculture. Food and Agriculture Organization of the United Nations

ICAR-IIWM (2023) Watershed and rubber dam intervention impact assessment report. Indian Council of Agricultural Research

IPCC (2014) Climate change 2014: impacts, adaptation, and vulnerability. Cambridge University Press

Kassam A et al (2009) The spread of conservation agriculture: justification and sustainability. Int J Agric Sustain 7(4):292–320

Lal R (2015) Restoring soil quality to mitigate soil degradation. Sustainability 7:5875–5895

Mandal D et al (2021) Soil erosion assessment in India. Environ Monit Assess 193:457

Mitsui Y et al (2010) Willow clones with high biomass yield in short rotation coppice. Biomass Bioenergy 34(4):467–473

Naresh Kumar S et al (2013) Vulnerability of rice production to climate change in India. Clim Chang 118:683–699

Naresh Kumar S et al (2014) Vulnerability of wheat production to climate change in India. Clim Res 59:173–187

Naresh Kumar S et al (2015) Climate change impact on potato in indo-Gangetic Plains. Int J Plant Prod 9(1):151–170

Nakahama N et al (2024) Utilization of abandoned land as cattle grazing restores biodiversity. Ecol Res 39(5):730–743

Shoyama K et al (2021) Community-based approaches to countering land abandonment in Japan. Environ Manag 68(3):353–365

Su G et al (2018) Spatial pattern of farmland abandonment in Japan. Sustainability 10(10):3676

Velayutham M et al (1999) Land resource inventory and planning. National Bureau of Soil Survey

MIX
Papier aus verantwortungsvollen Quellen
Paper from responsible sources
FSC® C105338

If you have any concerns about our products,
you can contact us on
ProductSafety@springernature.com

In case Publisher is established outside the EU,
the EU authorized representative is:
Springer Nature Customer Service Center GmbH
Europaplatz 3, 69115 Heidelberg, Germany

Printed by Libri Plureos GmbH
in Hamburg, Germany